Mathematics for Engineering, Technology and Computing Science

Mathematics for Engineering, Technology and Computing Science

BY

HEDLEY G. MARTIN,

B.A. (Keele), Ph.D. (Cantab,), F.I.M.A., A.Inst.P.

Principal Lecturer
Department of Mathematics, Science and Computing
Staffordshire College of Technology Stafford
(A Constituent College of North Staffordshire Polytechnic)

1966

PERGAMON PRESS

Oxford · London · Edinburgh · New York
Toronto · Sydney · Paris · Braunschweig

Pergamon Press Ltd., Headington Hill Hall, Oxford
4 & 5 Fitzroy Square, London W.1
Pergamon Press (Scotland) Ltd., 2 & 3 Teviot Place, Edinburgh 1
Pergamon Press Inc., Maxwell House, Fairview Park, Elmsford,
New York 10523
Pergamon of Canada Ltd., 207 Queen's Quay West, Toronto 1
Pergamon Press (Aust.) Pty. Ltd., 19a Boundary Street,
Rushcutters Bay, N.S.W. 2011, Australia
Pergamon Press S.A.R.L., 24 rue des Écoles, Paris 5ᵉ
Vieweg & Sohn GmbH, Burgplatz 1, Braunschweig

First edition 1970
Library of Congress Catalog Card No. 79–102403

Printed in Hungary

08 013960 4 (flexicover)
08 013961 2 (hard cover)

Contents

Preface

MANY of the courses in Engineering, Technology and Computing Science which contain a standard of Mathematics beyond GCE A-level have certain major sections on mathematics in common. Of these this book takes linear algebra, ordinary differential equations and vector analysis, together with line and multiple integrals, and develops the material in a manner and to a degree which should be suitable for the following courses:

> University and CNAA degrees in Electrical, Electronic, Electro-Mechanical and Mechanical Engineering.
> University and CNAA degrees in Computing Science.
> Higher National Diploma and Certificate Courses in Engineering, Electrical and Electronic Engineering, and in Mathematics, Statistics and Computing.

Most of the syllabus for the second, final year of the HNC course in Mathematics, Statistics and Computing is covered.

The author has lectured to students on these courses and the book is based on the lectures given. The standard of the material presented is mainly higher than that needed for the HNC Engineering courses, but the exposition should suit many students on these courses and encourage them to study some fields further.

Beaconside   H. G. MARTIN
Stafford

CHAPTER 1

Determinants and Linear Equations

1.1. INTRODUCTION

The studies to be made of determinants in this chapter and of matrices in the next follow courses which develop methods for solving sets of simultaneous linear equations. The determinant approach to such methods is the one more readily conceived but it is less easy to generalise and to use when many equations are involved. The matrix method is based on a deeper analysis and has a more general application to systems of many equations and to those where the number of equations and the number of unknowns are unequal; also it forms the basis for numerical methods suitable for automatic computation.

Distinctive features of the two approaches become apparent through the analysis of simple electrical circuits, for example. The determinant method is no more than a technique for solving a set of equations, which may relate various currents in a network. The matrix approach provides a similar technique, though rather more complicated. However, the nature of a matrix allows the characteristics of the components in a circuit, in most general terms its complex impedances, to be abstracted symbolically from the currents they affect and be expressed as a composite mathematical symbol, that is as a matrix. In this sense matrix analysis may be thought of as being concerned more with causes whereas determinants deal with effects. This distinction will become apparent as the studies progress.

A determinant is an orderly arrangement of a set of symbols, which may be or may represent numbers or functions. Such an arrangement implies that the symbols combine algebraically in clearly defined ways to produce a single result called the value of the determinant. Determinants are used mainly as convenient ways of representing complicated expressions and for detecting relations between numbers or functions which in other forms would probably be obscure. That all but two of the other chapters in this book use determinants is a measure of their importance and is the reason for placing the subject first.

1.2. NOTATION AND DEFINITIONS

The purpose of this section is to explain what determinants are and to give the definitions needed for showing, in the following section, how the value of a determinant is defined and how it may be obtained.

A determinant of order n is an arrangement of n^2 symbols, called elements, in n rows and an equal number of columns. The square array so formed is shown between a pair of vertical bars to indicate that the elements, in most cases numbers, are subject to certain algebraic rules of combination. The definitions given below are illustrated by determinants of order two and three but they apply also to those of higher order.

Column number

$$
\begin{array}{c}
\text{Row} \\
\text{number}
\end{array}
\quad
\begin{array}{c}
 \\
1 \\
2 \\
3
\end{array}
\begin{array}{ccc}
1 & 2 & 3 \\
\begin{vmatrix} a_{11} & a_{12} & a_{13} \\ a_{21} & a_{22} & a_{23} \\ a_{31} & a_{32} & a_{33} \end{vmatrix}
\end{array}
\qquad
\begin{vmatrix} a_{11} & a_{12} \\ a_{21} & a_{22} \end{vmatrix}
$$

$$
\qquad\qquad\qquad D_3 \qquad\qquad\qquad\qquad D_2
$$

Determinants of third and of second order, D_3 and D_2, are shown above; the position of an element a_{ij} is indicated by the digits in the subscript, i being the row number and j the column number. The rows and columns are numbered from the top left-hand element a_{11}, called the leading diagonal element. The *leading*, or principal, *diagonal* is drawn from a_{11} to the element in the lowest right-hand position, a_{33} and a_{22} for D_3 and D_2 respectively.

Associated with each element of a determinant of order n is a determinant of order $(n-1)$ formed by deleting the row and the column in which the element occurs. These determinants are called the *minors* of the respective elements in the original determinant. Thus the minor of the element a_{33} in D_3 is D_2 and the minor of a_{11} in D_2 is the single element a_{22} which, in this sense, can be regarded as a determinant of order one. When the minor of an element a_{ij} is multiplied by $(-1)^{i+j}$, the result is called the *cofactor* of that element and is denoted here by A_{ij}. Thus the cofactor of a_{33} in D_3 is $A_{33} = +D_2$ and the cofactors of the four elements in D_2 are $A_{11} = +a_{22}$, $A_{12} = -a_{21}$, $A_{21} = -a_{12}$ and $A_{22} = +a_{11}$.

Problems 1.2

1. What is the relation between the signs of the minor and the cofactor of any element in the leading diagonal of a determinant of any order?

2. Apply the question in **1** to other diagonal lines of elements parallel to the leading diagonal.

3. In the determinant $\begin{vmatrix} 3 & 4 \\ 7 & -5 \end{vmatrix}$ replace each element by its cofactor.

4. State the minors of the elements in the leading diagonal of the determinant $\begin{vmatrix} 1 & 2 & 3 \\ 7 & 6 & 5 \\ 8 & 9 & 4 \end{vmatrix}$.

5. For the determinant $\begin{vmatrix} 3 & -7 \\ 8 & 13 \end{vmatrix}$ multiply each element of the first row by its own cofactor and add the results. Repeat the process for the elements in the second row and for those in the two columns.

6. For the determinant $\begin{vmatrix} 14 & -9 \\ -5 & 8 \end{vmatrix}$ multiply each element of the first row by the cofactor of the respective element in the second row and add the results. Repeat the process in terms of columns.

1.3. EVALUATION OF DETERMINANTS

The value of a determinant is defined as the sum of the products of the elements in any one row or column with their respective cofactors. The evaluation of a determinant according to the process implied in this definition is called the expansion or development of the determinant by a specified row or column. The definition implies that there are four such ways in which a second-order determinant may be evaluated. For example, the value of D_2 may be obtained from development:

by the first row, giving

$$a_{11}A_{11} + a_{12}A_{12} = a_{11}a_{22} - a_{12}a_{21},$$

or the second row, giving

$$a_{21}A_{21} + a_{22}A_{22} = -a_{21}a_{12} + a_{22}a_{11},$$

or the first column, giving

$$a_{11}A_{11} + a_{21}A_{21} = a_{11}a_{22} - a_{21}a_{12},$$

or the second column, giving

$$a_{12}A_{12} + a_{22}A_{22} = -a_{12}a_{21} + a_{22}a_{11},$$

all of which give the same value for D_2. It is usual to visualise the expansion of a second-order determinant by the scheme

$$\begin{vmatrix} a_{11} & a_{12} \\ a_{21} & a_{22} \end{vmatrix} = \begin{vmatrix} a_{11} & a_{12} \\ a_{21} & a_{22} \end{vmatrix} = a_{11}a_{22} - a_{21}a_{12}.$$

rather than to consider a formal expansion according to the definition each time.

The value of D_3 from expansion by its first row is

$$a_{11}A_{11} + a_{12}A_{12} + a_{13}A_{13},$$

where each cofactor is a determinant of order two. Expansions by the other two rows and the three columns give the same value. An extension of these results shows that a determinant of order four could be evaluated in eight ways and that expansion by its first row would have the form

$$a_{11}A_{11} + a_{12}A_{12} + a_{13}A_{13} + a_{14}A_{14},$$

where each cofactor is a determinant of order three. Thus a determinant of any order can be reduced, in principle at least, to an expression which involves a sum of multiples of second-order determinants only. This conclusion does not imply that such a process is the only or even the best way of evaluating a determinant.

Example 3.1.

Obtain the value of the determinant

$$D = \begin{vmatrix} 4 & 2 & -3 \\ 2 & 4 & 1 \\ -3 & 7 & 5 \end{vmatrix}.$$

Expansion by the first row gives

$$D = 4(+1)\begin{vmatrix} 4 & 1 \\ 7 & 5 \end{vmatrix} + 2(-1)\begin{vmatrix} 2 & 1 \\ -3 & 5 \end{vmatrix} - 3(+1)\begin{vmatrix} 2 & 4 \\ -3 & 7 \end{vmatrix}$$

$$= 4(20-7) - 2(10+3) - 3(14+12) = -52.$$

Expansion by the second column gives

$$D = 2(-1)\begin{vmatrix} 2 & 1 \\ -3 & 5 \end{vmatrix} + 4(+1)\begin{vmatrix} 4 & -3 \\ -3 & 5 \end{vmatrix} + 7(-1)\begin{vmatrix} 4 & -3 \\ 2 & 1 \end{vmatrix}$$

$$= -2(10+3) + 4(20-9) - 7(4+6) = -52, \text{ as before.}$$

In these expansions the factors $(+1)$ and (-1) are $(-1)^{i+j}$, being part of each cofactor. After a little practice, D may be expanded mentally about the third column, say, as

$$D = -3\times26 - 1\times34 + 5\times12 = -52.$$

The sign of $(-1)^{i+j}$ in the cofactor of each element is usually selected by visualising the scheme

$$\begin{vmatrix} + & - & + \\ - & + & - \\ + & - & + \end{vmatrix}$$

for a third-order determinant; its extension for determinants of other orders is immediate.

Problems 1.3

Evaluate each of the following determinants in at least two ways.

1. $\begin{vmatrix} 10 & -7 \\ -3 & 8 \end{vmatrix}.$

2. $\begin{vmatrix} 1+2i & 3-4i \\ 3+4i & 1-2i \end{vmatrix}$ $(i^2 = -1).$

3. $\begin{vmatrix} 1 & -2 & 3 \\ -4 & 5 & -6 \\ 7 & -8 & 9 \end{vmatrix}.$

4. $\begin{vmatrix} 0 & -\sin\theta & \cos\theta \\ 1 & 0 & 0 \\ 0 & \cos\theta & \sin\theta \end{vmatrix}.$

5. $\begin{vmatrix} 4 & 5 & -2 & 1 \\ 0 & 2 & 3 & 5 \\ 7 & 8 & 6 & -1 \\ 9 & 0 & -4 & 3 \end{vmatrix}.$

1.4. PROPERTIES OF DETERMINANTS

Work with determinants can often be made easier and more satisfying by using a selection of their properties. These properties show how certain relations between elements or between rows or columns may affect the value of a determinant, and they often suggest, through inspection, the most economical method of evaluation.

(a) If each row of a determinant is interchanged with its respective column, or conversely, then the value of the determinant obtained equals that of the original determinant.

For example $\begin{vmatrix} 1 & 2 \\ 3 & 4 \end{vmatrix} = \begin{vmatrix} 1 & 3 \\ 2 & 4 \end{vmatrix} = -2.$

This operation is called transposition and the resulting determinant is called the *transpose*. This property implies that a general statement about the rows of a determinant applies also to the columns. All the properties listed below apply to rows and to columns but they are expressed in terms of rows only, it being understood that they apply equally well to columns.

(b) A factor which is common to each element in a row of a determinant may be extracted from that row to multiply the remaining determinant.

For example $\begin{vmatrix} 0 & 1 & 2 \\ 2 & 4 & 6 \\ 3 & 5 & 3 \end{vmatrix} = 2 \begin{vmatrix} 0 & 1 & 2 \\ 1 & 2 & 3 \\ 3 & 5 & 3 \end{vmatrix},$

where the factor 2 is extracted from the second row. This operation may be applied simultaneously to any number of rows and columns.

For example $\begin{vmatrix} 12 & 0 & 6 \\ -2 & 3 & 7 \\ 8 & 9 & -1 \end{vmatrix} = 6 \times 2 \times 3 \begin{vmatrix} 1 & 0 & 1 \\ -1 & 1 & 7 \\ 4 & 3 & -1 \end{vmatrix}$

where factors are taken from the first row and the first and second columns. The converse process applies also, namely that a multiple of a determinant may be included as the factor of any one row or any one column.

(c) If each element of a row is zero then the value of the determinant is zero.

The validity of this is evident from the expansion of such a determinant by its row of zero elements and from property (b) with zero as the common factor.

(d) If each element in a row is expressed as the sum of two terms then the determinant may be expressed as the sum of two determinants.

For example $\begin{vmatrix} 5 & 4 \\ -3 & 1 \end{vmatrix} = \begin{vmatrix} 2+3 & 4 \\ 1-4 & 1 \end{vmatrix} = \begin{vmatrix} 2 & 4 \\ 1 & 1 \end{vmatrix} + \begin{vmatrix} 3 & 4 \\ -4 & 1 \end{vmatrix}.$

The validity of this property is evident from expansion by the row or column whose elements are expressed as binomials. This property may be extended to elements which are expressed as sums of more than two terms.

(e) If any two rows are interchanged then the value of the resulting determinant is equal numerically but is opposite in sign to that of the original determinant.

For example $\begin{vmatrix} 2 & 3 & 7 \\ 4 & -1 & 2 \\ -3 & 5 & 4 \end{vmatrix} = - \begin{vmatrix} 7 & 3 & 2 \\ 2 & -1 & 4 \\ 4 & 5 & -3 \end{vmatrix} = 25,$

where the first and third columns are interchanged.

(f) If corresponding elements of two rows are equal or proportional then the value of the determinant is zero.

If respective elements are proportional, then by (b) a determinant can be formed in which they are equal. Clearly the interchange of two equal rows cannot affect the value of a determinant, but (e) asserts that the value is changed though in sign only. Hence the value is zero.

(g) If to the elements in any row there is added a constant multiple of the respective elements in any other row then the value of the determinant obtained equals that of the original determinant.

Consider $\begin{vmatrix} 1 & 8 & 5 \\ 6 & 2 & 7 \\ 4 & -1 & 3 \end{vmatrix}$ and add twice the third column to the first column to obtain $\begin{vmatrix} 1+10 & 8 & 5 \\ 6+14 & 2 & 7 \\ 4+6 & -1 & 3 \end{vmatrix},$

which by (b) and (d) may be expressed as

$$\begin{vmatrix} 1 & 8 & 5 \\ 6 & 2 & 7 \\ 4 & -1 & 3 \end{vmatrix} + 2 \begin{vmatrix} 5 & 8 & 5 \\ 7 & 2 & 7 \\ 3 & -1 & 3 \end{vmatrix}.$$

But by (f) the second determinant is zero and the result (g) follows. This property is probably the one used most often, mainly to simplify the evaluation of a determinant. Evaluation is easier if the expansion is by a row or a column in which some, preferably all but one, of the elements are zero.

Example 4.1.

Evaluate
$$\begin{vmatrix} 10 & 5 & 1 \\ 3 & 2 & 4 \\ 7 & 6 & 8 \end{vmatrix}.$$

The elements 10 and 5 in the first row, for example, may be converted to zeros by (g). To achieve this subtract twice the second column from the first column and then subtract five times the third column from the second column to obtain

$$\begin{vmatrix} 0 & 0 & 1 \\ -1 & -18 & 4 \\ -5 & -34 & 8 \end{vmatrix}$$

which, by expansion about the first row, gives the value -56 on inspection. It is good practice to show the operations performed on rows and columns by abbreviations such as $c_1' = c_1 - 2c_2$ and $c_2' = c_2 - 5c_3$, where a prime $'$ indicates the newly formed column c or row r.

There is a temptation to apply property (g) in a sense which often seems convenient but is incorrect. The form of (g) is

new row = old row + a multiple of another row

so the symbolic form must always start as

$$r_i' = r_i + \ldots \quad \text{or as} \quad c_j' = c_j + \ldots.$$

2*

Consider, for example, the evaluation of $D = \begin{vmatrix} 4 & 0 & 3 \\ 2 & 6 & -1 \\ 3 & 1 & 2 \end{vmatrix}$,

The leading diagonal element 4 can be reduced to zero by the operation $c_1' = 3c_1 - 4c_3$ giving $D' = \begin{vmatrix} 0 & 0 & 3 \\ 10 & 6 & 7 \\ 1 & 1 & 2 \end{vmatrix}$,

but this is not equivalent to D because (g) has not been applied literally due to the term $3c_1$ which, by (b), multiplies the value of D by 3. The error is easily corrected through the relation $D = D'/3$ and, on inspection, $D = 4$.

The mental manipulation of elements in a determinant is easily subject to error and great care should always be taken. There is little point in applying many manipulations to obtain a few zeros when a straightforward expansion may be quicker and less prone to error. The student is advised to look for ways of simplification but to achieve a balance between the elegance of an evaluation and the effort involved.

Example 4.2.

Evaluate $\qquad D = \begin{vmatrix} 3 & -16 & 30 \\ 3 & -5 & 4 \\ 1 & 2 & -7 \end{vmatrix}$.

$$D = \begin{vmatrix} -3 & -6 & 22 \\ 3 & -5 & 4 \\ 1 & 2 & -7 \end{vmatrix} = \begin{vmatrix} 0 & 0 & 1 \\ 3 & -5 & 4 \\ 1 & 2 & -7 \end{vmatrix} = 11.$$

$$\qquad r_1' = r_1 - 2r_2 \qquad\qquad r_1' = r_1 + 3r_3$$

Note that a prime indicates the most newly formed row or column at each stage. The arrow shows the row about which the final expansion is made. The manipulations shown could have been achieved by the single operation $r_1' = r_1 - 2r_2 + 3r_3$. Probably the most obvious operations are $r_1' = r_1 - 3r_3$ and $r_2' = r_2 - 3r_3$, which produce two zeros in the first column.

An important result, to be used later, is that if each element in any row is multiplied by the cofactor of the respective element in any other row, then the sum of the products is zero. This process is called expansion in terms of *alien cofactors* and the result should be verified for both rows and columns. See **6** of Problems 1.2.

The minors obtained by deletion of one row and one column are called the first minors of a determinant. Second minors of a determinant are formed by deletion of a pair of rows and a pair of columns; third minors are formed by deletion of three rows and three columns. Thus first, second, third,..., $(n-1)$th minors may be formed from a determinant of order n. The *rank* of a determinant is the order of its highest order non-zero minor(s). This is a very important concept to be used in the solution of linear equations by determinants and by matrix algebra.

The rank r of a determinant is evaluated most easily by converting the determinant into a form where the existence of a non-zero minor of order r can be seen clearly. As rank is unaffected by manipulation of rows or columns, such a conversion is achieved by the processes already described. The value of a determinant in which all the elements below, or all the elements above, the leading diagonal are zero is given by the product of the elements in the leading diagonal. This is evident when the expansion of such a determinant is formulated and it is the basis for the determination of rank shown by the following example.

Example 4.3.

Obtain the rank of $\begin{vmatrix} 1 & -2 & -5 & 0 \\ 1 & 0 & 1 & 2 \\ 1 & 2 & 7 & 4 \\ 1 & -1 & -2 & 1 \end{vmatrix}$.

First form a determinant in which all the elements below, or above, the leading diagonal are zero.

$$D = \begin{vmatrix} 1 & -2 & -5 & 0 \\ 0 & 2 & 6 & 2 \\ 0 & 4 & 12 & 4 \\ 0 & 1 & 3 & 1 \end{vmatrix} = \begin{vmatrix} 1 & -2 & -5 & 0 \\ 0 & 2 & 6 & 2 \\ 0 & 0 & 0 & 0 \\ 0 & 0 & 0 & 0 \end{vmatrix}.$$

$$r_i' = r_i - r_1 \qquad\qquad r_3' = r_3 - 2r_2,$$
$$(i = 2, 3, 4) \qquad\qquad r_4' = r_4 - \tfrac{1}{2}r_2$$

As $D = 0$ $r \neq 4$. Clearly a non-zero 3×3 minor cannot be formed so $r \neq 3$, but a non-zero 2×2 minor can be found so $r = 2$. If desired, by manipulation of columns, D may be expressed in a form such as

$$\begin{vmatrix} 1 & 0 & 0 & 0 \\ 0 & 2 & 0 & 0 \\ 0 & 0 & 0 & 0 \\ 0 & 0 & 0 & 0 \end{vmatrix}$$

which shows clearly that $r = 2$.

Sometimes an unknown quantity occurs in an equation involving a determinant which itself includes the unknown. Such an expression is called a determinantal equation and it may be solved by expansion of the determinant.

Example 4.4. Solve the determinantal equations

(a) $\begin{vmatrix} x & -1 \\ 3x+4 & x-2 \end{vmatrix} = 10,$ (b) $\begin{vmatrix} x & x+1 & 3 \\ 2 & x+2 & x \\ x+2 & 1 & x+1 \end{vmatrix} = 0.$

(a) Expansion gives $x^2 + x - 6 = 0$, whence $x = -3, 2$.

(b) Direct expansion leads to a cubic equation in x which may be hard to solve. By properties (b) and (g):

$$D = \begin{vmatrix} 2x+4 & x+1 & 3 \\ 2x+4 & x+2 & x \\ 2x+4 & 1 & x+1 \end{vmatrix} = 2(x+2) \begin{vmatrix} 0 & x & -x+2 \\ 0 & x+1 & -1 \\ 1 & 1 & x+1 \end{vmatrix}$$

$$c_1' = c_1 + c_2 + c_3 \qquad r_1' = r_1 - r_3, \quad r_2' = r_2 - r_3$$

$$= 2(x+2)(x^2 - 2x - 2) = 0,$$

whence $\qquad\qquad x = -2, \quad 1 \pm \sqrt{3}.$

Problems 1.4

1. Evaluate $\begin{vmatrix} -2 & 3 & -4 \\ 7 & -6 & 5 \\ -1 & 8 & -9 \end{vmatrix}$.

2. Evaluate $\begin{vmatrix} 41 & 37 & -6 \\ 43 & 39 & 5 \\ 44 & 40 & 17 \end{vmatrix}$.

3. Show by inspection that

$$\begin{vmatrix} 2 & -5 & -3 \\ 4 & -7 & 6 \\ -6 & 15 & 9 \end{vmatrix} = 0.$$

4. Show by extracting factors of rows and columns, and then by expanding one only second-order determinant, that the value of

$$\begin{vmatrix} 2 & 8 & 4 \\ 10 & 20 & 15 \\ 9 & 24 & 12 \end{vmatrix} \text{ is } 5!.$$

5. Evaluate $\begin{vmatrix} 2 & 3 & 4 & 5 \\ 4 & 8 & 6 & 1 \\ 5 & 4 & 7 & 2 \\ 3 & 4 & 10 & 9 \end{vmatrix}$

by an expansion which involves one only third-order determinant.

6. Evaluate
$$\begin{vmatrix} 2 & 7 & 3 & -2 \\ 3 & 4 & 2 & 4 \\ 3 & 3 & -3 & 0 \\ 4 & 10 & 4 & 5 \end{vmatrix}$$

by first reducing it effectively to a third-order determinant.

7. Evaluate
$$\begin{vmatrix} a+1 & a+2 & a+3 \\ a+5 & a+6 & a+4 \\ a+6 & a+9 & a+8 \end{vmatrix}.$$

8. Show that
$$\begin{vmatrix} a+b & a & b \\ a & a+c & c \\ b & c & b+c \end{vmatrix} = 4abc.$$

9. Show that
$$\begin{vmatrix} a+b & b+c & a+c \\ b+c & a+c & a+b \\ a+c & a+b & b+c \end{vmatrix} = 2 \begin{vmatrix} a & b & c \\ b & c & a \\ c & a & b \end{vmatrix}.$$

10. The letters a, b, c, d, e represent any five consecutive integers. Show that the value of

$$\begin{vmatrix} a^2 & b^2 & c^2 \\ b^2 & c^2 & d^2 \\ c^2 & d^2 & e^2 \end{vmatrix}$$

does not depend on the value of any one of the five integers.

11. Show by inspection that

$$\begin{vmatrix} 2a & a+b & a+c & a+d \\ b+a & 2b & b+c & b+d \\ c+a & c+b & 2c & c+d \\ d+a & d+b & d+c & 2d \end{vmatrix} = 0.$$

12. Show that
$$\begin{vmatrix} (a+3) & (a+4) & 1 & a+4 \\ (a+4) & (a+5) & 1 & a+5 \\ (a+5) & (a+6) & 1 & a+6 \end{vmatrix} = 2.$$

13. Evaluate
$$\begin{vmatrix} a & -a & a & a \\ b & b & b & b \\ c & c & -c & c \\ d & d & -d & -d \end{vmatrix}.$$

14. (a) Evaluate
$$\begin{vmatrix} a & b & c \\ d & e & f \\ g & h & i \end{vmatrix}$$

where $a, b, \ldots, h, i$ are any nine consecutive integers.

(b) Replace each element of the determinant in (a) by its square and show that the value of the resulting determinant is independent of whichever nine integers are chosen.

15. Show that $(a+b+c)$ is a factor of

$$\begin{vmatrix} a & a^2 & b+c \\ b & b^2 & a+c \\ c & c^2 & a+b \end{vmatrix}$$

and, by a suitable expansion, obtain the three other factors directly.

16. Show that

$$\begin{vmatrix} a^3 & a & a^2 & 1 \\ 1 & 1 & 1 & 1 \\ 3a-1 & a+1 & 2a & 2 \\ 3a^2 & 2a+1 & a^2+2a & 3 \end{vmatrix} = (a-1)^6.$$

17. Find the rank of each of the following determinants.

(a) $\begin{vmatrix} 2 & 0 & 2 & -5 \\ 3 & 3 & -5 & 4 \\ 1 & 2 & 2 & 2 \\ 4 & 3 & 7 & -10 \end{vmatrix}$, (b) $\begin{vmatrix} 3 & 0 & 1 & 2 \\ 0 & -4 & 5 & 6 \\ 3 & 1 & 4 & 9 \\ 1 & -3 & 2 & 1 \end{vmatrix}$,

(c) $\begin{vmatrix} 1 & -1 & -2 \\ 5 & 1 & -4 \\ 0 & 2 & 2 \end{vmatrix}$.

18. Obtain the value of x which satisfies the equation

$$\begin{vmatrix} 11-2x & 12 & 10 \\ 5-3x & 18 & 16 \\ 5-x & 15 & 13 \end{vmatrix} = 0.$$

19. Obtain the two values of x which satisfy the equation

$$\begin{vmatrix} 2x+14 & x & x+2 \\ x+7 & 2 & 3 \\ 4 & 2x+13 & 2x+15 \end{vmatrix} = 0.$$

20. Show that the equation

$$\begin{vmatrix} x-5 & 6 & -2 \\ 6 & x-4 & 4 \\ -2 & 4 & x \end{vmatrix} = 0$$

is satisfied by $x = 0$ and find the other roots.

1.5. SOLUTION OF LINEAR EQUATIONS

Determinant notation, though not in its present form, originated in the West with Leibnitz (1693) who used it to express the result of eliminating the n unknowns from $n+1$ simultaneous linear equations. A *linear equation*, or an equation of first degree, is a relation of equality between sums of terms, each term being either a constant or a constant multiple of an unknown of degree one. For example, the equation $x+2y = 3z-4$ is linear in the unknowns x, y and z.

It is instructive to consider first a single equation in one unknown,

$$ax = k. \tag{5.1}$$

Depending upon the nature of the constants, a and k, the solution is one of three distinct types:

(1) $a \neq 0$. There is a unique solution given by $x = k/a$.
(2) $a = 0$.

    (a) If $k \neq 0$ then there is no solution.
    (b) If $k = 0$ then every value of x is a solution.

Distinctions like these can be made about the solution of a system of equations, the number of which may or may not equal the number of unknowns, and the conditions corresponding to the *regular* case (1) and the *singular* case (2) may be expressed in terms of determinants.

If a system of equations has at least one solution then the equations are said to be *consistent*; if there is no solution they are said to be *inconsistent*.

Consider now the system of two linear equations, in the unknowns x and y, expressed in the form

$$\left. \begin{array}{l} a_1x + b_1y = k_1 \\ a_2x + b_2y = k_2 \end{array} \right\}, \tag{5.2}$$

where the letters with suffices represent constants. When at least one of the constants k_1, k_2 is not zero the system is said to be *inhomogeneous*; when both these constants are zero the system is *homogeneous*, because then every term has a similar form. These descriptions apply with similar meaning to a system of any number of equations. The elimination of y and of x between the equations gives the solutions

$$x = \frac{k_1b_2 - k_2b_1}{a_1b_2 - a_2b_1} \quad \text{and} \quad y = \frac{a_1k_2 - a_2k_1}{a_1b_2 - a_2b_1}$$

which may be expressed in the determinant forms

$$x = \begin{vmatrix} k_1 & b_1 \\ k_2 & b_2 \end{vmatrix} \bigg/ \Delta = \Delta_1/\Delta \quad \text{and} \quad y = \begin{vmatrix} a_1 & k_1 \\ a_2 & k_2 \end{vmatrix} \bigg/ \Delta = \Delta_2/\Delta, \tag{5.3}$$

where

$$\Delta \text{ (delta)} = \begin{vmatrix} a_1 & b_1 \\ a_2 & b_2 \end{vmatrix}$$

is the determinant formed from the coefficients of the unknowns when these are set out in the orderly way shown, that is with the x terms all in one column and the y terms all in another column. The numerator determinants Δ_1, Δ_2 in the solutions are each formed from Δ by replacing the column relating to the respective unknown by the column of constants on the right of the equations. Thus the pattern of the solutions in determinant form is remarkably simple and it enables them to be obtained by inspection. For example, it should be clear, from a mental evaluation of the determinants involved, that the solution of the system

$$\left.\begin{array}{l} x - y = -1 \\ x + 2y = 5 \end{array}\right\} \qquad (5.4)$$

is $x = 3/3 = 1$ and $y = 6/3 = 2$. This corresponds with the regular case (1) where the condition $a \neq 0$ is replaced by $\Delta \neq 0$.

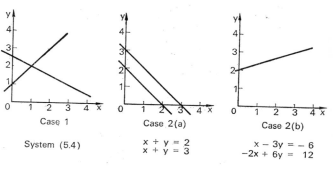

Case 1

System (5.4)

Case 2(a)

$x + y = 2$
$x + y = 3$

Case 2(b)

$x - 3y = -6$
$-2x + 6y = 12$

Fig. 1.1.

Each equation of the system (5.2) represents a straight line; a pair of numbers (x, y) is a solution of the system if the lines intersect at the point (x, y). There are three distinct arrangements of the lines corresponding with the three cases given for the solution of a single equation. These are shown in Fig. 1.1 where a specific example is given for each case.

Case 1 illustrates the general result that if any two lines in a plane are not parallel then their equations have a unique solution and are, therefore, consistent. The lines represented by equations (5.2) are not parallel provided the gradients $-a_1/b_1$ and $-a_2/b_2$ are not equal, that is if

$$a_1 b_2 - a_2 b_1 = \Delta \neq 0.$$

In case 2(a) the lines are parallel so $\Delta = 0$ and their equations are inconsistent in the sense that the relations between x and

y given by each can never be satisfied together. In the extreme case 2(b) the lines are parallel and coincident, there is an infinite set of solutions and, although the equations are consistent, either is redundant because each is merely a multiple of the other. Here $\Delta = 0$ and Δ_1, Δ_2 vanish also, by property (f). The solutions may be expressed in terms of a parameter t as $x = t$ and $y = 2 + t/3$, where t can take any value.

To summarise, the three cases to consider for the solution of the inhomogeneous system (5.2) are as follows:

(1) The regular case, $\Delta \neq 0$. There is a unique solution, given by (5.3).

(2) The singular case, $\Delta = 0$.

    (a) If Δ_1 and $\Delta_2 \neq 0$ then there is no solution.
    (b) If Δ_1 and $\Delta_2 = 0$ then there is an unlimited number of solutions.

It is worth taking this type of analysis one stage further to consider an inhomogeneous system of three equations in three unknowns, expressed as

$$\left. \begin{array}{l} a_1x + b_1y + c_1z = k_1 \\ a_2x + b_2y + c_2z = k_2 \\ a_3x + b_3y + c_3z = k_3 \end{array} \right\}. \tag{5.5}$$

As each equation represents a plane, the types of solution of the system may be associated with ways in which the planes intersect or fail to intersect. With an obvious extension of the notation used in (5.3) the unique solution of (5.5) arises in the regular case, $\Delta \neq 0$, and is given by

$$x = \Delta_1/\Delta, \quad y = \Delta_2/\Delta, \quad z = \Delta_3/\Delta, \tag{5.6}$$

where all four determinants are of order three. The singular case, $\Delta = 0$, may again be divided into two parts:

(a) If at least one of Δ_1, Δ_2 and $\Delta_3 \neq 0$ then there is no solution.

(b) If Δ_1, Δ_2 and $\Delta_3 = 0$ then there may or may not be solutions.

The distinction between consistency and inconsistency of the equations in system (5.5) can be made geometrically. The equations are consistent if the planes they represent have one point only in common, the regular case, if they have a common line of intersection or if all three coincide. The equations are inconsistent if the planes form a triangular prism, so that the three lines of intersection are parallel, if two are parallel and each is intersected by the third, if two coincide and are parallel to the third or if all three are parallel.

Example 5.1.

Show that the system of equations

$$x+2y+3z = 14,$$
$$2x-y-3z = -7,$$
$$3x+4y+z = 6$$

has a unique solution and then obtain it.

The determinant of the coefficients of x, y and z is

$$\Delta = \begin{vmatrix} 1 & 2 & 3 \\ 2 & -1 & -3 \\ 3 & 4 & 1 \end{vmatrix} = 22 \neq 0,$$

which is the condition for such a system to have a unique solution. Solving for x, the required numerator determinant is

$$\Delta_1 = \begin{vmatrix} 14 & 2 & 3 \\ -7 & -1 & -3 \\ 6 & 4 & 1 \end{vmatrix} = 66$$

and by (5.6) $x = \Delta_1/\Delta = 3$. Although values for y and z may be obtained similarly, a better method is to substitute $x = 3$

in any two of the equations, the first and second, say. These then become

$$2y + 3z = 11,$$
$$y + 3z = 13,$$

which upon subtraction give $y = -2$ and hence $z = 5$. Thus the solution of the system is

$$x = 3, \quad y = -2, \quad z = 5.$$

This should be checked by substitution in the third equation since it was not used for y and z.

Example 5.2.

Examine the solutions of the system

$$x - y + z = k,$$
$$2x - 3y + 4z = 0,$$
$$3x - 4y + 5z = 1$$

when $k \neq 1$ and when $k = 1$.

For this system $\Delta = 0$, so a unique solution does not exist. When $k \neq 1$ Δ_1, Δ_2 and $\Delta_3 \neq 0$, so there is no solution. When $k = 1$, Δ_1, Δ_2 and $\Delta_3 = 0$, so there may be solutions. Clearly the third equation is the sum of the first two, so it provides no extra information and can be discarded. By the same token any one of the equations may be discarded. The elimination of x and then of y between the first two equations gives $y = 2(z + 1)$ and $x = z + 3$. In terms of the parameter t the solutions may be expressed as

$$x = t + 3, \quad y = 2(t + 1), \quad z = t,$$

where t can take any value.

Results similar in form to those considered apply to systems of equations with more than three unknowns, and this indicates how determinants can be used to ascertain the properties of such systems as well as to simplify the derivation of the solutions. The results are now presented in a general form but they are limited to systems in which the number of equations equals the number of unknowns. Methods for solving systems in which these two quantities are not equal depend ultimately on the case of equality, but they will not be discussed here.

A system of n linear equations relating the n unknowns $x_1, x_2, \ldots, x_n$ may be expressed as

$$\left.\begin{array}{r}a_{11}x_1+a_{12}x_2+ \ldots +a_{1n}x_n = k_1 \\ a_{21}x_1+a_{22}x_2+ \ldots +a_{2n}x_n = k_2 \\ \cdot \qquad \cdot \quad \ldots \quad \cdot \qquad \cdot \\ a_{n1}x_1+a_{n2}x_2+ \ldots +a_{nn}x_n = k_n\end{array}\right\} . \tag{5.7}$$

The determinant Δ of all the coefficients a_{ij} is called the determinant of the system. To show that the solution of (5.7) may be expressed in a generalised form of (5.3) and (5.6), multiply equations (5.7) by the cofactors $A_{11}, A_{21}, \ldots, A_{n1}$ respectively and add all the resulting equations together. The result may be arranged in the form

$$\begin{aligned} (a_{11}A_{11} + a_{21}A_{21} + & \ldots + a_{n1}A_{n1})x_1 + \\ (a_{12}A_{11} + a_{22}A_{21} + & \ldots + a_{n2}A_{n1}) x_2 + \ldots \\ + (a_{1n}A_{11} + a_{2n}A_{21} + & \ldots + a_{nn}A_{n1}) x_n \\ = k_1A_{11} + k_2A_{21} + & \ldots + k_nA_{n1}. \end{aligned} \tag{5.8}$$

The coefficient of x_1 is the expansion of Δ by its first column; the sum of terms on the right of the equation is a similar expansion, Δ_1 say, except that each element a_{i1} in the first column of Δ is replaced by the respective constant k_i. The coefficients of all the other unknowns are zero because they are expansions of Δ in terms of alien cofactors. Thus (5.8) may be expressed as $\Delta x_1 = \Delta_1$ whence $x_1 = \Delta_1/\Delta$, provided $\Delta \neq 0$. The solution $x_2 = \Delta_2/\Delta$ follows similarly when equations (5.7) are multiplied by the cofactors $A_{12}, A_{22}, \ldots, A_{n2}$ respectively. The application of this process for each unknown leads to the unique solution of the system

$$x_1 = \Delta_1/\Delta, \quad x_2 = \Delta_2/\Delta, \ldots, \quad x_n = \Delta_n/\Delta; \quad \Delta \neq 0,$$

which is known as *Cramer's rule*.

In the singular case, $\Delta = 0$, if $\Delta_j \neq 0$ for at least one value of $j = 1, 2, \ldots, n$ then the system has no solution, but if $\Delta_j = 0$ for all j then the system may or may not have solutions.

Systems of linear equations which arise in practice usually have a unique solution due to the nature of the physical states which they describe, so in the great majority of applications the regular case is the only one of interest. The following section demonstrates the practical use of determinants in solving equations which arise in some problems concerning electrical circuits.

Problems 1.5

Solve the following systems of linear equations by determinant methods. When a system does not have a solution give a geometrical reason.

1. $3x + 4y = 1,$
$7x + 10y = -1.$

2. $5x - 7y = 2,$
$10x - 14y = 3.$

3. $x - 3y = 2,$
$3x - 9y = 6.$

4. $2x + 7y = 2,$
$3x + 11y = 1.$

5. $3x + 2y - 2z = 1,$
$4x - 3y + z = 1,$
$5x + 2y - 3z = 0.$

6. $x + 4y + 2z = 8,$
$3x + 7y + 5z = 14,$
$2x + 3y + 3z = 6.$

7. $6x + 3y + z = 10,$
$5x + 3y + z = 7,$
$x - y - z = 1.$

8. $3x - 2y + z = 4,$
$6x - 4y + 2z = 5,$
$9x - 6y + 3z = 6.$

9. $x + 4y - 3z = 2,$
$2x + 8y - 6z = 4,$
$3x + 12y + 9z = 6.$

10. $w + x + y + z = 2,$
$3w - 5x + 6y - 2z = -1,$
$7w + 8x + 8y + z = 3,$
$4w - 3x + y + 3z = 3.$

1.6. MESH CURRENT AND NODE VOLTAGE NETWORK ANALYSIS

The application of the two methods for network analysis named in the title of this section depends on two laws due to Kirchhoff. Mesh current analysis is based on the voltage law, which asserts that the algebraic sum of the potential differences in a closed circuit is zero. Figure 1.2 shows a circuit which

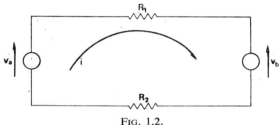

FIG. 1.2.

contains two sources of potential difference, v_a, v_b, and two resistances R_1, R_2. The direction of the current i shown is only assumed. When there are two or more sources in one circuit and not all their directions are the same, positive voltages are assigned to those in the same direction as the assumed current. Thus for the circuit shown the voltage law implies that

$$v_a - R_1 i - v_b - R_2 i = 0$$

or, if potential rises are equated to potential drops, equivalently that

$$v_a - v_b = R_1 i + R_2 i.$$

The mesh current method considers a network as a set of circuits in each of which a mesh, or loop, current is assumed to flow. A network analysed in this way is shown in Fig. 1.3, where there are three mesh currents. Application of the voltage law to each mesh gives an inhomogeneous system of three linear equations in the three unknown currents. The equations are

$$R_1 i_1 + (i_1 - i_2) R_2 = v_a,$$
$$(i_2 - i_1) R_2 + R_3 i_2 + (i_2 - i_3) R_4 = 0,$$
$$(i_3 - i_2) R_4 + R_5 i_3 = -v_b,$$

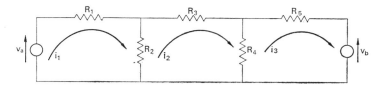

FIG. 1.3.

which when rearranged in the form of (5.5) become

$$\left. \begin{aligned} (R_1 + R_2) i_1 \quad &- \quad R_2 i_2 \quad &&= v_a \\ R_2 i_1 - (R_2 + R_3 + R_4) i_2 \quad &+ \quad R_4 i_3 &&= 0 \\ R_4 i_2 \quad &- (R_4 + R_5) i_3 &&= v_b \end{aligned} \right\} . \quad (6.1)$$

Note that when a resistance, such as R_2 or R_4, has two mesh currents flowing through it, the positive direction of current is associated with that current in the mesh to which the voltage law is being applied. The equations are now in a form suitable for solution by determinants.

Node voltage analysis is based on the current law, which asserts that the algebraic sum of all the currents which meet at a common junction is zero. Figure 1.4 shows a junction where five currents meet. If currents flowing towards the junction are considered positive and those flowing away negative, then the current law implies that

$$i_1 - i_2 + i_3 + i_4 - i_5 = 0.$$

An equivalent form of the law is that at any point of a circuit the sum of the inflowing currents equals the sum of the outflowing currents; thus

$$i_1 + i_3 + i_4 = i_2 + i_5.$$

The network in Fig. 1.3 is shown again in Fig. 1.5 suitably labelled for node analysis. A point in a network common to two or more circuit elements is called a node. If three or more elements join at a node then it is called a principal node or a

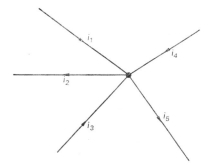

FIG. 1.4.

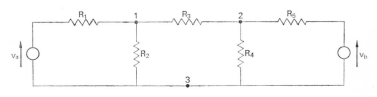

FIG. 1.5.

junction. The nodes marked 1, 2, 3 are all junctions, node 3 being a common point for four circuit elements. The voltage of a node is measured relative to one node in particular, called the reference node. If node 3 is the reference node then v_{13} denotes the voltage between nodes 1 and 3, v_{23} the voltage between nodes 2 and 3. Since in any one analysis the reference node remains the same, the suffix 3 in the example given may be omitted and v_1, v_2 are sufficient to describe the node voltages of nodes 1 and 2.

The node voltage method obtains the voltages at all the principal nodes relative to the reference node. To achieve this the current law is applied to each junction, other than the reference node, and equations for the unknown node voltages follow. Assuming that all branch currents flow away from the node, the current law applied to node 1 gives

$$\frac{v_1 - v_a}{R_1} + \frac{v_1}{R_2} + \frac{v_1 - v_2}{R_3} = 0,$$

and applied to node 2 gives

$$\frac{v_2 - v_1}{R_3} + \frac{v_2}{R_4} + \frac{v_2 - v_b}{R_5} = 0.$$

In terms of the conductance $G = 1/R$ and the form of (5.2), these equations become

$$\left.\begin{array}{rl} (G_1 + G_2 + G_3)\,v_1 \quad - \quad G_3 v_2 &= G_1 v_a \\ -G_3 v_1 + (G_3 + G_4 + G_5)\,v_2 &= G_5 v_b \end{array}\right\}. \qquad (6.2)$$

For both methods there are refinements which can be used to simplify the derivation of the equations. The analysis applies equally well to circuits with complex voltage sources and complex impedances.

Problems 1.6

1. Solve eqns. (6.1) with the values $R_1 = R_2 = R_3 = R_4 = R_5 = 1$ ohm, $v_a = 3$ and $v_b = 2$ V. Solve equations (6.2) also and verify that the solutions of the two sets of equations are mutually consistent.

2. Obtain expressions for the mesh currents I_1, I_2 shown in the network of Fig. 1.6. Evaluate the currents i_1, i_2, i_3 when $R_1 = 1$, $R_2 = 2$, $R_3 = 3$, $R_4 = 4$ ohm and $v = 7$ V.

3. Derive three equations which relate the mesh currents I_1, I_2, I_3 shown in the network of Fig. 1.7. For the values $R_1 = 2$, $R_2 = 4$, $R_3 = 3$, $R_4 = 10$, $R_5 = 6$, $R_6 = 8$ ohm determine the value of v required to produce a current of 7 A through R_4 in the direction of I_2.

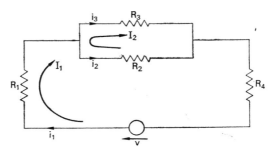

FIG. 1.6.

4. Taking the reference node to be the junction of R_2, R_4 and R_5, derive three equations which relate the voltages v_1, v_2, v_3 at the nodes shown in the network of Fig. 1.8. Evaluate these voltages when $R_1 = 4$, $R_2 = 3$, $R_3 = 2$, $R_4 = 6$, $R_5 = 8$, $R_6 = 1$ ohm and $v_a = 2$, $v_b = 1$ V. Verify that the values obtained imply that the algebraic sum of the currents at the reference node is zero.

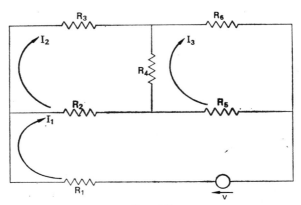

FIG. 1.7.

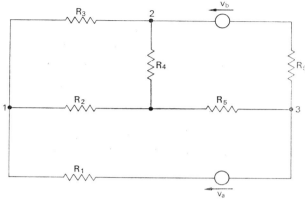

FIG. 1.8.

1.7. HOMOGENEOUS SYSTEMS OF EQUATIONS

So far attention has been confined to inhomogeneous systems, the homogeneous type having been mentioned only when defined. Homogeneous systems really require a separate treatment because, in their many applications, the singular case is far more significant than the regular case.

Consider a pair of homogeneous equations in two unknowns,

$$\left.\begin{array}{l} a_1x+b_1y = 0 \\ a_2x+b_2y = 0 \end{array}\right\}. \tag{7.1}$$

(1) The regular case, $\Delta \neq 0$. The application of Cramer's rule gives $\Delta x = \Delta_1 = 0$ and $\Delta y = \Delta_2 = 0$ whence the only solution is $x = 0$, $y = 0$. As $\Delta_1 = \Delta_2 = 0$ it follows that if the equations are known to have a non-zero solution, that is x and y are not both zero, then $\Delta = 0$.

(2) The singular case, $\Delta = a_1b_2 - a_2b_1 = 0$. Assuming that at least one coefficient is not zero, a_1 say, the condition $\Delta = 0$ implies $b_2 = a_2b_1/a_1$, whence the second equation may be expressed as $(a_2/a_1)(a_1x+b_1y) = 0$, which is merely a multiple

of the first equation. Thus there is effectively one equation only, $x = -b_1y/a_1$, and this has an unlimited number of solutions for each of which the ratio x/y is constant. In terms of the parameter t one possible form of solution is $x = -b_1t/a_1$, $y = t$, where t may take any value.

From the above reasoning follows the important theorem that $\Delta = 0$ is the necessary and sufficient condition for the system (7.1) to have non-zero solutions. Owing to its obvious and usually uninteresting nature, the solution $x = y = 0$ is called the trivial solution. Geometrically the regular case means that the lines represented by the equations are distinct and have a common point at the origin only, whereas in the singular case the lines coincide and have an infinite number of common points. It follows that, as homogeneous equations are always consistent, the important distinction is between zero and non-zero solutions.

If for a system of n homogeneous equations in n unknowns $\Delta \neq 0$, then the only solution is $x_1 = x_2 = \ldots = x_n = 0$. For such a system to have non-zero solutions the necessary and sufficient condition is $\Delta = 0$. An investigation into the nature of the solutions in the singular case is beyond the scope intended here, but the general result can be stated in terms of the rank r of Δ. It is that the n unknowns can be expressed in terms of $n-r$ independent parameters. Also $n-r$ rows, or columns, of Δ can be expressed as linear combinations of r linearly independent other rows, or columns.

Example 7.1.

Obtain the solutions of the following homogeneous systems of equations.

(a) $x - y + z = 0,$ (b) $x - y + 2z = 0,$
     $3x + y - 5z = 0,$ $2x + 5y - 3z = 0,$
     $4x - y - z = 0.$ $3x + 4y - z = 0.$

(c) $-x-y+2z = 0,$ (d) $8x+12y-7z = 0,$

$\quad 2x+2y-4z = 0,$ $\quad 2x+3y+4z = 0,$

$\quad 3x+3y-6z = 0.$ $\quad 4x+6y+8z = 0.$

(a) $\Delta = 4 \neq 0$ so the only solution is $x = y = z = 0$.

(b) $\Delta = 0$ so non-zero solutions exist. The rank r of Δ is 2. Omit $n-r = 3-2 = 1$ equation so that at least one non-zero $r \times r = 2 \times 2$ minor can be derived from the coefficients in the remaining two equations. In this example the omission of any one equation satisfies this condition. Omit the third equation, say, and express the first pair in the form

$$x- y =-2z,$$
$$2x+5y = 3z.$$

This implies that z is being treated as an arbitrary quantity, though this treatment could have been applied equally well to x or to y in this example. Now solve this pair of equations for x and for y in terms of z by Cramer's rule.

$$\Delta = 7, \quad \Delta x = \begin{vmatrix} -2z & -1 \\ 3z & 5 \end{vmatrix} = -7z, \quad \Delta y = \begin{vmatrix} 1 & -2z \\ 2 & 3z \end{vmatrix} = 7z$$

so $x = -z$ and $y = z$, which should be checked in the omitted equation. In terms of $n-r = 1$ independent parameter $z = t$, say, the solution of the system may be expressed as

$$x =-t, \quad y = t, \quad z = t.$$

(c) $\Delta = 0$ so non-zero solutions exist. The rank r of Δ is 1. Omit $n-r = 3-1 = 2$ equations so that the remaining equation has at least one non-zero coefficient. Clearly any pair may be omitted, say the second and third equations. This leaves the solutions $x+y = 2z$, which may be expressed in terms of $n-r = 2$ independent parameters $y = s$ and $z = t$, say, giving

$$x = 2t-s, \quad y = s, \quad z = t.$$

(d) $\Delta = 0$ so non-zero solutions exist. The rank r of Δ is 2. As in system (b), one equation may be omitted but, to satisfy the condition on the 2×2 minors, it may not be the first equation; suppose the third equation is omitted. The remaining pair may be solved by Cramer's rule but if z is treated as an arbitrary quantity, as it was in system (b), the singular case arises. It is preferable, therefore, to treat x or y in this way then, with y arbitrary,

$$\Delta = 46, \quad \Delta x = \begin{vmatrix} -12y & -7 \\ -3y & 4 \end{vmatrix} = 69y, \quad \Delta z = \begin{vmatrix} 8 & -12y \\ 2 & -3y \end{vmatrix} = 0,$$

so the solutions are $2x = -3y$ and $z = 0$ or, in terms of $y = t$,

$$2x = -3t, \quad y = t, \quad z = 0.$$

This formal treatment should not obscure the fairly obvious fact that the solutions to (c) and (d) may be derived by inspection.

Problems 1.7

Solve the following homogeneous systems of equations.

1. $3x - 4y + 5z = 0,$
    $2x + 3y - z = 0,$
    $2x - 3y + 2z = 0.$

2. $x + y + z = 0,$
    $3x + 5y - z = 0,$
    $2x - 4y + 14z = 0.$

3. $5x - 4y + 2z = 0,$
    $7x + 2y - 3z = 0,$
    $3x - 10y + 7z = 0.$

4. $8x - 9y + z = 0,$
    $3x + 2y - z = 0,$
    $6x + 4y - 2z = 0.$

5. $x - 3y + 5z = 0,$
    $-2x + 6y - 10z = 0,$
    $4x - 12y + 20z = 0.$

6. $3x - 2y + z = 0,$
    $6x - 4y - 5z = 0,$
    $12x - 8y + 4z = 0.$

1.8. DEPENDENCE BETWEEN LINEAR EQUATIONS

Two linear functions of two variables, such as

$$f_1 = a_1 x + b_1 y - k_1 \quad \text{and} \quad f_2 = a_2 x + b_2 y - k_2,$$

are said to be *dependent* if there are non-zero constants, l and m, such that the identity

$$lf_1 + mf_2 \equiv 0$$

is satisfied. If such an identity does not exist then the functions f_1, f_2 are said to be *independent*. Similarly, three linear functions f_1, f_2, f_3 are said to be dependent if the identity

$$lf_1 + mf_2 + nf_3 \equiv 0$$

can be satisfied by constants l, m, n which are *not all* zero. This definition may be extended to include any number of linear functions in any number of variables. The constant terms in f_1 and f_2 are associated with a minus sign merely to preserve a consistent notation between sections of this chapter.

Linear dependence is particularly important in relation to linear equations. The two linear equations

$$f_1 = a_1x + b_1y - k_1 = 0$$

and
$$f_2 = a_2x + b_2y - k_2 = 0$$

are said to be dependent or to be independent according to whether the functions f_1 and f_2 are dependent or independent. Similarly the three linear equations

$$f_1 = a_1x + b_1y - k_1 = 0,$$
$$f_2 = a_2x + b_2y - k_2 = 0,$$
$$f_3 = a_3x + b_3y - k_3 = 0$$

are dependent if the functions f_1, f_2, f_3 are dependent. The identity

$$lf_1 + mf_2 + nf_3 \equiv 0 \tag{8.1}$$

implies that the coefficient of x, the coefficient of y and the sum of the constant terms should all vanish. This set of three conditions is the homogeneous system of equations in l, m and n: —

$$a_1l + a_2m + a_3n = 0,$$
$$b_1l + b_2m + b_3n = 0, \tag{8.2}$$
$$-k_1l - k_2m - k_3n = 0.$$

Since at least one of l, m and n may not vanish, the system is required to have a non-zero solution and the condition for this is that

$$\begin{vmatrix} a_1 & a_2 & a_3 \\ b_1 & b_2 & b_3 \\ -k_1 & -k_2 & -k_3 \end{vmatrix} = 0$$

or, transposing and extracting the factor -1, that

$$\begin{vmatrix} a_1 & b_1 & k_1 \\ a_2 & b_2 & k_2 \\ a_3 & b_3 & k_3 \end{vmatrix} = 0.$$

It is important to realise that the dependence of three or more equations in two unknowns does not necessarily imply consistency. For example the equations

$$2x-3y = 1, \quad 4x-6y = 3 \quad \text{and} \quad 2x-3y = 2$$

are clearly inconsistent, because they represent parallel lines and therefore have no common solution, but they are linearly dependent through the relation $1st-2nd+3rd = 0$.

Example 8.1.

Show that the equations

$$2x+3y = -4,$$
$$3x-5y = 13,$$
$$4x+3y = -2$$

are linearly dependent and obtain the dependence.

The determinant

$$\begin{vmatrix} 2 & 3 & -4 \\ 3 & -5 & 13 \\ 4 & 3 & -2 \end{vmatrix}$$

has the value zero, so the equations are dependent. Assume

$n \neq 0$, let $l' = l/n$ and $m' = m/n$ then the identity (8.1) becomes

$$l'f_1 + m'f_2 + f_3 \equiv 0,$$

which is a more convenient form to use. In terms of l', m' and the given equations system (8.2) is

$$2l' + 3m' + 4 = 0,$$
$$3l' - 5m' + 3 = 0,$$
$$4l' - 13m' + 2 = 0,$$

whence $l' = -29/19$ and $m' = -6/19$. The dependence is therefore

$$29f_1 + 6f_2 = 19f_3,$$

where $f_1 = 2x + 3y + 4$, $f_2 = 3x - 5y - 13$ and $f_3 = 4x + 3y + 2$. Ratios such as l/n and m/n can be obtained equally well from (8.2) by keeping it as a homogeneous system and by using methods of the previous section to solve for l, m and n.

Problems 1.8

1. Show that the equations

$$3x + 4y = 1,$$
$$x + 2y = -3,$$
$$9x + 13y = -2$$

are dependent and obtain the dependence.

2. Determine the value of c for which the equations

$$3x - 7y - 4z = 8,$$
$$-2x + 6y + 11z = 21,$$
$$5x - 21y + 7z = 10c,$$
$$x + 23y + 13z = 41$$

are dependent and obtain the dependence.

1.9. CONSISTENCY OF EQUATIONS

Provided $\Delta \neq 0$, an inhomogeneous system of n equations in n unknowns has a unique solution, obtainable by means of Cramer's rule. If the number of equations exceeds the number n of unknowns, then a solution which satisfies all the equations may not exist. If such a solution does exist then the equations are said to be consistent, by definition, but not more than n of them can be independent. The purpose of this section is to establish some tests for the consistency of such systems. The analysis is rather involved and the student is warned that some texts over-simplify the problem and obtain conclusions which are not always valid.

Consider a system of three equations in two unknowns,

$$a_1x + b_1y = k_1,$$
$$a_2x + b_2y = k_2, \qquad (9.1)$$
$$a_3x + b_3y = k_3,$$

and assume that at least one pair of equations has a unique solution. This assumption implies that at least one of the determinants

$$\begin{vmatrix} a_1 & b_1 \\ a_2 & b_2 \end{vmatrix}, \quad \begin{vmatrix} a_1 & b_1 \\ a_3 & b_3 \end{vmatrix} \quad \text{and} \quad \begin{vmatrix} a_2 & b_2 \\ a_3 & b_3 \end{vmatrix}$$

does not vanish. Suppose, without loss of generality, that the first pair of equations has a unique solution, then, for the system to be consistent, this solution must satisfy the third equation also. Substitution of the values for x and y, derived from Cramer's rule, in the third equation gives

$$a_3 \begin{vmatrix} k_1 & b_1 \\ k_2 & b_2 \end{vmatrix} + b_3 \begin{vmatrix} a_1 & k_1 \\ a_2 & k_2 \end{vmatrix} = k_3 \begin{vmatrix} a_1 & b_1 \\ a_2 & b_2 \end{vmatrix}$$

which, as a single determinant E, is

$$E \equiv \begin{vmatrix} a_1 & b_1 & k_1 \\ a_2 & b_2 & k_2 \\ a_3 & b_3 & k_3 \end{vmatrix} = 0.$$

This result follows just as well from the assumption that either of the other two pairs of equations has a unique solution. The condition $E = 0$ is called the *eliminant* of the system, since it is derived from the elimination of the unknowns from the system, though under conditions of assumed consistency.

By analogy with the analysis of system (5.2) the above treatment of system (9.1) may be said to conform with the regular case, meaning that at least one of the three 2×2 minors derivable from the coefficients on the left-hand side of the system does not vanish. In the regular case the condition $E = 0$ is both necessary and sufficient for consistency.

As this treatment progresses, seven different types of system (9.1) will be illustrated by actual equations and a sketch of how the straight lines they represent are related. In the regular case there are two types of consistent systems.

(1) $2x - 3y = -4,$
$\quad\; 3x + 5y = 13,$
$\quad\; 4x - 3y = -2.$

$$E = \begin{vmatrix} 2 & -3 & -4 \\ 3 & 5 & 13 \\ 4 & -3 & -2 \end{vmatrix} = -4(-29) - 13(6) - 2(19) = 0.$$

E is expanded by its third column to show the values, in brackets, of the three 2×2 minors derivable from the coefficients of x and y. Here $E = 0$ and all three 2×2 minors $\neq 0$.

(2) $3x+4y = 9,$
$x+y = 2,$
$2x+2y = 4.$

Two coincident lines

$$E = \begin{vmatrix} 3 & 4 & 9 \\ 1 & 1 & 2 \\ 2 & 2 & 4 \end{vmatrix} = 9(0)-2(-2)+4(-1) = 0.$$

Here $E = 0$ and two of the 2×2 minors $\neq 0$.

In the regular case there are two types of inconsistent systems.

(3) $3x-2y = 4,$
$4x-y = 7,$
$x+y = 8.$

$$E = \begin{vmatrix} 3 & -2 & 4 \\ 4 & -1 & 7 \\ 1 & 1 & 8 \end{vmatrix} = 4(5)-7(5)+8(5) = 25.$$

Here $E \neq 0$ and all three 2×2 minors $\neq 0$.

(4) $3x+3y = 4,$
$2x+2y = 1,$
$x-2y = 3.$

$$E = \begin{vmatrix} 3 & 3 & 4 \\ 2 & 2 & 1 \\ 1 & -2 & 3 \end{vmatrix} = 4(-6)-1(-9)+3(0) = -15.$$

Here $E \neq 0$ and two of the 2×2 minors $\neq 0$.

Again by analogy with system (5.2), the singular case occurs when all three 2×2 minors are zero. The system (9.1) is inconsistent when not all the six 2×2 determinants correspond-ing with $\varDelta_1$ and $\varDelta_2$ vanish, as in (5) and (6).

(5) $x + y = 1,$
$\quad 2x + 2y = 3,$
$\quad 3x + 3y = 5.$

$$E = \begin{vmatrix} 1 & 1 & 1 \\ 2 & 2 & 3 \\ 3 & 3 & 5 \end{vmatrix} = 1(0) - 3(0) + 5(0) = 0.$$

Here $E = 0$, all the 2×2 minors $= 0$ and all six $\varDelta_1$, $\varDelta_2 \neq 0$.

(6) $2x - 3y = \quad 1,$
$\quad 6x - 9y = \quad 3,$
$\quad 4x - 6y = -2.$

Two coincident lines

$$E = \begin{vmatrix} 2 & -3 & 1 \\ 6 & -9 & 3 \\ 4 & -6 & -2 \end{vmatrix} = 1(0) - 3(0) - 2(0) = 0.$$

Here $E = 0$, all the 2×2 minors $= 0$ and four of the $\varDelta_1$, $\varDelta_2 \neq 0$.

When all the determinants corresponding with $\varDelta_1$ and $\varDelta_2$ vanish the system is consistent but there is an unlimited number of solutions, as in (7).

(7) $x - y = \quad 1,$
$\quad 2x - 2y = \quad 2,$
$\quad -x + y = -1.$

Three coincident lines

$$E = \begin{vmatrix} 1 & -1 & 1 \\ 2 & -2 & 2 \\ -1 & 1 & -1 \end{vmatrix} = 1(0) - 2(0) - 1(0) = 0.$$

Here $E = 0$, all the 2×2 minors $= 0$ and all the $\varDelta_1$, $\varDelta_2 = 0$.

This analysis shows that the system (9.1) is consistent and has a *unique* solution if $E = 0$ and at least one of the three 2×2 minors derivable from the coefficients of x and y does not vanish.

A complete analysis of consistency is given later by means of matrix algebra.

Problems 1.9

1. Obtain the two values of λ which satisfy the eliminant of the system

$$
\begin{aligned}
(9-\lambda)x - 2y + 3 &= 0, \\
3x + (2-\lambda)y + 9 &= 0, \\
5x - 10y + (23-\lambda) &= 0.
\end{aligned}
$$

Show that for both values the system is consistent and obtain the solution in each case, giving a graphical interpretation.

2. Obtain the two values of λ which satisfy the eliminant of the system

$$
\begin{aligned}
(9-\lambda)x - y + 4 &= 0, \\
9x + (3-\lambda)y + 12 &= 0, \\
6x - 2y + (11-\lambda) &= 0.
\end{aligned}
$$

Show that for one value only the system is consistent and obtain the solution in this case. Give graphical interpretations of both cases.

1.10. PERMUTATIONS

The expansion of the third-order determinant

$$
\begin{vmatrix}
a_1 & b_1 & c_1 \\
a_2 & b_2 & c_2 \\
a_3 & b_3 & c_3
\end{vmatrix}
$$

is

$$
a_1b_2c_3 - a_1b_3c_2 - a_2b_1c_3 + a_2b_3c_1 + a_3b_1c_2 - a_3b_2c_1.
$$

This form follows directly from expansion by the first column and it is preferred here because alphabetical order of the letters is preserved in each term. Consider the order of the number suffices in each term. In the first term the natural order 1, 2, 3 occurs. In the second term the order 1, 3, 2 can be made natural by interchanging 2 with 3; such a change is called an inversion. The suffices in the fourth term require two successive inversions to restore the natural order. When all six terms are considered in this way it is found that those associated with a plus sign require an even number of inversions, zero being taken as even, and those with a minus sign require an odd number of inversions to restore the natural order. The six arrangements of the three numbers are the 3! possible permutations of the numbers taken three at a time. Thus there are simple rules for defining the factors of each term and the sign to be associated with each term.

These considerations may be generalised and applied to a determinant of any order. Accordingly, in terms of an obvious extension of the above scheme, the value of a determinant of order n may be defined as the sum of all the $n!$ products which can be formed by keeping the n letters a, b, c, d, ... in alphabetical order and by attaching to them suffices corresponding to the permutations of the n numbers 1, 2, 3, 4, ..., n, each product being signed according to the number of inversions needed to restore the natural order. The properties of determinants and the methods of expansion follow from this definition.

1.11. OPERATIONS ON DETERMINANTS

FACTORISATION

Factors of a determinant may often be obtained from an inspection of its elements.

Example 11.1.

By inspection factorise $D \equiv \begin{vmatrix} 1 & s & s^2 \\ 1 & t & t^2 \\ 1 & u & u^2 \end{vmatrix}$.

4*

Suppose that $s = t$, then $r_1 = r_2$ and $D = 0$. This implies that $(s-t)$ must be a factor of D to ensure that D vanishes when $s = t$. Similar reasoning shows that $(t-u)$ and $(s-u)$ are factors of D. The expansion of D is of degree 3 in s, t, u and this is produced also by the product of the three factors, showing that $D = k(s-t)(t-u)(s-u)$, where k is a constant. When coefficients of the leading term tu^2, say, are compared, it is seen that $k = -1$ and hence that D may be expressed in the factorised form

$$D = (s-t)(t-u)(u-s).$$

MULTIPLICATION

Clearly the product of two determinants may be obtained as the product of their expansions; also determinants which are factors of a product need not be of the same order. The purpose now is to show how the product of two determinants may be expressed immediately in determinant form. Assume that the determinants are of the same order and consider the product

$$D_1 \times D_2 \equiv \begin{vmatrix} a_{11} & a_{12} \\ a_{21} & a_{22} \end{vmatrix} \times \begin{vmatrix} b_{11} & b_{12} \\ b_{21} & b_{22} \end{vmatrix} = D_3.$$

A determinant D_3 of order two equal to the product $D_1 \times D_2$ may be obtained in each of the four similar ways shown below.

Row into column

The ijth element of D_3, that is the element in the ith row and jth column, is formed by summing the products of the elements in the ith row of D_1 with the respective elements in the jth column of D_2. By this process

$$D_3 = \begin{vmatrix} a_{11}b_{11}+a_{12}b_{21} & a_{11}b_{12}+a_{12}b_{22} \\ a_{21}b_{11}+a_{22}b_{21} & a_{21}b_{12}+a_{22}b_{22} \end{vmatrix}.$$

This way of forming the product determinant is given first because it is the only one of the four which applies to the formation of matrix products and for this reason is the best one to practice. The other three ways, being similar in form, should be described sufficiently by their titles and the form of D_3 now given for each.

Row into row: $\qquad D_3 = \begin{vmatrix} a_{11}b_{11}+a_{12}b_{12} & a_{11}b_{21}+a_{12}b_{22} \\ a_{21}b_{11}+a_{22}b_{12} & a_{21}b_{21}+a_{22}b_{22} \end{vmatrix}.$

Column into row: $\quad D_3 = \begin{vmatrix} a_{11}b_{11}+a_{21}b_{12} & a_{11}b_{21}+a_{21}b_{22} \\ a_{12}b_{11}+a_{22}b_{12} & a_{12}b_{21}+a_{22}b_{22} \end{vmatrix}.$

Column into column: $\; D_3 = \begin{vmatrix} a_{11}b_{11}+a_{21}b_{21} & a_{11}b_{12}+a_{21}b_{22} \\ a_{12}b_{11}+a_{22}b_{21} & a_{12}b_{12}+a_{22}b_{22} \end{vmatrix}.$

To demonstrate the validity of D_3 express the first form given as the sum of four determinants by property (d).

$$D_3 = \begin{vmatrix} a_{11}b_{11} & a_{11}b_{12} \\ a_{21}b_{11} & a_{21}b_{12} \end{vmatrix} + \begin{vmatrix} a_{11}b_{11} & a_{12}b_{22} \\ a_{21}b_{11} & a_{22}b_{22} \end{vmatrix} + \begin{vmatrix} a_{12}b_{21} & a_{11}b_{12} \\ a_{22}b_{21} & a_{21}b_{12} \end{vmatrix}$$
$$+ \begin{vmatrix} a_{12}b_{21} & a_{12}b_{22} \\ a_{22}b_{21} & a_{22}b_{22} \end{vmatrix}.$$

By property (f) the first and last determinants are zero so, by applying properties (a) and (b) to the two remaining determinants, D_3 has the form

$$D_3 = (b_{11}b_{22}-b_{21}b_{12}) \begin{vmatrix} a_{11} & a_{12} \\ a_{21} & a_{22} \end{vmatrix} = D_2 \times D_1.$$

These methods may be applied to determinants of any order.

The product of two determinants whose orders are unequal may be obtained by a simple modification to the one of smaller order. For example the product determinant of

$$\begin{vmatrix} 1 & 3 & 2 \\ 7 & 4 & 5 \\ 3 & 2 & 1 \end{vmatrix} \times \begin{vmatrix} 8 & 3 \\ 7 & 4 \end{vmatrix}$$

may be obtained by any one of the four methods when the second-order determinant is replaced by either of the equivalent forms.

$$\begin{vmatrix} 8 & 3 & 0 \\ 7 & 4 & 0 \\ 0 & 0 & 1 \end{vmatrix} \quad \text{or} \quad \begin{vmatrix} 1 & 0 & 0 \\ 0 & 8 & 3 \\ 0 & 7 & 4 \end{vmatrix}.$$

Since the product of two determinants is really the product of two numbers, or expressions representing numbers, the process does not depend on the order in which the factor determinants are placed.

DIFFERENTIATION

If in a determinant of order n several or all elements are functions of a variable, then the derivative of the determinant with respect to that variable may be expressed as the sum of n determinants each of order n. These are similar to the original determinant except that each in turn has a different row, or column, containing the derivatives of the elements in the original row, or column. The form of such an expression follows from the way in which derivatives of products are formed.

Example 11.2.

Obtain the derivative of $D = \begin{vmatrix} x^2 & 1 & 2 \\ 3 & e^x & 1 \\ -1 & 0 & x \end{vmatrix}$.

By columns
$$\frac{dD}{dx} = \begin{vmatrix} 2x & 1 & 2 \\ 0 & e^x & 1 \\ 0 & 0 & x \end{vmatrix} + \begin{vmatrix} x^2 & 0 & 2 \\ 3 & e^x & 1 \\ -1 & 0 & x \end{vmatrix}$$

$$+ \begin{vmatrix} x^2 & 1 & 0 \\ 3 & e^x & 0 \\ -1 & 0 & 1 \end{vmatrix} = (x^3 + 3x^2 + 2)e^x - 3.$$

Problems 1.11

1. By inspection factorise $\begin{vmatrix} 1 & 1 & 1 \\ s & t & u \\ tu & us & st \end{vmatrix}$.

2. Multiply together, in the order shown, the determinants

$$\begin{vmatrix} 2 & 3 & 1 \\ -1 & 0 & 2 \\ 5 & 3 & 1 \end{vmatrix}, \quad \begin{vmatrix} 0 & 2 & 3 \\ 1 & 4 & 7 \\ 6 & 4 & 2 \end{vmatrix}$$

and evaluate the product determinant. Verify that a change in the order does not affect the value of the product.

3. Express the square of the determinant

$$D = \begin{vmatrix} 0 & b & a \\ a & 0 & c \\ b & c & 0 \end{vmatrix}$$

as one of order three. By inspection show that c is a factor of D and of D^2.

4. Obtain $\dfrac{d^2}{dx^2} \begin{vmatrix} xe^x & \sin x \\ \cos x & x \log x \end{vmatrix}$ by direct expansion first and also by differentiating the determinant twice.

5. Determine the value of c such that $\int_0^1 f(x)\, dx = 0$, where

$$f(x) = \begin{vmatrix} x^2 & x & c \\ 2 & 3 & 6 \\ 3 & 4 & 6 \end{vmatrix}.$$

CHAPTER 2

Matrix Algebra and Linear Equations

2.1. BASIC NOTATION AND DEFINITIONS

A matrix of order $m \times n$ (m by n) is an arrangement of mn elements in m rows and n columns. The rectangular or square array so formed is shown between a pair of curved or square brackets. Thus a typical matrix of general form is

$$\begin{pmatrix} a_{11} & a_{12} & \ldots & a_{1n} \\ a_{21} & a_{22} & \ldots & a_{2n} \\ \vdots & \vdots & \vdots & \vdots \\ a_{m1} & a_{m2} & \ldots & a_{mn} \end{pmatrix}.$$

The elements may be real or complex numbers; if they are real the matrix is called a *real matrix*.

There are several ways in which a full matrix array can be abbreviated, and the suitability of each depends on the context. The most common way is by a single letter, often **A**, **B**, **C** or **a**, **b**, **c**. Here this kind of symbolism always appears in bold type, though the practice is not common to all texts. Other ways show the general element a_{ij}, in the ith row and jth column, between square or curved brackets as $[a_{ij}]$, (a_{ij}) or, by omitting the suffices, simply as $[a]$, (a).

A matrix does not have a value in the sense that a determinant has a value. The value of a matrix is no more than the array of elements, there being no implication that these are combined in any algebraic way. Whereas work with determinants involves the familiar algebra of numbers, work with matrices is based on an algebra which defines how arrays are combined in the sum, difference, product and quotient senses. In this section many of the terms to be used are described and the algebra of matrix addition is defined.

A matrix for which $m = 1$, $n > 1$ has one row only and is called a *row matrix* or a *row vector*. $\mathbf{a} = [a_1\ a_2\ \ldots\ a_n]$ and $\mathbf{k} = (3\ 0\ 2\ -1)$ are row vectors. Likewise a matrix for which $n = 1$, $m > 1$ has one column only and is called a *column matrix* or a *column vector*. $\mathbf{a} = \begin{bmatrix} a_1 \\ a_2 \\ \vdots \\ a_m \end{bmatrix}$ and $\mathbf{k} = \begin{pmatrix} 4 \\ 1 \\ 2 \\ 3 \end{pmatrix}$ are column vectors, but in contexts where space can be saved such an array is shown often as a row of elements between curly brackets, as in {4 1 2 3}, the nature of the brackets implying that the matrix is of column type. When a single letter denotes a row or column matrix it is usually in lower case; $\mathbf{a}$, $\mathbf{b}$, $\mathbf{k}$, $\mathbf{x}$, $\mathbf{y}$ for example.

A matrix of order $m \times n$ in which every element is zero is called the *zero matrix* or *null matrix* of order $m \times n$ and is denoted by $[0]_{mn}$; in most contexts the symbol $\mathbf{0}$ is adequate.

An array where $m = n$, $m > 1$, is called a *square matrix* of order n, or an n-square matrix.

A *diagonal matrix* is a square matrix in which all elements not in the leading diagonal are zero. It is defined by $a_{ij} = 0$, $i \neq j$, and this does not restrict the values which may be taken by the elements in the leading diagonal.

The diagonal matrix of order n in which each element of the leading diagonal is unity is called the *unit* or *identity matrix* of order n and is denoted by $\mathbf{I}$, the order being apparent from the context.

A *triangular matrix* is a square matrix in which all the elements above, or below, the leading diagonal are zero.

The relation of equality can exist between two matrices only if they are of the same order. Two such matrices $A = (a_{ij})$ and $B = (b_{ij})$ of order $m \times n$ are *equal* when corresponding elements are equal, that is when

$$a_{ij} = b_{ij} \quad \text{for all} \quad i = 1, \ldots, m \quad \text{and} \quad j = 1, \ldots, n.$$

Equality of matrices is expressed by

$$A = B.$$

When two matrices are mutually suitable for combination by an algebraic operation they are said to be *conformable* for that operation. Two matrices are conformable for *addition* when their orders are equal. The operation of addition on two $m \times n$ matrices $A = (a_{ij})$ and $B = (b_{ij})$ produces their *sum* which is defined as the $m \times n$ matrix $C = (c_{ij})$ whose elements are

$$c_{ij} = a_{ij} + b_{ij}, \quad i = 1, \ldots, m \quad \text{and} \quad j = 1, \ldots, n.$$

The matrix sum is expressed as

$$C = A + B.$$

Multiplication of a matrix $A = (a_{ij})$ by a number c is defined to give the product matrix (ca_{ij}) which is denoted by cA or Ac, each element of A being multiplied by c. The *difference* D of two matrices A and B conformable for addition is defined by

$$D = A + (-1)B = A - B = (a_{ij} - b_{ij}).$$

Example 1.1.

(a) Let $A = \begin{pmatrix} -3 & 0 & 1 \\ 2 & 5 & -4 \end{pmatrix}$ and $B = \begin{pmatrix} 3 & 1 & -1 \\ 7 & 0 & 2 \end{pmatrix}$,

then $A + B = \begin{pmatrix} 0 & 1 & 0 \\ 9 & 5 & -2 \end{pmatrix}$ and $A - B = \begin{pmatrix} -6 & -1 & 2 \\ -5 & 5 & -6 \end{pmatrix}$.

(b) Let $\mathbf{A} = \begin{pmatrix} 1 & 2 \\ 3 & 4 \end{pmatrix}$ and $\mathbf{B} = \begin{pmatrix} 7 & -5 \\ 3 & -2 \end{pmatrix}$,

then $7\mathbf{A} - 2\mathbf{B} = \begin{pmatrix} 7 & 14 \\ 21 & 28 \end{pmatrix} - \begin{pmatrix} 14 & -10 \\ 6 & -4 \end{pmatrix} = \begin{pmatrix} -7 & 24 \\ 15 & 32 \end{pmatrix}$.

As the addition of matrices is based directly on the addition of their elements, which are numbers, the laws that apply to addition in ordinary algebra apply also to the addition of matrices. Thus if $\mathbf{A}$, $\mathbf{B}$, $\mathbf{C}$ are any matrices of the same order and a, b are any constant numbers, then the validity of the following relations should be evident.

$$\mathbf{A} + \mathbf{B} = \mathbf{B} + \mathbf{A} \qquad \text{(commutation)},$$
$$\mathbf{A} + (\mathbf{B} + \mathbf{C}) = (\mathbf{A} + \mathbf{B}) + \mathbf{C} \qquad \text{(association)},$$
$$a(\mathbf{A} + \mathbf{B}) = a\mathbf{A} + a\mathbf{B} \qquad \text{(distribution)},$$
$$(a + b)\mathbf{A} = a\mathbf{A} + b\mathbf{A} \qquad \text{(distribution)},$$
$$a(b\mathbf{A}) = (ab)\mathbf{A} \qquad \text{(association)}.$$

An important operation is transposition. The *transpose* of an $m \times n$ matrix $\mathbf{A} = (a_{ij})$ is defined as the $n \times m$ matrix (a_{ji}) and is denoted by $\mathbf{A}'$. Thus the operation simply interchanges rows with columns, as with determinants.

A square matrix $\mathbf{A} = (a_{ij})$ which is equal to its transpose is said to be *symmetric*, that is

$$\mathbf{A} = \mathbf{A}' \quad \text{or} \quad a_{ij} = a_{ji}, \qquad i, j = 1, \ldots, n.$$

A square matrix $\mathbf{A} = (a_{ij})$ is said to be *skew-symmetric* if

$$\mathbf{A} = -\mathbf{A}' \quad \text{or} \quad a_{ij} = -a_{ji}, \qquad i, j = 1, \ldots, n,$$

which implies that the elements in the leading diagonal are zero. Any square matrix $\mathbf{A}$ may be expressed as the sum $\mathbf{S} + \mathbf{U}$ of a symmetric matrix $\mathbf{S}$ and a skew-symmetric matrix $\mathbf{U}$ where these are defined by

$$\mathbf{S} = \tfrac{1}{2}(\mathbf{A} + \mathbf{A}') \quad \text{and} \quad \mathbf{U} = \tfrac{1}{2}(\mathbf{A} - \mathbf{A}').$$

Problems 2.1

1. Show that $(\mathbf{A}+\mathbf{B})' = \mathbf{A}'+\mathbf{B}'$, $(a\mathbf{A})' = a\mathbf{A}'$ and $(\mathbf{A}')' = \mathbf{A}$.

2. Given $\mathbf{A} = \begin{pmatrix} -1 & 2 & 0 \\ 3 & 5 & -4 \end{pmatrix}$, $\mathbf{B} = \begin{pmatrix} 6 & 1 & -7 \\ 4 & 5 & 2 \end{pmatrix}$,

$\mathbf{C} = \begin{pmatrix} -15 & 4 & 14 \\ 1 & 5 & -16 \end{pmatrix}$,

obtain (a) $\mathbf{A}+\mathbf{B}+\mathbf{C}$, (b) $\mathbf{A}-\mathbf{B}'+\mathbf{C}$, (c) $\frac{1}{2}(5\mathbf{A}+\mathbf{C})'$, (d) $(\mathbf{A}'+2\mathbf{B}')'$, (e) $3\mathbf{A}-2\mathbf{B}-\mathbf{C}$. Give the reason if any one of the suggested combinations is not defined.

3. According to the definitions is a null square matrix (a) diagonal, (b) triangular, (c) symmetric, (d) skew-symmetric?

4. Let $\mathbf{A} = \begin{pmatrix} 1 & 0 & 2 \\ -3 & 1 & 4 \\ -2 & 6 & 9 \end{pmatrix}$, $\mathbf{B} = \begin{pmatrix} 3 & 5 & -2 \\ 0 & 1 & 4 \\ 7 & 2 & -3 \end{pmatrix}$, $\mathbf{C} = \begin{pmatrix} 2 & -1 & 0 \\ 6 & 2 & 4 \\ 5 & -2 & -6 \end{pmatrix}$.

(a) Obtain $\mathbf{A}+\mathbf{B}'+\mathbf{C}$.
(b) Represent $\mathbf{C}$ as the sum of two triangular matrices.
(c) Express $\mathbf{A}$ as the sum of a symmetric matrix and a skew-symmetric matrix.
(d) Express $(\mathbf{A}+\mathbf{C})$ as the sum of $3\mathbf{I}$ and two triangular matrices, where $\mathbf{I}$ is the 3×3 unit matrix.
(e) Obtain a triangular matix, with zero elements above the leading diagonal, which when added to $\mathbf{B}$ produces a symmetric matrix in which all elements on the leading diagonal have the value 2.

5. Determine constants a, b, c such that

$$aA + bB + cC = \begin{pmatrix} 1 & 13 & -2 \\ -21 & -3 & 0 \\ -3 & 16 & 21 \end{pmatrix},$$

where $\mathbf{A}$, $\mathbf{B}$, $\mathbf{C}$ are defined in **4**.

2.2. MATRIX PRODUCTS

The way in which two numbers are multiplied together does not suggest immediately how the product of two number arrays should be formed. The algebra of matrices is designed to describe, in a concise form, systems of linear equations and relations between them. A study of how two such systems can be related in terms of number algebra should indicate the basis for a consistent relation in terms of matrix algebra.

Suppose the positions of points in a plane are defined by three different coordinate systems called the x_1, x_2 system, the y_1, y_2 system and the z_1, z_2 system. Let the systems be related linearly by

$$x_1 = a_{11}y_1 + a_{12}y_2 \atop x_2 = a_{21}y_1 + a_{22}y_2 \Bigr\} \qquad (2.1)$$

and

$$y_1 = b_{11}z_1 + b_{12}z_2 \atop y_2 = b_{21}z_1 + b_{22}z_2 \Bigr\} . \qquad (2.2)$$

Equations (2.2) relate a given point, defined in terms of the z system as (z_1, z_2), to the y system in which the point is (y_1, y_2). Similarly equations (2.1) relate the same point (y_1, y_2) to the x system, in which it is (x_1, x_2). The substitution of (2.2) in (2.1) eliminates the y system explicitly and the result expresses the x system in terms of the z system directly by

$$x_1 = c_{11}z_1 + c_{12}z_2 \atop x_2 = c_{21}z_1 + c_{22}z_2 \Bigr\} , \qquad (2.3)$$

where

$$c_{11} = a_{11}b_{11} + a_{12}b_{21}, \quad c_{12} = a_{11}b_{12} + a_{12}b_{22} \atop c_{21} = a_{21}b_{11} + a_{22}b_{21}, \quad c_{22} = a_{21}b_{12} + a_{22}b_{22} \Bigr\} . \qquad (2.4)$$

Let the column vectors $\mathbf{x}$, $\mathbf{y}$, $\mathbf{z}$ be $\{x_1\ x_2\}$, $\{y_1\ y_2\}$, $\{z_1\ z_2\}$ respectively and let $\mathbf{A}$, $\mathbf{B}$, $\mathbf{C}$ be the 2×2 matrices formed from the coefficients in (2.1), (2.2), (2.3) so that

$$\mathbf{A} = \begin{pmatrix} a_{11} & a_{12} \\ a_{21} & a_{22} \end{pmatrix}, \quad \mathbf{B} = \begin{pmatrix} b_{11} & b_{12} \\ b_{21} & b_{22} \end{pmatrix}, \quad \mathbf{C} = \begin{pmatrix} c_{11} & c_{12} \\ c_{21} & c_{22} \end{pmatrix}. \quad (2.5)$$

Given a suitable definition for the product of matrices, systems (2.1), (2.2) and (2.3) may be expressed in concise matrix form as

$$\mathbf{x} = \mathbf{A}\mathbf{y}, \quad \mathbf{y} = \mathbf{B}\mathbf{z} \quad \text{and} \quad \mathbf{x} = \mathbf{C}\mathbf{z}$$

whence the elimination of **y** through

$$\mathbf{x} = \mathbf{Ay} = \mathbf{A(Bz)} = (\mathbf{AB})\,\mathbf{z} = \mathbf{Cz},$$

shows that the product **AB** should be identifiable with **C** if the definition is to be consistent with result (2.4) derived from the algebra of numbers. These matrix equations agree completely with the expanded forms above if multiplication of a matrix by a matrix is defined according to the "row into column" process described for determinants in section 1.11. These ideas are now placed on a formal basis.

Let **x**, **y** be two vectors with components $x_1, x_2, \ldots, x_n$ and $y_1, y_2, \ldots, y_n$. The sum of products

$$x_1 y_1 + x_2 y_2 + \ldots + x_n y_n,$$

denoted by $\mathbf{x} \cdot \mathbf{y}$, is called the *scalar product* of the two vectors and it is defined only when the vectors have the same number of components.

Two matrices **A** and **B** are conformable for the product **AB**, in this order, when the number of columns in **A** equals the number of rows in **B**.

The *product* **AB** is defined as the matrix whose element in the ith row and jth column is the scalar product of the ith row of **A** with the jth column of **B**. The matrix **AB** has as many rows as **A** and as many columns as **B**.

The matrices **A** and **B** in (2.5) may be expressed in the forms

$$\mathbf{A} = \begin{pmatrix} \mathbf{a}_1 \\ \mathbf{a}_2 \end{pmatrix} \quad \text{and} \quad \mathbf{B} = (\mathbf{b}_1 \quad \mathbf{b}_2),$$

where $\mathbf{a}_1 = (a_{11}\ a_{12})$ and $\mathbf{a}_2 = (a_{21}\ a_{22})$ are row vectors representing the rows of **A** and $\mathbf{b}_1 = \{b_{11}\ b_{21}\}$, $\mathbf{b}_2 = \{b_{12}\ b_{22}\}$ are column vectors representing the columns of **B**. **A** and **B** are conformable for the product **AB** and by definition this is

$$\mathbf{AB} = \begin{pmatrix} \mathbf{a}_1 \\ \mathbf{a}_2 \end{pmatrix} (\mathbf{b}_1 \quad \mathbf{b}_2) = \begin{pmatrix} \mathbf{a}_1 \cdot \mathbf{b}_1 & \mathbf{a}_1 \cdot \mathbf{b}_2 \\ \mathbf{a}_2 \cdot \mathbf{b}_1 & \mathbf{a}_2 \cdot \mathbf{b}_2 \end{pmatrix}, \equiv \mathbf{C}$$

by (2.4) and (2.5).

Example 2.1.

Let $$\mathbf{A} = \begin{pmatrix} 2 & 3 & -1 \\ 4 & 0 & 5 \end{pmatrix}, \qquad \mathbf{B} = \begin{pmatrix} 3 & -2 & 1 \\ 0 & 5 & -1 \\ 4 & -6 & 7 \end{pmatrix},$$

$$\mathbf{c} = (1 \quad 4 \quad 5) \quad \text{and} \quad \mathbf{d} = \begin{pmatrix} 3 \\ -1 \\ 4 \end{pmatrix}.$$

Then $$\mathbf{AB} = \begin{pmatrix} 2 & 17 & -8 \\ 32 & -38 & 39 \end{pmatrix}, \qquad \mathbf{Ad} = \begin{pmatrix} -1 \\ 32 \end{pmatrix},$$

$$\mathbf{BB} = \mathbf{B}^2 = \begin{pmatrix} 13 & -22 & 12 \\ -4 & 31 & -12 \\ 40 & -80 & 59 \end{pmatrix}, \qquad \mathbf{Bd} = \begin{pmatrix} 15 \\ -9 \\ 46 \end{pmatrix},$$

$$\mathbf{cB} = (23 \quad -12 \quad 32), \quad \mathbf{cd} = 19, \quad \mathbf{dc} = \begin{pmatrix} 3 & 12 & 15 \\ -1 & -4 & -5 \\ 4 & 16 & 20 \end{pmatrix}.$$

The products $\mathbf{A}^2$, $\mathbf{Ac}$, $\mathbf{BA}$, $\mathbf{Bc}$, $\mathbf{cA}$, $\mathbf{c}^2$, $\mathbf{dA}$, $\mathbf{dB}$ and $\mathbf{d}^2$ are not defined. Note that $\mathbf{cd} \neq \mathbf{dc}$.

Matrix multiplication conforms with the following laws, which apply also to the algebra of numbers.

$$(\mathbf{AB})\mathbf{C} = \mathbf{A}(\mathbf{BC}) \qquad \text{(association)}.$$

This product may be expressed as $\mathbf{ABC}$.

$$\mathbf{A}(\mathbf{B}+\mathbf{C}) = \mathbf{AB}+\mathbf{AC} \quad \text{(distribution)},$$
$$(\mathbf{A}+\mathbf{B})\,\mathbf{C} = \mathbf{AC}+\mathbf{BC} \quad \text{(distribution)}.$$

These statements imply that $\mathbf{A}$, $\mathbf{B}$, $\mathbf{C}$ are conformable for the products indicated and that the order in which the various factors appear is preserved.

There are two respects in which matrix multiplication and its implications differ from the multiplication of numbers.

If **A** and **B** are conformable for the products **AB** and **BA**, then in general

$$\mathbf{AB} \neq \mathbf{BA},$$

that is the law of commutation does not apply. To distinguish between these two products, in the first **A** is said to pre-multiply **B** whereas in the second it is said to post-multiply **B**. The relation

$$\mathbf{AB} = \mathbf{0}$$

does not imply that $\mathbf{A} = \mathbf{0}$ or that $\mathbf{B} = \mathbf{0}$. Similarly, $\mathbf{A}^2 = \mathbf{0}$ does not imply that **A** is a null matrix.

The transpose of a product of matrices in a certain order equals the product of the transposed matrices in the opposite order, that is

$$(\mathbf{ABC})' = \mathbf{C}'\mathbf{B}'\mathbf{A}'.$$

The determinant of a square matrix **A** is defined to have the same respective elements as **A** and is denoted by $|\mathbf{A}|$ or det **A**; if it is zero then the matrix is said to be *singular*, otherwise non-singular or regular. The following relations apply to the product of two square matrices of order n.

$$|\mathbf{AB}| = |\mathbf{A}| \times |\mathbf{B}|, \quad |\mathbf{BA}| = |\mathbf{B}| \times |\mathbf{A}|. \qquad (2.6)$$

As $|\mathbf{A}|$ and $|\mathbf{B}|$ are numbers the commutative law does apply to their product, so $|\mathbf{A}| \times |\mathbf{B}| = |\mathbf{B}| \times |\mathbf{A}|$ showing that

$$|\mathbf{AB}| = |\mathbf{BA}|.$$

If k is a number then

$$|k\mathbf{A}| = k^n |\mathbf{A}|,$$

by property (b) of determinants.

Problems 2.2

1. Let $\mathbf{A} = \begin{pmatrix} 1 & 2 \\ -1 & 3 \end{pmatrix}$, $\mathbf{B} = \begin{pmatrix} 2 & 3 & 1 \\ 0 & -1 & 2 \end{pmatrix}$, $\mathbf{c} = (4 \quad 1 \quad -2)$, $\mathbf{d} = \begin{pmatrix} 5 \\ -2 \end{pmatrix}$.
State which of the products $\mathbf{A}^2$, **AB**, **Ac**, **Ad**, **B′A**, **BB′**, **Bc′**, **B′d**, **cA**, **cB′**, **cc′**, **c′d′**, **d′A**, **d′B**, **dc**, **dd′** are not defined and obtain those that are.

2. Evaluate the matrix given by the triple product

$$\begin{pmatrix} 5 & 1 \\ 2 & 0 \\ 3 & -1 \end{pmatrix} \begin{pmatrix} 4 & 0 \\ -1 & 2 \end{pmatrix} \begin{pmatrix} -1 & 3 & 2 \\ 4 & 0 & 1 \end{pmatrix}.$$

3. Obtain $A^3 - 2A^2 + A - I$ as a single matrix when

$$A = \begin{pmatrix} 1 & 1 & 2 \\ 1 & 1 & 1 \\ 2 & 1 & 1 \end{pmatrix}.$$

4. Express the linear transformations

$$\begin{aligned} x_1 &= 3y_1 - 2y_2 \\ x_2 &= 5y_1 + y_2 \end{aligned} \quad \text{and} \quad \begin{aligned} y_1 &= z_1 - z_2 \\ y_2 &= 4z_1 + z_2 \end{aligned}$$

in matrix form and obtain the transformation which expresses x_1, x_2 in terms of z_1, z_2.

5. Give a geometrical interpretation of the transformation $\mathbf{x} = \mathbf{Ay}$ for each form of

$$A = \begin{pmatrix} 1 & 0 \\ 0 & -1 \end{pmatrix}, \quad \begin{pmatrix} 1 & 0 \\ 0 & 1 \end{pmatrix}, \quad \begin{pmatrix} -1 & 0 \\ 0 & -1 \end{pmatrix}, \quad \begin{pmatrix} 0 & -1 \\ 1 & 0 \end{pmatrix}, \quad \begin{pmatrix} -1 & 0 \\ 0 & 1 \end{pmatrix},$$

$$\begin{pmatrix} \cos\theta & -\sin\theta \\ \sin\theta & \cos\theta \end{pmatrix}.$$

(For example, the first form represents reflection in the y_1 axis.)

6. (a) Show that $(AB)' = B'A'$.

(b) Show that $AI = IA = A$, where A is a square matrix and I is the unit matrix of the same order.

7. Expand $(A+B)^2$ and check the validity of the expansion for

$$A = \begin{pmatrix} 4 & -3 \\ -1 & 7 \end{pmatrix} \quad \text{and} \quad B = \begin{pmatrix} -3 & 2 \\ 1 & -6 \end{pmatrix}.$$

8. Show that $\mathbf{ab}$ is singular when $\mathbf{a}$ is a column vector and $\mathbf{b}$ is a row vector of the same order.

2.3. THE INVERSE OF A SQUARE MATRIX

By the definition of matrix multiplication system (2.1) may be expressed in the matrix form

$$\begin{pmatrix} a_{11} & a_{12} \\ a_{21} & a_{22} \end{pmatrix} \begin{pmatrix} y_1 \\ y_2 \end{pmatrix} = \begin{pmatrix} x_1 \\ x_2 \end{pmatrix}$$

or as

$$\mathbf{Ay} = \mathbf{x}. \tag{3.1}$$

For a given value of $\mathbf{x} \neq \mathbf{0}$ equation (3.1) is the matrix representation of an inhomogeneous system of two linear equations in the two unknowns y_1, y_2. Generally it may be regarded as a system of n equations in n unknowns. Such a system has a unique solution if, and only if, $\Delta = \det \mathbf{A} \neq 0$ and this is given by Cramer's rule, in the notation of section 1.5, as

$$y_1 = \Delta_1/\Delta, \ldots, y_n = \Delta_n/\Delta.$$

If Δ_1 is expanded by its first column, Δ_2 by its second column, ..., and Δ_n by its nth column, then the solution of the system assumes the form

$$\left. \begin{array}{l} y_1 = (x_1 A_{11} + \ldots + x_n A_{n1})/\det \mathbf{A} \\ y_2 = (x_1 A_{12} + \ldots + x_n A_{n2})/\det \mathbf{A} \\ \vdots \\ y_n = (x_1 A_{1n} + \ldots + x_n A_{nn})/\det \mathbf{A} \end{array} \right\}, \tag{3.2}$$

where, by the notation of section 1.2, A_{ij} is the cofactor of the element a_{ij} in $\det \mathbf{A}$. The set of relations (3.2) may be expressed in the matrix form

$$\mathbf{y} = \mathbf{Rx} \tag{3.3}$$

where

$$\mathbf{R} = \frac{1}{\det \mathbf{A}} \begin{pmatrix} A_{11} & A_{21} & \ldots & A_{n1} \\ A_{12} & A_{22} & \ldots & A_{n2} \\ \cdot & \cdot & \ldots & \cdot \\ A_{1n} & A_{2n} & \ldots & A_{nn} \end{pmatrix} \equiv \frac{1}{\det \mathbf{A}} \operatorname{adj} \mathbf{A}.$$

The matrix $\operatorname{adj} \mathbf{A}$, called the *adjugate* (or adjoint) of $\mathbf{A}$, is defined as shown, that is by replacing each element of $\mathbf{A}$ by its cofactor in $\det \mathbf{A}$ and transposing the result. Conversely the transpose of $\mathbf{A}$ may be taken before each element is replaced by its cofactor, relative to the transposed position of the element.

Substitution of (3.3) in (3.1) gives

$$x = Ay = A(Rx) = (AR)x,$$

showing that AR is the identity matrix of order n. The converse substitution gives

$$y = Rx = R(Ay) = (RA)y,$$

showing that RA is the same identity matrix. Thus A and R are related by

$$AR = RA = I$$

and, by virtue of this relation, R is called the reciprocal or *inverse* of A and is denoted by A^{-1}. Thus

$$AA^{-1} = A^{-1}A = I \qquad (3.4)$$

and by (2.6)

$$\det(AA^{-1}) = \det A \det A^{-1} = 1,$$

which is true only if both $\det A \neq 0$ and $\det A^{-1} \neq 0$. The solution y in (3.3) of system (3.1) was defined to be unique through the condition $\det A \neq 0$. Since $x \neq 0$ has a given value it follows that A^{-1} is unique. These results are now collected as a formal definition.

The inverse A^{-1} of a square matrix A exists only when A is non-singular. It is unique and is defined as adj $A/\det A$.

Example 3.1.

By matrix algebra obtain the solution of the system

$$2x - 5y - 7z = 0,$$
$$3x - y - z = 4,$$
$$8x - 3y - 2z = 7.$$

In matrix form the system is

$$\begin{pmatrix} 2 & -5 & -7 \\ 3 & -1 & -1 \\ 8 & -3 & -2 \end{pmatrix} \begin{pmatrix} x \\ y \\ z \end{pmatrix} = \begin{pmatrix} 0 \\ 4 \\ 7 \end{pmatrix}$$

or

$$\mathbf{Ax} = \mathbf{k}, \tag{3.5}$$

where $\mathbf{x} = \{x\,y\,z\}$ and $\mathbf{k} = \{0\,4\,7\}$. Now $\varDelta = \det \mathbf{A} = 15 \neq 0$, so the system has a unique solution and $\mathbf{A}^{-1}$ exists.

$$\text{adj } \mathbf{A} = \begin{pmatrix} -1 & -2 & -1 \\ 11 & 52 & -34 \\ -2 & -19 & 13 \end{pmatrix}' = \begin{pmatrix} -1 & 11 & -2 \\ -2 & 52 & -19 \\ -1 & -34 & 13 \end{pmatrix}.$$

Pre-multiply (3.5) by $\mathbf{A}^{-1}$ then

$$\mathbf{A}^{-1}\mathbf{A}\mathbf{x} = \mathbf{Ix} = \mathbf{x} = \mathbf{A}^{-1}\mathbf{k},$$

so the solution is given by

$$\begin{pmatrix} x \\ y \\ z \end{pmatrix} = \frac{1}{15} \begin{pmatrix} -1 & 11 & -2 \\ -2 & 52 & -19 \\ -1 & -34 & 13 \end{pmatrix} \begin{pmatrix} 0 \\ 4 \\ 7 \end{pmatrix} = \frac{1}{15} \begin{pmatrix} 30 \\ 75 \\ -45 \end{pmatrix} = \begin{pmatrix} 2 \\ 5 \\ -3 \end{pmatrix}.$$

Thus $x = 2$, $y = 5$ and $z = -3$.

If $\mathbf{B}$ is a matrix with inverse $\mathbf{B}^{-1}$ then $\mathbf{BB}^{-1} = \mathbf{I}$. Suppose that $\mathbf{B} = \mathbf{A}^{-1}$; then $\mathbf{A}^{-1}(\mathbf{A}^{-1})^{-1} = \mathbf{I}$ which by (3.4) shows that $(\mathbf{A}^{-1})^{-1} = \mathbf{A}$. Thus two successive inversions of a matrix restore the original matrix. In (3.4) replace $\mathbf{A}$ by the product $\mathbf{BC}$; then

$$(\mathbf{BC})\,(\mathbf{BC})^{-1} = \mathbf{I},$$

which pre-multiplied by $\mathbf{B}^{-1}$ gives

$$(\mathbf{B}^{-1}\mathbf{B})\,\mathbf{C}(\mathbf{BC})^{-1} = \mathbf{B}^{-1}$$

or

$$\mathbf{C}(\mathbf{BC})^{-1} = \mathbf{B}^{-1}.$$

Pre-multiplication of this result by $\mathbf{C}^{-1}$ shows that

$$(\mathbf{BC})^{-1} = \mathbf{C}^{-1}\mathbf{B}^{-1}$$

which illustrates the general result that the inverse of a product of matrices equals the product of their inverses in reverse order.

Problems 2.3

1. Obtain the inverse of each matrix and, by showing (3.4) to be satisfied, verify that it is correct.

$$\begin{pmatrix} 3 & 2 \\ 7 & 5 \end{pmatrix}, \quad \begin{pmatrix} 2 & 0 & 1 \\ 0 & 3 & 0 \\ 1 & -1 & 4 \end{pmatrix}, \quad \begin{pmatrix} 1 & 0 & 1 \\ 0 & 1 & 0 \\ 1 & 0 & 1 \end{pmatrix}, \quad \begin{pmatrix} 0 & 0 & 1 \\ 0 & 1 & 0 \\ 1 & 0 & 0 \end{pmatrix}, \quad \begin{pmatrix} \cos\theta & \sin\theta \\ -\sin\theta & \cos\theta \end{pmatrix}.$$

2. By matrix algebra obtain the solution of the system

$$3x + y - z = 3,$$
$$x + 3y - z = 5,$$
$$6x + y - 2z = 0.$$

3. Obtain the converse of the linear transformation

$$X = x + y + z,$$
$$Y = x - y + 2z,$$
$$Z = 3x - 2y + z,$$

that is, a set of three linear equations with x, y and z as their respective subjects.

4. Show that the operations of transposition and inversion are commutative:
$$(A')^{-1} = (A^{-1})'.$$

5. Show that if A is singular then

$$A \text{ adj } A = (\text{adj } A) A = 0.$$

6. Show that the inverse of a non-singular diagonal matrix is the diagonal matrix in which each non-zero element is the reciprocal of the original.

2.4. SUBMATRICES AND RANK

Systems of equations where the number of equations equals the number of unknowns have been studied in the previous chapter by means of determinants and, when possible, their solutions were obtained by Cramer's rule. The previous section has shown how the inverse of a matrix is connected with the processes implied by Cramer's rule, although in practice the matrix and determinant methods for solving equations seem hardly to be related. When the number of equations and unknowns is small, say not greater than five, the application of Cramer's rule is usually easier than the use of an inverse matrix. The solution of many equations would probably be arranged as an automatic computation and for this a matrix approach is preferable.

The next two sections consider systems where the number of equations may not equal the number of unknowns. The essential features of such systems are analysed more easily in terms of matrices rather than determinants, although the latter may be preferred for actually obtaining the solution if one exists. The matrix analysis depends on the concept of a submatrix and the rank of a matrix, terms which are similar in meaning to a minor and the rank of a determinant.

A matrix which is derived from a given matrix **A** by the omission of some rows or some columns, possibly both, is called a *submatrix* of **A**. At one extreme each element of **A** is a submatrix and, at the other, **A** is a submatrix of itself. For instance, the matrix

$$\begin{pmatrix} a & b & c \\ d & e & f \end{pmatrix}$$

contains the three 2×2 submatrices

$$\begin{pmatrix} a & b \\ d & e \end{pmatrix}, \quad \begin{pmatrix} a & c \\ d & f \end{pmatrix} \quad \text{and} \quad \begin{pmatrix} b & c \\ e & f \end{pmatrix},$$

the two 1×3 submatrices $(a\ b\ c)$ and $(d\ e\ f)$, the three 2×1 submatrices $\{a\ d\}$, $\{b\ e\}$ and $\{c\ f\}$, the six 1×2 submatrices $(a\ b)$, $(a\ c)$, $(b\ c)$, $(d\ e)$, (df), (ef) and the six 1×1 submatrices a, b, c, d, e, f. Including itself, the matrix has twenty-one submatrices.

An $m \times n$ matrix **A** is of *rank r* if it contains a non-singular square submatrix of order r and if every square submatrix of order greater than r, which might be contained in **A**, is singular. The rank of a matrix equals the maximum number of linearly independent row vectors or column vectors and is unaffected by transposition. The value of r cannot exceed the smaller of the numbers m and n. If $r = 0$ then $\mathbf{A}$ is a null matrix.

The rank of a matrix is unchanged when rows and columns are manipulated by the elementary operations described for determinants so, if desired, the rank of a matrix may be found by the method given in section 1.4 for determinants. To make this quite clear, an elementary operation on a matrix is one which interchanges two rows or which multiplies a row by a number ($\neq 0$) or which adds a row to another row; similarly for columns. Two matrices $\mathbf{A}$ and $\mathbf{B}$ are said to be *equivalent*, $\mathbf{A} \sim \mathbf{B}$, if each can be transformed into the other by a sequence of elementary operations. Note that equivalence, in this sense, does not imply equality.

Example 4.1.

Determine the rank of $\begin{pmatrix} 2 & 1 & 0 & 1 \\ 4 & 2 & 1 & 3 \\ 6 & 3 & 4 & 7 \end{pmatrix}$.

$$\begin{pmatrix} 2 & 1 & 0 & 1 \\ 4 & 2 & 1 & 3 \\ 6 & 3 & 4 & 7 \end{pmatrix} \qquad \sim \qquad \begin{pmatrix} 0 & 1 & 0 & 0 \\ 0 & 2 & 1 & 1 \\ 0 & 3 & 4 & 4 \end{pmatrix}$$

$$c_1' = c_1 - 2c_2$$
$$c_4' = c_4 - c_2$$

$$\sim \qquad \begin{pmatrix} 0 & 1 & 0 & 0 \\ 0 & 2 & 1 & 0 \\ 0 & 3 & 4 & 0 \end{pmatrix}.$$

$$c_4' = c_4 - c_3$$

This form shows that a non-singular 3×3 submatrix does not exist, so $r \neq 3$. Clearly a non-singular 2×2 submatrix exists, so $r = 2$. By further operations on rows and columns the equivalent matrix

$$\begin{pmatrix} 1 & 0 & 0 & 0 \\ 0 & 1 & 0 & 0 \\ 0 & 0 & 0 & 0 \end{pmatrix}$$

may be obtained, showing clearly that $r = 2$.

2.5. GENERAL HOMOGENEOUS SYSTEMS

A limited study of homogeneous systems of linear equations was made in section 1.7 and there the number of equations was taken to equal the number of unknowns. This restriction is now removed and a more general treatment is given in terms of matrices.

A homogeneous system of m equations in n unknowns has the form

$$\left.\begin{aligned} a_{11}x_1 + a_{12}x_2 + \ldots + a_{1n}x_n &= 0 \\ a_{21}x_1 + a_{22}x_2 + \ldots + a_{2n}x_n &= 0 \\ \vdots \qquad \vdots \qquad \vdots \qquad \vdots \qquad \vdots \\ a_{m1}x_1 + a_{m2}x_2 + \ldots + a_{mn}x_n &= 0 \end{aligned}\right\} . \tag{5.1}$$

Such a system has the zero solution $x_1 = x_2 = \ldots = x_n = 0$ in all cases. The object is to formulate conditions for non-zero solutions to exist and to show how these may be found. Let $\mathbf{A}$ be the $m \times n$ matrix formed from the coefficients a_{ij} and suppose its rank is r, so that is has a non-singular submatrix $\mathbf{B}$ of order r. To preserve simple symbolic forms, suppose that the equations and unknowns are so arranged in (5.1) that the elements of $\mathbf{B}$ occupy the upper left-hand region of $\mathbf{A}$. Thus without loss in generality $\mathbf{B}$ has the easily identified form

$$\mathbf{B} = \begin{pmatrix} a_{11} & a_{12} & \ldots & a_{1r} \\ a_{21} & a_{22} & \ldots & a_{2r} \\ . & . & \ldots & . \\ a_{r1} & a_{r2} & \ldots & a_{rr} \end{pmatrix} .$$

If $x_1, \ldots, x_n$ are (non-zero) solutions of the first r equations in system (5.1) then they satisfy the other $m-r$ equations of the system. This is stated without proof. Accordingly the first r equations only are sufficient to provide a solution of the system and these are expressed as

$$\left.\begin{array}{l} a_{11}x_1 + a_{12}x_2 + \ldots + a_{1r}x_r = -(a_{1,\,r+1}x_{r+1} + \ldots + a_{1n}x_n) \\ a_{21}x_1 + a_{22}x_2 + \ldots + a_{2r}x_r = -(a_{2,\,r+1}x_{r+1} + \ldots + a_{2n}x_n) \\ \quad\cdot \qquad \cdot \qquad \cdots \qquad \cdot \qquad\qquad \cdot \qquad\qquad \cdots \qquad \cdot \\ a_{r1}x_1 + a_{r2}x_2 + \ldots + a_{rr}x_r = -(a_{r,\,r+1}x_{r+1} + \ldots + a_{rn}x_n) \end{array}\right\}$$

$$(5.2)$$

to conform with the type of arrangement used in section 1.7 for the application of Cramer's rule. If $r = n$ then system (5.2) is homogeneous, and since **B** is non-singular this system, and therefore system (5.1), has zero solutions only. If $r < n$ then the unknowns $x_{r+1}, \ldots, x_n$ may be treated as arbitrary quantities and the system may be solved by Cramer's rule or by any other suitable method. Thus if $r < n$ the system always has non-zero solutions. The condition $r < n$ is bound to apply if $m < n$, for then r cannot exceed m, so a system (5.1) with more unknowns than equations always has non-zero solutions.

The stages of this analysis are now summarised in an application to a specific system. A solution which includes a relation between two or more unknowns is called the general solution of the system. A particular solution arises from the general solution when suitable values are given to the unknowns.

Example 5.1.

Obtain the general solution of the homogeneous system

$$\begin{array}{rcl} 2w + x \quad\;\; + z &=& 0, \\ 4w + 2x + y + 3z &=& 0, \\ 6w + 3x + 4y + 7z &=& 0. \end{array}$$

As $m < n$, the system has non-zero solutions. The coefficient matrix is

$$\mathbf{A} = \begin{pmatrix} 2 & 1 & 0 & 1 \\ 4 & 2 & 1 & 3 \\ 6 & 3 & 4 & 7 \end{pmatrix}.$$

In Example 4.1 this was found to have rank $r = 2$. Omit $m - r = 1$ equation chosen so that a non-singular 2×2 matrix can be formed from the remaining coefficients. The omission of any one equation satisfies the required condition in this case, so suppose the third is omitted. Arrange the remaining equations in the form of (5.2) so that the matrix corresponding with $\mathbf{B}$ has rank 2. Such an arrangement could be

$$2w + z = -x,$$
$$4w + 3z = -2x - y.$$

By Cramer's rule $\Delta = \det \mathbf{B} = 2$,

$$\Delta_1 = \begin{vmatrix} -x & 1 \\ -(2x+y) & 3 \end{vmatrix} = y - x \text{ and } \Delta_2 = \begin{vmatrix} 2 & -x \\ 4 & -(2x+y) \end{vmatrix} = 2y,$$

so the general solution of the system may be expressed as

$$2w = y - x, \quad z = -y.$$

Problems 2.5

Obtain, where possible, general solutions of the systems below.

1. $x + y - z = 0,$
   $5x + 2y - 3z = 0.$

2. $-5w + x + y + 10z = 0,$
   $2w + 4x + 6y + 17z = 0,$
   $x + y + 5z = 0.$

3. $5w - x - y - 10z = 0,$
   $2w + 4x + 6y + 17z = 0,$
   $7w + 3x + 5y + 7z = 0.$

4. $\quad w - x + 2y + z = 0,$
$\quad\quad 3w + 5x + 3y + 3z = 0,$
$\quad\quad 7w + x - 3y + 7z = 0.$

5. $\quad\quad\quad 2y + 9z = 0,$
$\quad\quad 2x + y + z = 0,$
$\quad\quad 3x + 2y - 7z = 0,$
$\quad\quad 5x + y - 3z = 0,$
$\quad\quad 8x + 6y - z = 0.$

6. $\quad v + w + x + y + z = 0,$
$\quad\quad 3v + 2w + 5x + 2y + 3z = 0,$
$\quad\quad 7v - 3w + 8x + y + 5z = 0,$
$\quad\quad 4v - w - 3x + 3y + 2z = 0.$

2.6. GENERAL INHOMOGENEOUS SYSTEMS

An inhomogeneous system of m equations in n unknowns has the form of (5.1) with constants $k_1, k_2, \ldots, k_m$ replacing the zeros on the right-hand side, that is

$$\left.\begin{array}{c} a_{11}x_1 + \ldots + a_{1n}x_n = k_1 \\ \cdot \quad\quad \cdots \quad\quad \cdot \quad\quad \cdot \\ a_{m1}x_1 + \ldots + a_{mn}x_n = k_m \end{array}\right\}, \quad (6.1)$$

where $k_i \neq 0$ for at least one value of $i = 1, \ldots, m$. The $m \times n$ matrix formed from the coefficients a_{ij} is again denoted by **A**.

By introducing a fictitious unknown $x_{n+1} = -1$, system (6.1) may be converted into the homogeneous system

$$\left.\begin{array}{c} a_{11}x_1 + \ldots + a_{1n}x_n + k_1 x_{n+1} = 0 \\ \cdot \quad\quad \cdots \quad\quad \cdot \quad\quad \cdot \\ a_{m1}x_1 + \ldots + a_{mn}x_n + k_m x_{n+1} = 0 \end{array}\right\}, \quad (6.2)$$

the coefficients a_{ij} and k_i in which are said to constitute the *augmented matrix*, denoted here by **G**. For this conversion to be valid, (6.2) must have $x_{n+1} = -1$ as a particular solution.

If (6.2) is satisfied only when $x_{n+1} = 0$, and not by any other value of x_{n+1}, then the equations in (6.1) are inconsistent so the system does not have a solution. For example consider the system

$$u - 2v + 3w = 4,$$
$$3u - 6v + 9w = 8$$

which clearly is inconsistent and, with x_{n+1} denoted by x, express it in the form of (6.2) as

$$u - 2v + 3w + 4x = 0,$$
$$3u - 6v + 9w + 8x = 0,$$

where $x = -1$. The general solution of this homogeneous system is $u = 2v - 3w$, $x = 0$ which confirms that, as x cannot take the value -1, the original inhomogeneous system has no solution. A non-zero value of x_{n+1} which satisfies the homogeneous system (6.2) can always be multiplied by a factor which changes it to -1. For instance the system

$$u + v - 2w + 5x = 0,$$
$$2u + v + w + 3x = 0$$

is satisfied by $u = -3$, $v = -6$, $w = 3$, $x = 3$ and by any multiple, say $-1/3$, of these values. These simple examples show that if x_{n+1} is arbitrary then the equations in (6.1) are consistent.

A systematic way of solving (6.1) is implied by the following important theorem, which provides a general test for the consistency of equations.

The necessary and sufficient condition for the general inhomogeneous system (6.1) to have solutions is that the coefficient matrix **A** and the augmented matrix **G** have the same rank.

This theorem may be established by the following reasoning, first for necessity and then for sufficiency. The matrix **A**, of rank r, contains a non-singular submatrix **B** of order r and, since **B** is a submatrix of **G** and **G** has an extra column, the rank of **G** is at least r and at most $r+1$. If the system does have a solution then **G** may be expressed as

$$\mathbf{G} = \begin{pmatrix} a_{11} & \cdots & a_{1n} & (a_{11}x_1 + \ldots + a_{1n}x_n) \\ \cdot & \cdots & \cdot & \cdot & \cdots & \cdot \\ a_{m1} & \cdots & a_{mn} & (a_{m1}x_1 + \ldots + a_{mn}x_n) \end{pmatrix},$$

where the ith element in the $(n+1)$th column is a sum of multiples of all the other elements which occur in the ith row. Now a square submatrix of **G** which is not a submatrix of **A** must contain a column composed entirely of elements in the $(n+1)$th column of **G**. By properties of determinants such a submatrix must be singular, so the rank of **G** cannot exceed the rank of **A**. Thus the consequence of assuming the system to be consistent is that the ranks of **A** and **G** are equal.

Now start by assuming the ranks of **A** and **G** to be equal and see what is implied. Let the equations and unknowns of the system be so arranged that the elements of a non-singular matrix **B** of order r occupy the upper left-hand region in the coefficient matrix **A** of (6.1), as in the previous section. By the properties of homogeneous systems, (6.2) has a solution for which the unknowns $x_{r+1}, \ldots, x_n, x_{n+1}$ are arbitrary. This means that to x_{n+1} may be assigned the value -1, or any multiple thereof, and therefore a solution of (6.1) exists.

Although the above analysis has considered the constants k_i to be multiplied by a fictitious unknown x_{n+1}, this device is not necessary for the practical solution of a system.

Example 6.1.

Obtain the general solution of the system

$$\begin{aligned}
2u + v - w - 3x + 4y - 3z &= -1, \\
3u - v + 2w - 4x + 8y - 7z &= 5, \\
7u + 3v - 6w + x - 9y + 6z &= 0, \\
8u + v - 3w \qquad - 5y + 2z &= 6, \\
8u + 6v - 10w - x - 9y + 7z &= -7.
\end{aligned}$$

The coefficient matrix and the augmented matrix have the same rank, $r = 3$, so the equations are consistent. Omit $m - r = 2$ equations chosen so that a 3×3 non-singular matrix can be formed from the coefficients in the remaining equations. By omitting the last two equations and by treating x, y and z as arbitrary quantities the equations to be solved are

$$2u + v - w = -1 + 3x - 4y + 3z,$$
$$3u - v + 2w = 5 + 4x - 8y + 7z,$$
$$7u + 3v - 6w = - x + 9y - 6z.$$

For solution by Cramer's rule $\Delta = \det \mathbf{B} = 16$ and

$$\Delta_1 = \begin{vmatrix} (-1 + 3x - 4y + 3z) & 1 & -1 \\ (5 + 4x - 8y + 7z) & -1 & 2 \\ (-x + 9y - 6z) & 3 & -6 \end{vmatrix} = 11x + 15(1 - y + z)$$

whence $\qquad 16u = 11x + 15(1 - y + z).$

Similarly $\qquad 16v = -57 + 83x - 151y + 103z$

and $\qquad 16w = -11 + 57x - 117y + 85z.$

Problems 2.6

Obtain, where possible, general solutions of the systems below.

1. $\begin{aligned} x - 2y + z &= -6, \\ x + y + z &= 0. \end{aligned}$

2. $\begin{aligned} 2x + 2y + z &= 11, \\ 5x + 3y - 2z &= 2. \end{aligned}$

3. $\begin{aligned} 3x - 2y &= 2, \\ 4x - 3y &= 1, \\ 2x + y &= 13, \\ -3x + 4y &= 8. \end{aligned}$

4. $\begin{aligned} 3x - 2y + 6z &= 1, \\ 6x - 4y + 12z &= 3. \end{aligned}$

5. $\begin{aligned} x - 2y + z &= 1, \\ 3x - 6y + 3z &= 3. \end{aligned}$

6. $\begin{aligned} w + x + y - z &= 3, \\ 2w - 7x + y - 3z &= 1, \\ 3w + 5x + 6y - 8z &= 0. \end{aligned}$

7.
$$w+ \ x+ \ y+ \ z = 0,$$
$$6w- \ 5x- \ 7y+ \ 7z = 2,$$
$$3w+ \ 3x+ \ 8y+ \ 5z = 4.$$

8.
$$w+ \ x+ \ y+ \ 6z = \ 4,$$
$$3w- \ 8x+ \ 3y+ \ z = \ 1,$$
$$w- 10x+ \ y- 11z = -7.$$

9.
$$w- \ x+ \ 2y- \ z = 2,$$
$$3w- \ 4x+ \ 5y- \ 2z = 0,$$
$$7w- \ 7x+ \ 3y- \ z = 6,$$
$$2w- \ x- \ 6y+ \ 3z = 1.$$

10.
$$x+ \ y+ \ z = \ 2,$$
$$2x+ \ 3y- \ 2z = -3,$$
$$3x- \ 5y+ \ 5z = \ 2,$$
$$-x+ \ 6y- \ 3z = \ 1,$$
$$5x+ \ 2y+ \ 3z = \ 3.$$

2.7. NETWORK ANALYSIS

In section 1.6 two methods are described, in a limited way, for the analysis of currents and voltages associated with electrical networks. Each method produces a set of simultaneous equations and these may be solved either by determinants or by matrices. The present section is concerned not so much with techniques of solution but rather with how a matrix, by its very nature, may be used to represent the characteristics of a network. By means of matrix notation the elements of a network can be separated from the currents and voltages associated with it. How a given network "affects" a certain current and voltage is found by operating on these quantities with the matrix of the network, the operation being matrix multiplication.

A *four-terminal network* is usually represented by a rectangle to which four connections are attached, as shown in Fig. 2.1. The rectangular box may contain a variety of electrical components, such as resistances, capacitors and inductances connected in a certain way. The four terminals are treated as two pairs, with each of which is associated a voltage v and a current i as shown. The pair to the left is usually associated with an input signal and that to the right with an output signal, so the contents of the box transform electrically the input into the output. A network is assumed to be linear, which implies that the four quantities i_1, i_2, v_1, v_2 are connected by relations such as

$$\left. \begin{array}{l} v_1 = av_2 + bi_2 \\ i_1 = cv_2 + di_2 \end{array} \right\} , \qquad (7.1)$$

Fig. 2.1.

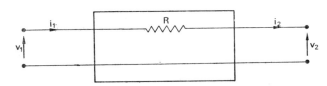

Fig. 2.2.

where a, b, c, d are constants which describe the features of a given network. A typical network problem is to evaluate any two of the unknowns from given values of the other two, most likely the output pair from the input pair.

The simple treatment given here considers networks to be composed of resistances only, though the forms obtained are valid for general impedances. Consider the simple network, shown in Fig. 2.2, which consists of a single series resistance. In the form of (7.1) the network equations are

$$v_1 = v_2 + Ri_2,$$
$$i_1 = i_2.$$

In matrix form this system is

$$\begin{pmatrix} v_1 \\ i_1 \end{pmatrix} = \begin{pmatrix} 1 & R \\ 0 & 1 \end{pmatrix} \begin{pmatrix} v_2 \\ i_2 \end{pmatrix},$$

which shows how a matrix, in this case the 2×2 matrix, can represent the features of a network. Similarly, a network composed of a single shunt resistance, shown in Fig. 2.3, is described by the equations

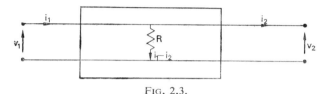

FIG. 2.3.

FIG. 2.4.

$$v_1 = v_2,$$

$$i_1 = \frac{1}{R} v_2 + i_2,$$

which in matrix form are

$$\begin{pmatrix} v_1 \\ i_1 \end{pmatrix} = \begin{pmatrix} 1 & 0 \\ 1/R & 1 \end{pmatrix} \begin{pmatrix} v_2 \\ i_2 \end{pmatrix},$$

and again the network is featured by a 2×2 matrix.

These two simple networks are the elements of some more complicated networks. Take, for example, the network shown in Fig. 1.3, remove the voltage sources v_a, v_b and show the resistances as a sequence of simple networks, as in Fig. 2.4.

HGM· – METCS 6

Whatever other circuitry may be joined to either end of the network is of no interest; indeed one end might be short-circuited. Suppose the values of i_1, v_1 for an input signal are known and the values of i_6, v_6 for the corresponding output signal are required. The matrix equations for the first two basic networks are

$$\begin{pmatrix} v_1 \\ i_1 \end{pmatrix} = \begin{pmatrix} 1 & R_1 \\ 0 & 1 \end{pmatrix} \begin{pmatrix} v_2 \\ i_2 \end{pmatrix} \quad \text{and} \quad \begin{pmatrix} v_2 \\ i_2 \end{pmatrix} = \begin{pmatrix} 1 & 0 \\ 1/R_2 & 1 \end{pmatrix} \begin{pmatrix} v_3 \\ i_3 \end{pmatrix},$$

whence, by the elimination of $\{v_2\ i_2\}$,

$$\begin{pmatrix} v_1 \\ i_1 \end{pmatrix} = \begin{pmatrix} 1 & R_1 \\ 0 & 1 \end{pmatrix} \begin{pmatrix} 1 & 0 \\ 1/R_2 & 1 \end{pmatrix} \begin{pmatrix} v_3 \\ i_3 \end{pmatrix}.$$

This elimination process simply takes the output of one network to be the input of the next and, by virtue of matrix notation, it is achieved merely by substitution for the column matrices. Treatment of the complete network in this way gives the required relation

$$\begin{pmatrix} v_1 \\ i_1 \end{pmatrix} = \begin{pmatrix} 1 & R_1 \\ 0 & 1 \end{pmatrix} \begin{pmatrix} 1 & 0 \\ 1/R_2 & 1 \end{pmatrix} \begin{pmatrix} 1 & R_3 \\ 0 & 1 \end{pmatrix} \begin{pmatrix} 1 & 0 \\ 1/R_4 & 1 \end{pmatrix} \begin{pmatrix} 1 & R_5 \\ 0 & 1 \end{pmatrix} \begin{pmatrix} v_6 \\ i_6 \end{pmatrix}.$$

It is left as an exercise to show that this reduces to the form

$$\begin{pmatrix} v_1 \\ i_1 \end{pmatrix} = \mathbf{A} \begin{pmatrix} v_6 \\ i_6 \end{pmatrix},$$

where

$$\mathbf{A} = \begin{pmatrix} (1+R_1/R_2)\,(1+R_3/R_4)+R_1/R_4 \\ (1+R_3/R_4)/R_2+1/R_4 \\ \end{pmatrix}$$
$$\{(1+R_1/R_2)\,(1+R_2/R_4)+R_1/R_4\}\,R_5+(1+R_1/R_2)\,R_3+R_1 \\ \{(1+R_3/R_4)/R_2+1/R_4\}\,R_5+R_3/R_2+1 \end{pmatrix}.$$

The output signal is given by

$$\begin{pmatrix} v_6 \\ i_6 \end{pmatrix} = \mathbf{A}^{-1} \begin{pmatrix} v_1 \\ i_1 \end{pmatrix}.$$

The values of the resistances are substituted in **A**, or preferably in the five constituent matrices, before the inverse is taken.

As a simple example let all the resistances have unit value and suppose the input values are $v_1 = 3$, $i_1 = 13/8$. Then

$$\mathbf{A} = \begin{pmatrix} 5 & 8 \\ 3 & 5 \end{pmatrix}, \quad \mathbf{A}^{-1} = \begin{pmatrix} 5 & -8 \\ -3 & 5 \end{pmatrix}$$

and

$$\begin{pmatrix} v_6 \\ i_6 \end{pmatrix} = \begin{pmatrix} 5 & -8 \\ -3 & 5 \end{pmatrix} \begin{pmatrix} 3 \\ 13/8 \end{pmatrix} = \begin{pmatrix} 2 \\ -7/8 \end{pmatrix}.$$

This result is consistent with the details in **1** of Problems 1.6 and its solution.

Problems 2.7

1. The resistances shown in the network of Fig. 2.5 are measured in ohms. (a) Evaluate the input i_1, v_1 if the output is $i_2 = 1$ A, $v_2 = 1$ V. (b) Evaluate the output i_2, v_2 if the input is $i_1 = 1$ A, $v_1 = 4$ V.

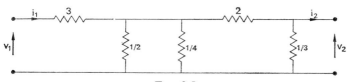

FIG. 2.5.

2. Consider the network shown in Fig. 2.6 and a network of similar form but with the order of resistances reversed, the first being R_4 in shunt, the second R_3 in series and so on. Derive an expression which relates R_1, R_2, R_3 and R_4 if, for a given input i_1, v_1, both networks produce the same output i_2, v_2. If $R_1 = R_3 = 2$ ohm evaluate R_2 and interpret the result physically. What can be said about R_4 in this case?

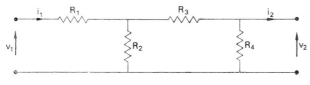

FIG. 2.6.

6*

2.8. PARTITIONED MATRICES

A matrix is said to be partitioned when it is divided into two or more submatrices by lines drawn between rows or between columns, possibly both, of its elements. Each submatrix formed by such partitions can be denoted by one symbol, rather than by the array of its elements, and this can often simplify algebraic operations. In some applications the submatrices into which a matrix is partitioned may represent distinct aspects of the whole operation implied by the complete matrix.

Let **A** and **B** be matrices which are conformable for an operation and suppose this is to be performed with **A** and **B** in partitioned forms. The partitions must be selected in ways which ensure that the submatrices of **A** and **B**, so defined, are themselves conformable for the operation. For example suppose that

$$\mathbf{A} = \begin{pmatrix} 1 & 0 & 2 \\ 3 & 0 & 1 \\ 2 & 1 & 0 \end{pmatrix} \quad \text{and} \quad \mathbf{B} = \begin{pmatrix} 1 & 0 & 1 & 1 \\ 2 & 0 & 1 & 3 \\ 1 & 3 & 1 & 2 \end{pmatrix};$$

these are conformable for the product **AB** because **A** has as many columns as **B** has rows. Suitable partitioning for the evaluation of **AB** is

$$\mathbf{A} = \left(\begin{array}{cc|c} 1 & 0 & 2 \\ 3 & 0 & 1 \\ \hline 2 & 1 & 0 \end{array} \right) \equiv \begin{pmatrix} \mathbf{A}_1 & \mathbf{A}_2 \\ \mathbf{A}_3 & \mathbf{A}_4 \end{pmatrix} \quad \text{and}$$

$$\mathbf{B} = \left(\begin{array}{ccc|c} 1 & 0 & 1 & 1 \\ 2 & 0 & 1 & 3 \\ \hline 1 & 3 & 1 & 2 \end{array} \right) \equiv \begin{pmatrix} \mathbf{B}_1 & \mathbf{B}_2 \\ \mathbf{B}_3 & \mathbf{B}_4 \end{pmatrix},$$

in terms of which the product is

$$\mathbf{AB} = \begin{pmatrix} \mathbf{A_1B_1 + A_2B_3} & \mathbf{A_1B_2 + A_2B_4} \\ \mathbf{A_3B_1 + A_4B_3} & \mathbf{A_3B_2 + A_4B_4} \end{pmatrix}$$

$$= \begin{pmatrix} \begin{pmatrix} 1 & 0 \\ 3 & 0 \end{pmatrix}\begin{pmatrix} 1 & 0 & 1 \\ 2 & 0 & 1 \end{pmatrix} + \begin{pmatrix} 2 \\ 1 \end{pmatrix}(1 \quad 3 \quad 1) & \begin{pmatrix} 1 & 0 \\ 3 & 0 \end{pmatrix}\begin{pmatrix} 1 \\ 3 \end{pmatrix} + \begin{pmatrix} 2 \\ 1 \end{pmatrix}(2) \\ (2 \quad 1)\begin{pmatrix} 1 & 0 & 1 \\ 2 & 0 & 1 \end{pmatrix} + (0)(1 \quad 3 \quad 1) & (2 \quad 1)\begin{pmatrix} 1 \\ 3 \end{pmatrix} + (0)(2) \end{pmatrix}$$

$$= \begin{pmatrix} \begin{pmatrix} 1 & 0 & 1 \\ 3 & 0 & 3 \end{pmatrix} + \begin{pmatrix} 2 & 6 & 2 \\ 1 & 3 & 1 \end{pmatrix} & \begin{pmatrix} 1 \\ 3 \end{pmatrix} + \begin{pmatrix} 4 \\ 2 \end{pmatrix} \\ (4 \quad 0 \quad 3) + (0 \quad 0 \quad 0) & (5) + (0) \end{pmatrix} = \begin{pmatrix} 3 & 6 & 3 & 5 \\ 4 & 3 & 4 & 5 \\ 4 & 0 & 3 & 5 \end{pmatrix}.$$

This example serves only to illustrate the meaning and a use of matrix partition; it does not suggest that the product is obtained more easily than by the direct method.

A square matrix which, by suitable partitions, has one or more simple submatrices, such as a unit or null matrix, may often be inverted through partitioning much more easily than by the direct method which requires its adjugate and determinant. This is specially true for such matrices of order greater than 3×3, so, as an example, consider the 4×4 matrix $\mathbf{A}$ partitioned as shown.

$$\mathbf{A} = \begin{pmatrix} 6 & 2 & 2 & 1 \\ 3 & 4 & 1 & 1 \\ \hline 1 & 1 & 1 & 0 \\ 3 & 1 & 0 & 1 \end{pmatrix} \equiv \begin{pmatrix} \mathbf{A_1} & \mathbf{A_2} \\ \mathbf{A_3} & \mathbf{I} \end{pmatrix}.$$

Let the required inverse $\mathbf{A}^{-1}$ be represented by the 4×4 matrix $\mathbf{B}$, partitioned in the same way as $\mathbf{A}$ so that $\mathbf{B} = \begin{pmatrix} \mathbf{B_1} & \mathbf{B_2} \\ \mathbf{B_3} & \mathbf{B_4} \end{pmatrix}$. Now

$$\mathbf{AA}^{-1} = \mathbf{AB} = \begin{pmatrix} \mathbf{A_1B_1 + A_2B_3} & \mathbf{A_1B_2 + A_2B_4} \\ \mathbf{A_3B_1 + B_3} & \mathbf{A_3B_2 + B_4} \end{pmatrix} = \begin{pmatrix} \mathbf{I} & \mathbf{0} \\ \mathbf{0} & \mathbf{I} \end{pmatrix},$$

where I and 0 are the 2×2 unit and null matrices. This matrix equation is equivalent to the four equations

$$A_1B_1+A_2B_3 = I, \quad A_1B_2+A_2B_4 = 0,$$
$$A_3B_1+B_3 = 0, \quad A_3B_2+B_4 = I.$$

From these the four submatrix elements of B may each be expressed in terms of the four submatrix elements of A and it is left as a simple exercise to show that

$$A^{-1} = B = \begin{pmatrix} B_1 & B_2 \\ B_3 & B_4 \end{pmatrix}$$
$$= \begin{pmatrix} (A_1-A_2A_3)^{-1} & -(A_1-A_2A_3)^{-1}A_2 \\ -A_3(A_1-A_2A_3)^{-1} & I+A_3(A_1-A_2A_3)^{-1}A_2 \end{pmatrix}.$$

By inspection $(A_1-A_2A_3) = \begin{pmatrix} 1 & -1 \\ -1 & 2 \end{pmatrix}$, $B_1 = \begin{pmatrix} 2 & 1 \\ 1 & 1 \end{pmatrix}$,

$B_2 = -B_1A_2 = -\begin{pmatrix} 5 & 3 \\ 3 & 2 \end{pmatrix}$, $B_3 = -A_3B_1 = -\begin{pmatrix} 3 & 2 \\ 7 & 4 \end{pmatrix}$ and

$B_4 = I-A_3B_2 = \begin{pmatrix} 9 & 5 \\ 18 & 12 \end{pmatrix}$ so $A^{-1} = \begin{pmatrix} 2 & 1 & -5 & -3 \\ 1 & 1 & -3 & -2 \\ -3 & -2 & 9 & 5 \\ -7 & -4 & 18 & 12 \end{pmatrix}.$

Problems 2.8

1. By means of suitable partitions obtain the inverse of

(a) $\begin{pmatrix} 4 & 3 & -2 & 1 \\ -2 & -1 & -5 & 2 \\ 2 & 3 & 0 & 0 \\ 1 & 2 & 0 & 0 \end{pmatrix}$, (b) $\begin{pmatrix} 0 & 0 & 0 & 1 \\ 0 & 0 & -1 & 4 \\ 0 & 1 & 5 & 4 \\ 1 & -3 & 7 & 1 \end{pmatrix}.$

2. Invert $\begin{pmatrix} 1 & 0 & 0 & 2 & 3 & -1 \\ 0 & 1 & 0 & 4 & 7 & 0 \\ 0 & 0 & 1 & 5 & 8 & -2 \\ 1 & -3 & -3 & 0 & 0 & 0 \\ 1 & 9 & 4 & 0 & 0 & 0 \\ 0 & 5 & 3 & 0 & 0 & 0 \end{pmatrix}.$

2.9. EIGENVALUES AND EIGENVECTORS

Let a square matrix $\mathbf{A}$ of order n multiply a column vector $\mathbf{k}$ which has n elements. In general there is nothing special about the product $\mathbf{Ak}$, a column vector, though for every square matrix there does exist at least one special $\mathbf{k} \neq \mathbf{0}$ such that $\mathbf{Ak}$ is proportional to $\mathbf{k}$. This distinctive case is represented by the equation

$$\mathbf{Ak} = \lambda\mathbf{k}, \tag{9.1}$$

where λ is the standard symbol used to denote the constant of proportionality. Equations of the form (9.1) arise in many applications and are known as eigenvalue problems. In practice values for λ, the eigenvalues, have to be determined before the corresponding values for $\mathbf{k}$, the eigenvectors, can be obtained. Numerical methods are available which enable an eigenvalue and its associated eigenvector to be evaluated simultaneously by successive approximation.

In order to develop an analytic method for the evaluation of eigenvalues, express (9.1) in the form

$$(\mathbf{A} - \lambda\mathbf{I})\mathbf{k} = \mathbf{0}, \tag{9.2}$$

where $\mathbf{I}$ is the unit matrix of order n. Equation (9.2) is a homogeneous system of n linear equations in the n unknown elements of $\mathbf{k}$ and it has a non-zero solution if, and only if,

$$\det(\mathbf{A} - \lambda\mathbf{I}) = 0. \tag{9.3}$$

This determinantal equation is a polynomial of degree n in λ and it is called the *characteristic equation* of $\mathbf{A}$. Its n roots, which are not necessarily distinct, are the *eigenvalues* of $\mathbf{A}$, synonymous terms being proper values, characteristic roots, latent roots or simply roots. A column vector $\mathbf{k}$ which, through (9.1), corresponds with an eigenvalue λ is an *eigenvector* of $\mathbf{A}$, other terms being characteristic vector and latent vector.

Example 9.1.

Determine the eigenvalues of the matrix

$$\mathbf{A} = \begin{pmatrix} 4 & -2 \\ 1 & 1 \end{pmatrix}$$

and obtain associated eigenvectors.

The characteristic equation is

$$\det(\mathbf{A} - \lambda\mathbf{I}) = \begin{vmatrix} 4-\lambda & -2 \\ 1 & 1-\lambda \end{vmatrix} = 0$$

which, upon expansion of the determinant, becomes

$$\lambda^2 - 5\lambda + 6 = (\lambda - 2)(\lambda - 3) = 0,$$

so the eigenvalues are $\lambda_1 = 2$ and $\lambda_2 = 3$. Let $\mathbf{k} = \{x_1\ x_2\}$ then for $\lambda = 2$ the expanded form of equation (9.2) is the system

$$2x_1 - 2x_2 = 0,$$
$$x_1 - x_2 = 0,$$

which has the solution $x_2 = x_1$. Thus the eigenvalue $\lambda_1 = 2$ has the associated eigenvector

$$\mathbf{k}_1 = \{x_1\ \ x_1\} = x_1\{1\ \ 1\},$$

where x_1 is arbitrary. This result is consistent with (9.1) since the condition is still valid when $\mathbf{k}$ is multiplied by an arbitrary constant. Usually an arbitrary quantity such as x_1 is omitted and $\mathbf{k}_1$ is expressed as $\mathbf{k}_1 = \{1\ \ 1\}$ or, more correctly, as $\mathbf{k}_1 \propto \{1\ \ 1\}$. Sometimes it is useful to *normalise* an eigenvector, that is to divide it by the square root of the sum of the squares of its elements. This ensures that the product of the vector with its transpose, the scalar product, is unity. In normalised form $\mathbf{k}_1 = \{1\ \ 1\}/\sqrt{2}$. By similar reasoning the eigenvalue $\lambda_2 = 3$ is seen to have the associated eigenvector $\mathbf{k}_2 \propto \{2\ \ 1\}$, which in normalised form is $\mathbf{k}_2 = \{2\ \ 1\}/\sqrt{5}$. For numerical analysis normalisation is defined in a slightly different way. The validity of each eigenvalue and eigenvector should be confirmed by substitution in (9.1).

In more detail equation (9.3) is

$$\begin{vmatrix} a_{11}-\lambda & a_{12} & a_{13} & . & a_{1n} \\ a_{21} & a_{22}-\lambda & a_{23} & . & a_{2n} \\ a_{31} & a_{32} & a_{33}-\lambda & . & a_{3n} \\ . & . & . & & . \\ a_{n1} & a_{n2} & a_{n3} & . & a_{nn}-\lambda \end{vmatrix} = 0, \qquad (9.4)$$

where the a_{ij} are the elements of **A**. To illustrate certain points more simply suppose that $n = 3$, then by the properties of determinants, (d) in particular, (9.4) may be expressed as

$$-\lambda \left\{ \begin{vmatrix} a_{22}-\lambda & a_{23} \\ a_{32} & a_{33}-\lambda \end{vmatrix} + \begin{vmatrix} a_{11} & a_{13} \\ a_{31} & a_{33}-\lambda \end{vmatrix} + \begin{vmatrix} a_{11} & a_{12} \\ a_{21} & a_{22} \end{vmatrix} \right\}$$
$$+ \begin{vmatrix} a_{11} & a_{12} & a_{13} \\ a_{21} & a_{22} & a_{23} \\ a_{31} & a_{32} & a_{33} \end{vmatrix} = 0.$$

From this form the polynomial of degree 3 in λ is seen to be

$$\lambda^3 - (a_{11}+a_{22}+a_{33})\,\lambda^2$$
$$+ (a_{11}a_{22}+a_{11}a_{33}+a_{22}a_{33}-a_{12}a_{21}-a_{23}a_{32}-a_{13}a_{31})\,\lambda$$
$$-\det \mathbf{A} = 0.$$

The polynomial of degree n in λ arising from the expansion of (9.4) is

$$\lambda^n - (a_{11}+a_{22}+ \ldots +a_{nn})\,\lambda^{n-1}+ \ldots (-1)^n \det \mathbf{A} = 0, \quad (9.5)$$

terms of little interest having been omitted. If the roots of this polynomial are $\lambda_1,\ \lambda_2,\ \lambda_3,\ \ldots,\ \lambda_n$, the eigenvalues of **A**, then

$$(\lambda-\lambda_1)(\lambda-\lambda_2)(\lambda-\lambda_3) \ldots (\lambda-\lambda_n) = 0. \qquad (9.6)$$

By the properties of polynomials the coefficient of λ^{n-1} in (9.6) is $-\sum_{r=1}^{n} \lambda_r$ and the constant term is $(-1)^n \prod_{r=1}^{n} \lambda_r$. The sum of the elements in the leading diagonal of a square matrix **A** is called the *trace*, or spur, of **A** and is denoted by tr **A**. Comparison of the coefficients of λ^{n-1} in (9.5) and (9.6) shows that

$$\text{the sum of the eigenvalues equals tr } \mathbf{A}. \tag{9.7}$$

Comparison of the constant terms shows that

$$\text{the product of the eigenvalues equals det } \mathbf{A}. \tag{9.8}$$

This pair of interesting results is useful as a check on the correctness of eigenvalues; see Example 9.1. It follows from (9.8) that a singular matrix has at least one zero eigenvalue.

Eigenvectors may be found by a way more systematic than that used in Example 9.1. Recall the result, from 5 of Problems 2.3, that the product of a singular matrix and its adjugate is a null matrix. Accordingly, if λ is an eigenvalue of **A** then $(\mathbf{A} - \lambda \mathbf{I})$ is singular and

$$(\mathbf{A} - \lambda \mathbf{I}) \text{ adj } (\mathbf{A} - \lambda \mathbf{I}), = [\text{adj } (\mathbf{A} - \lambda \mathbf{I})] (\mathbf{A} - \lambda \mathbf{I}), = \mathbf{0}.$$

Comparison of this relation with (9.2) shows that **k** may be chosen as any one column of adj $(\mathbf{A} - \lambda \mathbf{I})$ or, more simply, as the cofactors of the elements in any one row of det $(\mathbf{A} - \lambda \mathbf{I})$.

Example 9.2.

Obtain the eigenvalues and eigenvectors of

$$\mathbf{A} = \begin{pmatrix} 0 & 1 & 0 \\ 1 & 1 & -1 \\ 0 & -1 & 0 \end{pmatrix}.$$

$$\det (\mathbf{A} - \lambda \mathbf{I}) = \begin{vmatrix} -\lambda & 1 & 0 \\ 1 & 1-\lambda & -1 \\ 0 & -1 & -\lambda \end{vmatrix} = \begin{vmatrix} -\lambda & 1 & 0 \\ 0 & 1-\lambda & -1 \\ -\lambda & -1 & -\lambda \end{vmatrix}$$

$$= \lambda \begin{vmatrix} 1 & 1 & 0 \\ 0 & 1-\lambda & 1 \\ 1 & -1 & \lambda \end{vmatrix} = \lambda \begin{vmatrix} 1 & 0 & 0 \\ 0 & 1-\lambda & 1 \\ 1 & -2 & \lambda \end{vmatrix}$$

$$= -\lambda(\lambda+1)(\lambda-2) = 0,$$

so the eigenvalues are -1, 0, 2. The cofactors of, say, the elements in the first row of $\det (\mathbf{A} - \lambda \mathbf{I})$ are

$$(\lambda^2 - \lambda - 1), \quad \lambda, \quad -1.$$

Substitution of the eigenvalues in these cofactors gives

for $\lambda = -1$ the eigenvector $\mathbf{k} \propto \{ \; 1 \; -1 \; -1 \}$,

for $\lambda = \;\; 0$ the eigenvector $\mathbf{k} \propto \{ -1 \;\;\;\; 0 \; -1 \}$ and

for $\lambda = \;\; 2$ the eigenvector $\mathbf{k} \propto \{ \; 1 \;\;\;\; 2 \; -1 \}$.

These eigenvectors may be reformed slightly, to minimise the number of minus signs, and expressed in the normalised forms

$$\{-1 \quad 1 \quad 1\}/\sqrt{3}, \quad \{1 \quad 0 \quad 1\}/\sqrt{2}, \quad \{1 \quad 2 \quad -1\}/\sqrt{6}.$$

When all the eigenvalues of a matrix are distinct, the associated eigenvectors are linearly independent. To prove this let matrix $\mathbf{A}$ of order n have the eigenvalues $\lambda_1, \lambda_2, \ldots, \lambda_n$, all distinct, and the related eigenvectors $\mathbf{k}_1, \mathbf{k}_2, \ldots, \mathbf{k}_n$; suppose that these vectors are not linearly independent but are related by

$$c_1 \mathbf{k}_1 + c_2 \mathbf{k}_2 + \ldots + c_n \mathbf{k}_n = \mathbf{0}, \tag{9.9}$$

where $c_1, c_2, \ldots, c_n$ are constants. For any root λ_i and its related vector $\mathbf{k}_i$

$$\mathbf{A} \mathbf{k}_i = \lambda_i \mathbf{k}_i.$$

From each side of this equation subtract $\lambda_j \mathbf{k}_i$, where λ_j is another eigenvalue, then

$$\mathbf{A}\mathbf{k}_i - \lambda_j \mathbf{k}_i = \lambda_i \mathbf{k}_i - \lambda_j \mathbf{k}_i$$

or

$$(\mathbf{A} - \lambda_j \mathbf{I})\mathbf{k}_i = (\lambda_i - \lambda_j)\mathbf{k}_i.$$

Now multiply (9.9) by $(\mathbf{A} - \lambda_j \mathbf{I})$ and observe that the vector $\mathbf{k}_j$ is eliminated because $(\mathbf{A} - \lambda_j \mathbf{I})\mathbf{k}_j = \mathbf{0}$. The result is therefore a linear combination of the remaining $n-1$ vectors. For example, suppose that $n = 3$ and $j = 3$, then

$$(\mathbf{A} - \lambda_3 \mathbf{I})\,(9.9) = c_1(\lambda_1 - \lambda_3)\mathbf{k}_1 + c_2(\lambda_2 - \lambda_3)\mathbf{k}_2 = \mathbf{0}.$$

This process may be applied repeatedly for $n-1$ different values of j until the result is, say,

$$c_1(\lambda_1 - \lambda_2)\,(\lambda_1 - \lambda_3) \ldots (\lambda_1 - \lambda_n)\mathbf{k}_1 = \mathbf{0} \qquad (9.10)$$

or a similar result in which any one of the eigenvectors remains. But the eigenvalues are all distinct so (9.10) and the equations similar to it are all satisfied by $c_1 = c_2 = \ldots = c_n = 0$ only, which implies that the supposition (9.9) is incorrect.

Result (9.10) shows that if two eigenvalues are equal then the coefficients of the eigenvectors in (9.9) are not necessarily zero, so eigenvectors associated with equal eigenvalues may or may not be linearly independent. These conclusions are now illustrated by specific matrices.

Example 9.3.

Obtain linearly independent eigenvectors of the matrix

$$\mathbf{A} = \begin{pmatrix} 4 & 3 & 3 \\ 2 & 3 & 2 \\ 1 & 1 & 2 \end{pmatrix}$$

given that its eigenvalues are 1, 1, 7.

The cofactors of the elements in the first row of det $(\mathbf{A} - \lambda\mathbf{I})$ are

$$(\lambda-1)(\lambda-4), \quad 2(\lambda-1), \quad (\lambda-1).$$

Substitution of the eigenvalue $\lambda = 7$ gives the eigenvector $\mathbf{k} \propto \{18 \quad 12 \quad 6\} \propto \{3 \quad 2 \quad 1\}$. Substitution of the repeated eigenvalue $\lambda = 1$ gives a zero eigenvector, which is of no interest. These results follow also from substitution in the cofactors of the elements in the second and third rows of det $(\mathbf{A} - \lambda\mathbf{I})$, the rank r of $(\mathbf{A} - \lambda\mathbf{I})$ being 1 when $\lambda = 1$. Let $\mathbf{k} = \{x_1 \quad x_2 \quad x_3\}$; then for $\lambda = 1$ the system $(\mathbf{A} - \lambda\mathbf{I})\mathbf{k} = \mathbf{0}$ is effectively one equation only, namely

$$x_1 + x_2 + x_3 = 0.$$

Thus any one element of $\mathbf{k}$ may be expressed in terms of the two other elements, which are arbitrary. With x_1 and x_2 arbitrary $\mathbf{k}$ has the form $\{x_1 \quad x_2 \quad -(x_1+x_2)\}$ which, by inspection, may be expressed as a linear combination of two independent eigenvectors,

$$\{x_1 \quad x_2 \quad -(x_1+x_2)\} \equiv x_1\{1 \quad 0 \quad -1\} + x_2\{0 \quad 1 \quad -1\},$$

the first having $x_2 = 0$ and the second $x_1 = 0$. Although this matrix $\mathbf{A}$ has a repeated eigenvalue it does have three independent eigenvectors, $\{3 \quad 2 \quad 1\}$, $\{1 \quad 0 \quad -1\}$ and $\{0 \quad 1 \quad -1\}$.

Example 9.4.

Obtain linearly independent eigenvectors of the matrix

$$\mathbf{A} = \begin{pmatrix} 5 & 3 & -5 \\ 5 & 7 & -7 \\ 6 & 6 & -7 \end{pmatrix}$$

given that its eigenvalues are 1, 2, 2.
The cofactors of the elements in the first row of det $(\mathbf{A} - \lambda\mathbf{I})$ are

$$\lambda^2 - 7, \quad 5\lambda - 7, \quad 6(\lambda - 2),$$

so $\lambda = 1$ gives the eigenvector $\mathbf{k} \propto \{3 \quad 1 \quad 3\}$ and the repeated root $\lambda = 2$ gives the eigenvector $\mathbf{k} \propto \{1 \quad -1 \quad 0\}$. Let $\mathbf{k} = \{x_1 \ x_2 \ x_3\}$; then for $\lambda = 2$ the system $(\mathbf{A} - \lambda\mathbf{I})\mathbf{k} = \mathbf{0}$ is effectively two equations, the rank r of $(\mathbf{A} - 2\mathbf{I})$ being 2, and its solutions are $x_2 = -x_1$, $x_3 = 0$. In this case there is just one arbitrary element of $\mathbf{k}$ and the repeated eigenvalue can be associated with one linearly independent eigenvector only.

If $\mathbf{A}$ is non-singular then its characteristic equation $\det (\mathbf{A} - \lambda\mathbf{I}) = 0$ can be multiplied by $\det (\mathbf{A}^{-1})\lambda^{-n}$, giving

$$\det (\mathbf{A}^{-1}) \, \lambda^{-n} \det (\mathbf{A} - \lambda\mathbf{I}) = 0.$$

By (2.6) and property (b) of determinants this equation is equivalent to

$$\det \{\mathbf{A}^{-1} - (1/\lambda) \, \mathbf{I}\} = 0,$$

which shows that $1/\lambda$ is an eigenvalue of the inverse $\mathbf{A}^{-1}$. Now $\mathbf{A}\mathbf{k} = \lambda\mathbf{k}$ and pre-multiplication by $(1/\lambda)\mathbf{A}^{-1}$ gives

$$(1/\lambda) \, \mathbf{k} = \mathbf{A}^{-1}\mathbf{k}$$

which, since $(1/\lambda)$ is an eigenvalue of $\mathbf{A}^{-1}$, shows that both $\mathbf{A}$ and its inverse have the same eigenvectors.

An important result, stated here without proof, is the Cayley–Hamilton theorem. This asserts that every square matrix satisfies its own characteristic equation. Thus if this equation of an $n \times n$ matrix is

$$\lambda^n + a_{n-1}\lambda^{n-1} + a_{n-2}\lambda^{n-2} + \ldots + a_1\lambda + a_0 = 0,$$

where $a_0, a_1, \ldots, a_{n-1}$ are constants, then the matrix equation

$$\mathbf{A}^n + a_{n-1}\mathbf{A}^{n-1} + a_{n-2}\mathbf{A}^{n-2} + \ldots + a_1\mathbf{A} + a_0\mathbf{I} = \mathbf{0} \quad (9.11)$$

is satisfied. By means of this theorem the inverse of a non-singular matrix may be expressed as a sum of multiples of powers of the matrix, a suitable rearrangement of (9.11) being

$$\mathbf{A}^{-1} = \frac{-1}{a_0} (\mathbf{A}^{n-1} + a_{n-1}\mathbf{A}^{n-2} + a_{n-2}\mathbf{A}^{n-3} + \ldots + a_1\mathbf{I}),$$

$$a_0 = (-1)^n \det \mathbf{A} \neq 0.$$

Problems 2.9a

Obtain the eigenvalues and normalised linearly independent eigenvectors of the following matrices.

1. $\begin{pmatrix} -2 & -1 \\ 4 & 3 \end{pmatrix}$.

2. $\begin{pmatrix} 1 & 1 \\ 0 & 1 \end{pmatrix}$.

3. $\begin{pmatrix} 0 & 0 \\ 0 & 0 \end{pmatrix}$.

4. $\begin{pmatrix} 4 & 0 \\ 0 & 0 \end{pmatrix}$.

5. $\begin{pmatrix} -1 & 2 \\ 0 & 1 \end{pmatrix}$.

6. $\begin{pmatrix} 3 & 0 \\ 0 & 3 \end{pmatrix}$.

7. $\begin{pmatrix} 12 & -25 \\ 4 & -8 \end{pmatrix}$.

8. $\begin{pmatrix} 7 & -4 \\ 6 & -7 \end{pmatrix}$.

9. $\begin{pmatrix} 1 & 0 & 0 \\ 0 & 2 & 0 \\ 0 & 0 & 3 \end{pmatrix}$.

10. $\begin{pmatrix} 5 & -1 & 0 \\ 2 & 0 & 8 \\ 1 & 0 & 0 \end{pmatrix}$.

11. $\begin{pmatrix} 3 & 1 & 1 \\ -4 & -2 & -1 \\ 4 & 4 & 3 \end{pmatrix}$.

12. $\begin{pmatrix} 2 & 0 & 0 \\ 0 & 5 & 0 \\ 0 & 0 & 2 \end{pmatrix}$.

13. $\begin{pmatrix} -3 & 0 & -6 \\ -4 & 1 & -6 \\ 2 & 0 & 4 \end{pmatrix}$.

14. $\begin{pmatrix} 1 & 0 & 0 \\ 4 & 3 & 1 \\ -12 & -9 & -3 \end{pmatrix}$.

15. $\begin{pmatrix} 3 & 0 & 0 & 0 \\ 4 & -7 & 0 & 0 \\ 2 & 5 & -10 & 0 \\ -1 & 12 & 8 & 21 \end{pmatrix}$.

16. Show that a square matrix and its transpose have identical eigenvalues.

17. Show that the elements in the leading diagonal of a triangular matrix are its eigenvalues.

18. Prove that tr $(\mathbf{AB})$ = tr $(\mathbf{BA})$.

19. Show that the eigenvalues of $c\mathbf{A}$, where c is a constant, are c times those of $\mathbf{A}$.

20. Show that the eigenvalues of $\mathbf{A}^c$, where c is a positive integer, are those of $\mathbf{A}$ raised to the power c.

21. Show that the characteristic equation of an n-square matrix of rank r has $n-r$ roots which are zero.

22. A symmetric matrix A has eigenvalues 2, 3, 6 and eigenvectors $\{1 \quad -1 \quad 0\}$, $\{1 \quad 1 \quad -1\}$ associated with the roots 2, 3 respectively. Obtain A.

23. If λ_1, λ_2 are the eigenvalues of a 2×2 matrix A, show that

$$(A - \lambda_1 I)(A - \lambda_2 I) = 0.$$

24. By means of the Cayley–Hamilton theorem derive the inverse of the matrix

$$A = \begin{pmatrix} -1 & 1 & 0 \\ 1 & 2 & -1 \\ -2 & 1 & 1 \end{pmatrix}.$$

Obtain the eigenvalues and eigenvectors of A. State the eigenvalues and eigenvectors of A^{-1} and verify their correctness.

25. By using the Cayley–Hamilton theorem show that

$$190A^{-1} = A^4 - 194A + 383I,$$

where

$$A = \begin{pmatrix} 2 & 2 & 1 \\ 1 & 3 & 1 \\ 1 & 2 & 2 \end{pmatrix},$$

and verify the result.

Matrices which arise from applications are often symmetric. An eigenvalue problem associated with a symmetric matrix may require a slightly more detailed analysis than that already given for the general case. Some properties peculiar to symmetric matrices are now examined and a way is explained in which independent eigenvectors, specially related to each other, may be obtained.

Two column vectors $x = \{x_1, x_2, \ldots, x_n\}$ and $y = \{y_1, y_2, \ldots, y_n\}$ are said to be *orthogonal* if they satisfy the condition

$$x'y = y'x = x_1 y_1 + x_2 y_2 + \ldots + x_n y_n = 0.$$

Real orthogonal vectors which have two elements represent perpendicular lines when their elements are taken to define points referred to a pair of perpendicular axes and these points are joined to the origin. In this way Fig. 2.7 shows the pair of orthogonal vectors $\{2 \quad -1\}$, $\{2 \quad 4\}$ and the pair $\{-3 \quad -1\}$, $\{3/2 \quad -9/2\}$. This interpretation of orthogonality is easily extended to three dimensions for vectors with three elements and the concept is associated with orthogonal vectors which have any number of elements.

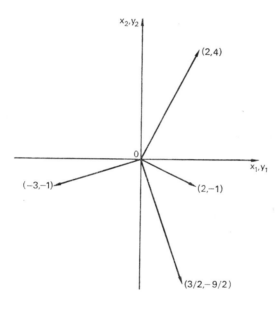

FIG. 2.7.

Two properties of symmetric matrices are now stated and proved.

(a) The eigenvectors associated with distinct eigenvalues of a symmetric matrix are mutually orthogonal.

Let $\mathbf{A}$ be a symmetric matrix and let λ_1, λ_2 be any two of its eigenvalues which are distinct, associated with eigenvectors

HGM: – METCS 7

k_1, k_2 respectively. By definition

$$\mathbf{Ak}_1 = \lambda_1 \mathbf{k}_1$$

which, upon transposition of each side, becomes

$$(\mathbf{Ak}_1)' = (\lambda_1 \mathbf{k}_1)' \quad \text{or} \quad \mathbf{k}_1'\mathbf{A} = \lambda_1 \mathbf{k}_1',$$

since $\mathbf{A}' = \mathbf{A}$. Post-multiplication by k_2 gives

$$\mathbf{k}_1'\mathbf{Ak}_2 = \lambda_1 \mathbf{k}_1'\mathbf{k}_2. \tag{i}$$

Again by definition

$$\mathbf{Ak}_2 = \lambda_2 \mathbf{k}_2$$

and pre-multiplication by $\mathbf{k}_1'$ gives

$$\mathbf{k}_1'\mathbf{Ak}_2 = \lambda_2 \mathbf{k}_1'\mathbf{k}_2,$$

which when subtracted from (i) yields

$$(\lambda_1 - \lambda_2)\, \mathbf{k}_1'\mathbf{k}_2 = 0.$$

But λ_1 is distinct from λ_2, so $\mathbf{k}_1'\mathbf{k}_2 = 0$, that is the eigenvectors k_1, k_2 are orthogonal. The above reasoning applies to any pair of distinct eigenvalues and implies therefore that all the eigenvectors associated with the distinct eigenvalues of $\mathbf{A}$ are mutually orthogonal.

(b) The eigenvalues of a real symmetric matrix are real.

The notation of (a) is used without suffices and a complex conjugate is denoted by a bar above a symbol. By definition

$$\mathbf{Ak} = \lambda \mathbf{k}, \tag{i}$$

the complex conjugate of which is

$$\mathbf{A}\overline{\mathbf{k}} = \overline{\lambda}\overline{\mathbf{k}},$$

since $\overline{\mathbf{A}} = \mathbf{A}$. Transposition of each side and post-multiplication by k gives

$$\overline{\mathbf{k}}'\mathbf{Ak} = \overline{\lambda}\overline{\mathbf{k}}'\mathbf{k}. \tag{ii}$$

Pre-multiplication of (i) by $\overline{\mathbf{k}}'$ gives

$$\overline{\mathbf{k}}'\mathbf{Ak} = \lambda\overline{\mathbf{k}}'\mathbf{k}$$

which when subtracted from (ii) yields

$$(\bar{\lambda}-\lambda)\overline{\mathbf{k}}'\mathbf{k} = 0.$$

But $\overline{\mathbf{k}}'\mathbf{k}$ is a sum of positive real terms so $\bar{\lambda} = \lambda$, which is true only if λ is real.

Suppose a symmetric matrix $\mathbf{A}$ of order n has an eigenvalue λ_i which occurs q times as a root of the characteristic equation det $(\mathbf{A}-\lambda\mathbf{I}) = 0$. It can be shown that the rank r of $(\mathbf{A}-\lambda_i\mathbf{I})$ is $n-q$; this means that the equation $(\mathbf{A}-\lambda_i\mathbf{I})\mathbf{k} = \mathbf{0}$ has a non-zero solution $\mathbf{k}$ which contains q arbitrary elements. Such a solution provides q linearly independent eigenvectors. This was seen to be the case for the non-symmetric matrix in Example 9.3 where $r = 1 = n-q$, but it did not apply to the non-symmetric matrix in Example 9.4 for which $r = 2 \neq n-q$. Any symmetric matrix of order n has the special property of having n linearly independent eigenvectors, whether or not its eigenvalues are distinct. Independent eigenvectors associated with repeated eigenvalues are not necessarily orthogonal, although orthogonal eigenvectors can be derived from them. As will appear later, it may be desirable to have eigenvectors of a symmetric matrix which are all mutually orthogonal. A way of achieving this is now shown.

For example let $\mathbf{A}$ be a 3×3 symmetric matrix which has a non-repeated eigenvalue λ_1, associated with the vector $\mathbf{k}_1$, and an eigenvalue λ_2 which occurs twice. The rank of $(\mathbf{A}-\lambda_2\mathbf{I})$ is $n-q = 1$, which implies that the rows of this matrix are proportional to each other; let any one row be $(a\ b\ c)$. As in Example 9.3 let $\mathbf{k}_2 = \{x_1\ x_2\ x_3\}$ then the system $(\mathbf{A}-\lambda_2\mathbf{I})\mathbf{k} = \mathbf{0}$ is effectively the single equation

$$(a \quad b \quad c)\{x_1 \quad x_2 \quad x_3\} = 0.$$

By inspection $\mathbf{k}_2$ may be chosen as

$$\{0 \quad c \quad -b\}, \quad \{c \quad 0 \quad -a\} \quad \text{or} \quad \{b \quad -a \quad 0\}$$

7*

but, as these three are linearly dependent, $\mathbf{k}_2$ is restricted to be either member of any one pair. Suppose the first two forms of $\mathbf{k}_2$ are chosen. By property (a) these are orthogonal to $\mathbf{k}_1$ but they are not necessarily orthogonal to each other. Any linear combination of the two forms will serve as a vector $\mathbf{k}_2$ and it will be orthogonal to $\mathbf{k}_1$. Take $\mathbf{k}_2$ to be $\{0 \quad c \quad -b\}$ and let a linear combination of the two forms be

$$\{0 \quad c \quad -b\} + p\{c \quad 0 \quad -a\}$$

denoted by $\tilde{\mathbf{k}}_2$, p being an undetermined constant. For $\mathbf{k}_2$ and $\tilde{\mathbf{k}}_2$ to be orthogonal

$$\mathbf{k}_2' \tilde{\mathbf{k}}_2, \ = \mathbf{k}_2' \mathbf{k}_2 + p\mathbf{k}_2'\{c \quad 0 \quad -a\}, \ = 0.$$

This condition determines p as

$$p = \frac{-\mathbf{k}_2' \mathbf{k}_2}{\mathbf{k}_2'\{c \quad 0 \quad -a\}}. \tag{9.12}$$

In this way three mutually orthogonal eigenvectors $\mathbf{k}_1$, $\mathbf{k}_2$, $\tilde{\mathbf{k}}_2$ may be selected for the symmetric matrix $\mathbf{A}$.

Example 9.5.

Obtain three mutually orthogonal eigenvectors of the symmetric matrix

$$\mathbf{A} = \begin{pmatrix} 7 & -2 & 1 \\ -2 & 10 & -2 \\ 1 & -2 & 7 \end{pmatrix}.$$

given that its eigenvalues are 6, 6, 12.

The cofactors of the elements in the first row of det $(\mathbf{A} - \lambda\mathbf{I})$ are

$$(\lambda - 6)(\lambda - 11), \quad 2(6 - \lambda), \quad (\lambda - 6)$$

and, for the unrepeated root $\lambda = 12$, these give an eigenvector proportional to $\mathbf{k}_1 = \{1 \ -2 \ 1\}$. For the repeated root

$\lambda = 6$ a typical row of the matrix $(\mathbf{A} - \lambda\mathbf{I})$ is $(1 \ -2 \ 1)$, $\equiv (a \ b \ c)$. In terms of the general theory let $\mathbf{k}_2 = \{0 \ 1 \ 2\}$, $\equiv \{0 \ c \ -b\}$, and let $\tilde{\mathbf{k}}_2 = \mathbf{k}_2 + p \{1 \ \ 0 \ -1\}$, $\equiv \mathbf{k}_2 + p \{c \ \ 0 \ -a\}$, then by (9.12)

$$p = \frac{-\mathbf{k}_2'\mathbf{k}_2}{\mathbf{k}_2'\{1 \ \ \ 0 \ -1\}} = \frac{5}{2}$$

whence $\tilde{\mathbf{k}}_2 = \{\frac{5}{2} \ 1 \ -\frac{1}{2}\}$ or, doubling this, $\tilde{\mathbf{k}}_2 = \{5 \ 2 \ -1\}$. Thus $\mathbf{A}$ has $\mathbf{k}_1 = \{1 \ -2 \ 1\}$, $\mathbf{k}_2 = \{0 \ 1 \ 2\}$ and $\tilde{\mathbf{k}}_2 = \{5 \ 2 \ -1\}$ as mutually orthogonal eigenvectors. It is left as an exercise to show that these are orthogonal and independent.

Problems 2.9b

1. Prove that the eigenvalues of a real skew-symmetric matrix are imaginary.

2. Obtain three mutually orthogonal eigenvectors of the matrix

$$\begin{pmatrix} 8 & 1 & 4 \\ 1 & 8 & -4 \\ 4 & -4 & -7 \end{pmatrix}.$$

(The third eigenvector may be obtained more easily by a vector product, defined in section 8.4, than by the method shown in Example 9.5.)

2.10. APPLICATIONS OF EIGENVALUES

Before some uses of eigenvalues can be shown, several definitions and results are needed. The concept of orthogonality is extended from row or column matrices to square matrices. A real square matrix $\mathbf{A}$ for which

$$\mathbf{A}' = \mathbf{A}^{-1},$$

that is the transpose equals the inverse, is called an *orthogonal matrix*. Let $\mathbf{A}$ be an orthogonal matrix of order n with column vectors $\mathbf{a}_1, \ldots, \mathbf{a}_n$ then by definition

$$\mathbf{A}^{-1}\mathbf{A} = \mathbf{A}'\mathbf{A} = \begin{pmatrix} \mathbf{a}'_1 \\ \mathbf{a}'_2 \\ \vdots \\ \mathbf{a}'_n \end{pmatrix} (\mathbf{a}_1, \mathbf{a}_2, \ldots, \mathbf{a}_n)$$

$$= \begin{pmatrix} \mathbf{a}'_1\mathbf{a}_1 & \mathbf{a}'_1\mathbf{a}_2 & \ldots & \mathbf{a}'_1\mathbf{a}_n \\ \mathbf{a}'_2\mathbf{a}_1 & \mathbf{a}'_2\mathbf{a}_2 & \ldots & \mathbf{a}'_2\mathbf{a}_n \\ . & . & \ldots & . \\ \mathbf{a}'_n\mathbf{a}_1 & \mathbf{a}'_n\mathbf{a}_2 & \ldots & \mathbf{a}'_n\mathbf{a}_n \end{pmatrix} = \mathbf{I},$$

which implies that

$$\mathbf{a}'_i\mathbf{a}_j = \begin{cases} 0 & \text{when} \quad i \neq j, \\ 1 & \text{when} \quad i = j. \end{cases}$$

This proves the important result that the column vectors of an orthogonal matrix are unit vectors which are mutually orthogonal, the same being true of the row vectors. Conversely if an orthogonal matrix is to be constructed from a set of column, or row, vectors then these must be normalised and be or, if possible, made to be mutually orthogonal. These requirements give purpose to the process of normalisation and to the attention given to the eigenvectors of symmetric matrices. Several properties of orthogonal matrices are now stated.

(a) The transpose and the inverse of an orthogonal matrix are orthogonal.
(b) The product of orthogonal matrices is an orthogonal matrix.
(c) The determinant of an orthogonal matrix equals ± 1.
(d) The eigenvalues, which may be complex, of an orthogonal matrix have unit modulus.
(e) If λ is an eigenvalue of an orthogonal matrix then $1/\lambda$ is an eigenvalue.

To make comprehension easier, the letter $\mathbf{H}$ will be used to denote an orthogonal matrix.

The main use of orthogonal matrices is in the transformation of coordinate systems. Before this is considered as an application, some results connected with the idea of a transformation are presented. Let $\mathbf{x}$, $\mathbf{y}$, $\mathbf{u}$ and $\mathbf{v}$ be column vectors, each of n elements, and let $\mathbf{A}$, $\mathbf{B}$ be square matrices of order n. Suppose that

$$\mathbf{y} = \mathbf{Ax}, \quad \mathbf{x} = \mathbf{Bu} \quad \text{and} \quad \mathbf{y} = \mathbf{Bv},$$

then

$$\mathbf{y} = \mathbf{Ax} = \mathbf{ABu} \quad \text{and, since} \quad \mathbf{y} = \mathbf{Bv},$$
$$\mathbf{v} = \mathbf{B^{-1}ABu} \quad (\det \mathbf{B} \neq 0).$$

Thus the elimination of $\mathbf{x}$ and $\mathbf{y}$ gives a relation between $\mathbf{u}$ and $\mathbf{v}$, in which $\mathbf{u}$ is said to be transformed into $\mathbf{v}$ by the transformation matrix

$$\mathbf{T} = \mathbf{B^{-1}AB}.$$

More generally, for this type of product, $\mathbf{T}$ is called the *transform* of $\mathbf{A}$ by $\mathbf{B}$ and, as will appear, it is the basis for some applications. A special case arises when $\mathbf{A}$ is symmetric and $\mathbf{B}$ is orthogonal for then

$$\mathbf{T} = \mathbf{H^{-1}AH} = \mathbf{H'AH},$$

which is a form simpler to use. Also the transform $\mathbf{T}$ is symmetric since

$$\mathbf{T'} = (\mathbf{H'AH})' = \mathbf{H'AH} = \mathbf{T}.$$

Let λ be an eigenvalue and $\mathbf{k}$ be an associated eigenvector of a matrix $\mathbf{A}$, so that $\mathbf{Ak} = \lambda\mathbf{k}$ and $\det(\mathbf{A} - \lambda\mathbf{I}) = 0$. Since $\mathbf{B}$ is non-singular the equation

$$\det(\mathbf{B^{-1}}) \det(\mathbf{A} - \lambda\mathbf{I}) \det \mathbf{B} = 0$$

is valid and this may be expressed as

$$\det\{\mathbf{B^{-1}(A} - \lambda\mathbf{I})\,\mathbf{B}\} = 0$$

which can be reduced to

$$\det(\mathbf{B^{-1}AB} - \lambda\mathbf{I}) = 0,$$

showing that λ is also an eigenvalue of the transform $\mathbf{T} = \mathbf{B}^{-1}\mathbf{AB}$. This result applies to all the eigenvalues of $\mathbf{A}$, showing that both the trace and the determinant of a matrix $\mathbf{A}$ are preserved in the transform $\mathbf{B}^{-1}\mathbf{AB}$. Furthermore as $\mathbf{Ak} = \lambda\mathbf{k}$ it follows that

$$\mathbf{ABB}^{-1}\mathbf{k} = \lambda\mathbf{k}$$

which, upon pre-multiplication by $\mathbf{B}^{-1}$, gives

$$\mathbf{B}^{-1}\mathbf{ABB}^{-1}\mathbf{k} = \lambda\mathbf{B}^{-1}\mathbf{k} \quad \text{or} \quad \mathbf{T}(\mathbf{B}^{-1}\mathbf{k}) = \lambda(\mathbf{B}^{-1}\mathbf{k}),$$

showing that $\mathbf{B}^{-1}\mathbf{k}$ is an eigenvector of the transform $\mathbf{T}$, associated with the eigenvalue λ.

DIAGONAL TRANSFORMS

If $\mathbf{A}$ is a square matrix of order n having n linearly independent eigenvectors $\mathbf{k}_1, \mathbf{k}_2, \ldots, \mathbf{k}_n$, and if $\mathbf{B}$ is the matrix which has these vectors for its columns, then the transform $\mathbf{B}^{-1}\mathbf{AB}$ is diagonal. This is now proved. By definition

$$\mathbf{Ak}_i = \lambda_i\mathbf{k}_i \quad (i = 1, 2, \ldots, n)$$

and

$$\mathbf{AB} = \mathbf{A}(\mathbf{k}_1, \mathbf{k}_2, \ldots, \mathbf{k}_n) = (\lambda_1\mathbf{k}_1, \lambda_2\mathbf{k}_2, \ldots, \lambda_n\mathbf{k}_n),$$

which may be expressed as the product

$$\mathbf{AB} = (\mathbf{k}_1, \mathbf{k}_2, \ldots, \mathbf{k}_n) \begin{pmatrix} \lambda_1 & 0 & \ldots & 0 \\ 0 & \lambda_2 & \ldots & 0 \\ \cdot & \cdot & \ldots & \cdot \\ 0 & 0 & \ldots & \lambda_n \end{pmatrix}.$$

This diagonal matrix, with the eigenvalues of $\mathbf{A}$ in the leading diagonal, is often denoted by $\mathbf{\Lambda}$ (capital lambda) or by diag λ. With the former notation the equation becomes

$$\mathbf{AB} = \mathbf{B\Lambda}$$

which, upon pre-multiplication by $\mathbf{B}^{-1}$, gives the required diagonal transform

$$\mathbf{B}^{-1}\mathbf{AB} = \mathbf{\Lambda}.$$

Note that the sequence of eigenvalues in diag λ corresponds with that of their related eigenvectors in **B**. If **A** is symmetric and **B** is an orthogonal matrix **H** constructed from normalised orthogonal eigenvectors of **A**, then the diagonal transform is simply

$$\mathbf{H'AH} = \mathbf{\Lambda}.$$

The algebra of diagonal matrices is very simple and in this lies the advantage of a transformation to diagonal form.

POWERS OF A MATRIX

Suppose a square matrix **A** of order n is raised to a power P, a positive integer, so high that the evaluation of $\mathbf{A}^P$ by successive products is impracticable. Provided **A** has n independent eigenvectors, the value of $\mathbf{A}^P$ can be obtained quite easily by means of a transformation to diagonal form. Let **B** have n independent eigenvectors of **A** for its n columns, then

$$\mathbf{B}^{-1}\mathbf{AB} = \mathbf{\Lambda}$$

which, upon pre-multiplication by **B** and post-multiplication by $\mathbf{B}^{-1}$, becomes

$$\mathbf{A} = \mathbf{B\Lambda B}^{-1}.$$

In terms of this form $\mathbf{A}^P$ can be expressed as

$$\mathbf{A}^P = (\mathbf{B\Lambda B}^{-1})(\mathbf{B\Lambda B}^{-1}) \ldots (\mathbf{B\Lambda B}^{-1}),$$

the factor $(\mathbf{B\Lambda B}^{-1})$ occurring P times, and by rearranging brackets

$$\mathbf{A}^P = \mathbf{B\Lambda}(\mathbf{B}^{-1}\mathbf{B})\,\mathbf{\Lambda}(\mathbf{B}^{-1}\mathbf{B}) \ldots (\mathbf{B}^{-1}\mathbf{B})\mathbf{\Lambda B}^{-1}.$$

Since $\mathbf{B}^{-1}\mathbf{B} = \mathbf{I}$,

$$\mathbf{A}^P = \mathbf{B\Lambda I\Lambda I} \ldots \mathbf{I\Lambda B}^{-1}$$

or

$$\mathbf{A}^P = \mathbf{B\Lambda\Lambda} \ldots \mathbf{\Lambda B}^{-1} = \mathbf{B\Lambda}^P\mathbf{B}^{-1},$$

where

$$\mathbf{\Lambda}^P = \{\text{diag}\,(\lambda_1, \lambda_2, \ldots, \lambda_n)\}^P$$

is simply

$$\text{diag}\,(\lambda_1^P, \lambda_2^P, \ldots, \lambda_n^P).$$

Example 10.1.

Obtain $\mathbf{A}^{101}$, where $\mathbf{A} = \begin{pmatrix} 3 & -2 \\ 4 & -3 \end{pmatrix}$.

The eigenvalues of $\mathbf{A}$ are -1, 1 and respective eigenvectors are $\{1 \; 2\}$, $\{1 \; 1\}$ so $\mathbf{B} = \begin{pmatrix} 1 & 1 \\ 2 & 1 \end{pmatrix}$ and $\mathbf{B}^{-1} = \begin{pmatrix} -1 & 1 \\ 2 & -1 \end{pmatrix}$.

$$\mathbf{\Lambda}^{101} = \begin{pmatrix} (-1)^{101} & 0 \\ 0 & (1)^{101} \end{pmatrix} = \begin{pmatrix} -1 & 0 \\ 0 & 1 \end{pmatrix}$$

so

$$\mathbf{A}^{101} = \mathbf{B}\mathbf{\Lambda}^{101}\mathbf{B}^{-1} = \begin{pmatrix} 1 & 1 \\ 2 & 1 \end{pmatrix} \begin{pmatrix} -1 & 0 \\ 0 & 1 \end{pmatrix} \begin{pmatrix} -1 & 1 \\ 2 & -1 \end{pmatrix} = \begin{pmatrix} 3 & -2 \\ 4 & -3 \end{pmatrix}$$
$$= \mathbf{A}.$$

This rather surprising result, $\mathbf{A}^{101} = \mathbf{A}$, has arisen because $\mathbf{A}$ satisfies the uncommon condition $\mathbf{A}^2 = \mathbf{I}$.

REDUCTION OF QUADRATIC FORMS

A real quadratic form Q in the n variables x_1, x_2, ..., x_n is a homogeneous polynomial which has the general form

$$Q = \sum_{i=1}^{n} \sum_{j=1}^{n} c_{ij}x_i x_j,$$

where the coefficients c_{ij} are real. By considering the variables as a column vector $\mathbf{x} = \{x_1, x_2, ..., x_n\}$ the form Q may be expressed compactly as the matrix product

$$Q = \mathbf{x}'\mathbf{A}\mathbf{x}.$$

For a given quadratic form the matrix $\mathbf{A}$ is not unique but it can be chosen to be symmetric. This symmetric matrix is called the matrix of the quadratic form and is the only form of $\mathbf{A}$ to be used in the following application.

The way in which **A** is constructed is most easily seen by an example. Using x, y, z, rather than x_1, x_2, x_3, a typical quadratic form in three variables is

$$Q = 5x^2 + 7y^2 + 8z^2 + 8xy - 6xz - 2yz.$$

In symmetric matrix form this is

$$(x \quad y \quad z) \begin{pmatrix} 5 & 4 & -3 \\ 4 & 7 & -1 \\ -3 & -1 & 8 \end{pmatrix} \begin{pmatrix} x \\ y \\ z \end{pmatrix}.$$

The elements in the leading diagonal are the coefficients of the squared variables in Q. The symmetric elements 4, -3 and -1 are half the coefficients which appear in the other terms of Q.

A quadratic form would be much simpler if it consisted of terms in the squared variables only, that is if $c_{ij} = 0$ when $i \neq j$ for all i and j in the general form. To achieve such simplicity the form would have to be defined in terms of another coordinate system, say the X, Y, Z system for three variables. The transformation to this system from the x, y, z system is now obtained. Because **A** is symmetric it may be expressed in the form $\mathbf{A} = \mathbf{H\Lambda H'}$, where **H** is orthogonal, and the quadratic form is

$$Q = \mathbf{x'Ax} = \mathbf{x'H\Lambda H'x}.$$

Let $\mathbf{X} = \{X\ Y\ Z\}$ and suppose **x** is related to **X** by $\mathbf{x} = \mathbf{HX}$. Pre-multiplication by $\mathbf{H'}$ shows that $\mathbf{X} = \mathbf{H'x}$ which, upon transposition, becomes $\mathbf{X'} = \mathbf{x'H}$. These relations between **X** and **x** enable Q to be expressed as

$$Q = \mathbf{X'\Lambda X} = \lambda_1 X^2 + \lambda_2 Y^2 + \lambda_3 Z^2$$

which is of the required simple form but in terms of a new set of variables.

Example 10.2.

Reduce to a sum of squares the quadratic form
$$Q = 7x^2 + 10y^2 + 7z^2 - 4xy + 2xz - 4yz.$$

The matrix of the form is $\mathbf{A} = \begin{pmatrix} 7 & -2 & 1 \\ -2 & 10 & -2 \\ 1 & -2 & 7 \end{pmatrix}$ which, in

Example 9.5, was found to have eigenvalues $6, 6, 12$ associated with orthogonal eigenvectors $\{0 \ 1 \ 2\}$, $\{5 \ \ 2 \ -1\}$, $\{1 \ -2 \ 1\}$ respectively. A reduced form of Q is therefore
$$Q = 6X^2 + 6Y^2 + 12Z^2,$$

where $\mathbf{X} = \mathbf{H}'\mathbf{x}$, $\mathbf{H}$ being the orthogonal matrix with the normalised eigenvectors of $\mathbf{A}$ as its columns. Thus

$$\mathbf{H} = \begin{pmatrix} 0 & 5/\sqrt{30} & 1/\sqrt{6} \\ 1/\sqrt{5} & 2/\sqrt{30} & -2/\sqrt{6} \\ 2/\sqrt{5} & -1/\sqrt{30} & 1/\sqrt{6} \end{pmatrix}$$
$$= \frac{1}{\sqrt{30}} \begin{pmatrix} 0 & 5 & \sqrt{5} \\ \sqrt{6} & 2 & -2\sqrt{5} \\ 2\sqrt{6} & -1 & \sqrt{5} \end{pmatrix}$$

and, as a set of three equations, the transformation $\mathbf{X} = \mathbf{H}'\mathbf{x}$ is
$$X = (y+2z)/\sqrt{5}, \quad Y = (5x+2y-z)/\sqrt{30},$$
$$Z = (x-2y+z)/\sqrt{6}.$$

The student may like to verify that $\mathbf{H}'\mathbf{A}\mathbf{H} = \text{diag}\ (6, 6, 12)$.

Example 10.3.

Obtain a pair of rectangular axes which coincide with the axes of the ellipse $7x^2 + 4y^2 + 4xy = 1$.

This equation is of the form $Q = 1$, where Q is a quadratic form in two variables. The symmetric matrix of the form is

$$A = \begin{pmatrix} 7 & 2 \\ 2 & 4 \end{pmatrix},$$

which has the eigenvalues 3, 8 with related eigenvectors $\{1 \; -2\}$, $\{2 \; 1\}$. An orthogonal matrix H which satisfies $H'AH = \text{diag}\,(3, 8)$ is

$$H = \frac{1}{\sqrt{5}} \begin{pmatrix} 1 & 2 \\ -2 & 1 \end{pmatrix}$$

and the original x, y coordinate system is related to the required X, Y system by $X = H'x$, which gives the pair of equations

$$X = (x-2y)/\sqrt{5}, \quad Y = (2x+y)/\sqrt{5}.$$

In terms of the new system the ellipse has the equation

$$3X^2 + 8Y^2 = 1$$

and its axes are coincident with the lines $X = 0$ and $Y = 0$ which, in terms of the old system, are the lines $2y = x$ and $y = -2x$. The ellipse and the two coordinate systems are shown in Fig. 2.8.

This kind of application can be made for a quadratic form Q in three variables, in which case the equation $Q = 1$ represents the surface defined by a central quadric. If Q is the form in Example 10.2 then $Q = 1$ is the equation of an ellipsoid. The reduced form obtained shows that the ellipsoid has semi-axes of length $1/\sqrt{6}, 1/\sqrt{6}$ and $1/2\sqrt{3}$. The principal planes of the ellipsoid are the planes $X = 0, Y = 0, Z = 0$ which, in terms of the original system, are defined by the equations

$$y+2z = 0, \quad 5x+2y-z = 0, \quad x-2y+z = 0.$$

The axes of the ellipsoid coincide with the three lines in which pairs of these planes meet.

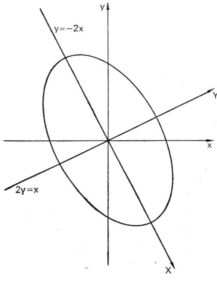

FIG. 2.8.

SIMULTANEOUS DIFFERENTIAL EQUATIONS

Students who have not studied differential equations may need to become familiar with Chapters 3 and 4 in order to understand this application. Consider the pair of simultaneous equations

$$\ddot{y}_1 = 5y_1 + 4y_2,$$
$$\ddot{y}_2 = y_1 + 8y_2,$$

where the dependent variables y_1, y_2 are functions of t and a dot denotes d/dt. Let

$$\mathbf{y} = \{y_1 \quad y_2\} \quad \text{and} \quad \mathbf{A} = \begin{pmatrix} 5 & 4 \\ 1 & 8 \end{pmatrix},$$

then the system can be expressed in the matrix form

$$\ddot{\mathbf{y}} = \mathbf{A}\mathbf{y}.$$

Suppose the system is satisfied by $\mathbf{y} = \mathbf{k}e^{mt}$, where $\mathbf{k}$ is a constant column vector and m is a constant number. Substitution of this assumed solution in the system gives

$$m^2\mathbf{k}\,e^{mt} = \mathbf{A}\mathbf{k}\,e^{mt}$$

which, since $e^{mt} \neq 0$, is effectively

$$\mathbf{A}\mathbf{k} = m^2\mathbf{k}.$$

For a non-zero solution to exist m^2 must be an eigenvalue of $\mathbf{A}$ and $\mathbf{k}$ an eigenvector. The eigenvalues of $\mathbf{A}$ are 4, 9 and related eigenvectors are $\{4 \ -1\}$, $\{1 \ \ 1\}$. Solutions of the system are therefore

$$\mathbf{y} = \{4 \ -1\}\,e^{\pm 2t} \quad \text{and} \quad \mathbf{y} = \{1 \ \ 1\}\,e^{\pm 3t},$$

each of which may be multiplied by an arbitrary constant. The general solution of the system is a linear combination of these four independent solutions and therefore may be expressed as

$$\mathbf{y} = A'\{4 \ -1\}\,e^{2t} + B'\{4 \ -1\}e^{-2t} + C'\{1 \ \ 1\}\,e^{3t} + D'\{1 \ \ 1\}\,e^{-3t},$$

where A', B', C', D' are arbitrary constants. In terms of the identities $e^{\pm mt} = \cosh mt \pm \sinh mt$ the solution has the form

$$\mathbf{y} = \{4 \ -1\}\,[(A'+B')\cosh 2t + (A'-B')\sinh 2t]$$
$$+ \{1 \ \ 1\}\,[(C'+D')\cosh 3t + (C'-D')\sinh 3t].$$

Now replace the four sums of arbitrary constants by the arbitrary constants A, B, C, D, simply for elegance, and separate y_1 from y_2 in $\mathbf{y}$ to obtain the general solution in the form

$$y_1 = 4(A\cosh 2t + B\sinh 2t) + C\cosh 3t + D\sinh 3t,$$
$$y_2 = -(A\cosh 2t + B\sinh 2t) + C\cosh 3t + D\sinh 3t,$$

the validity of which is evident on inspection.

Problems 2.10

1. The eigenvalues of a 3×3 symmetric matrix $\mathbf{A}$ are 0, 1, 4 and the associated eigenvectors are proportional to $\mathbf{k}_1 = \{1 \ -1 \ \ 0\}$, $\mathbf{k}_2 = \{0 \ \ 0 \ \ 1\}$, $\mathbf{k}_3$ respectively. Obtain $\mathbf{k}_3$ and form the matrix $\mathbf{A}$ by evaluating the product of three matrices, each suitably defined in terms of the given eigenvalues or the related eigenvectors.

2. Reconsider **22** of Problems 2.9a and use a more elegant method.

3. Use the Cayley–Hamilton theorem to derive the inverse of

$$A = \begin{pmatrix} 3 & -1 \\ 5 & -3 \end{pmatrix}$$

and show that $A^{13} = 2^{12}A$.

4. Obtain the eigenvalues and three linearly independent eigenvectors of the matrix

$$A = \begin{pmatrix} 4 & 0 & 2 \\ -6 & 1 & -4 \\ -6 & 0 & -3 \end{pmatrix}.$$

Express the matrix in the form

$$A = B(\text{diag } \lambda)B^{-1},$$

check the validity of this form and show that $A^m = A$ where m is a positive integer.

5. Show that the inverse of a non-singular matrix A is given by

$$A^{-1} = B(\text{diag } 1/\lambda)B^{-1}$$

and use this form to invert

$$A = \begin{pmatrix} 13 & 6 \\ -16 & -7 \end{pmatrix}.$$

6. Obtain A^P as a single 3×3 matrix, involving P, where

$$A = \begin{pmatrix} 1 & 1 & -2 \\ -1 & 2 & 1 \\ 0 & 1 & -1 \end{pmatrix}$$

and P is an even integer. Use the result to derive, in the most simplified form, the square of the inverse of A.

7. Reduce to a sum of squares the quadratic form

$$x^2 + 2y^2 - 3z^2 - 12xy - 8xz + 4yz$$

and give, as a set of three equations, the related transformation from the x, y, z system to the X, Y, Z system.

8. Find the lengths of the semi-axes of the ellipsoid

$$(4x + 5z)(5x + 4z) = 9(4 + xz - y^2)$$

and obtain the equations to planes which define its principal axes.

9. By means of an orthogonal matrix $\mathbf{H}$ obtain the linear transformation, as a set of three equations which relates a new coordinate system (X, Y, Z) to the (x, y, z) system, for the conversion of the quadratic form

$$Q = 2x^2 + 5y^2 + 2z^2 - 4xy + 2xz - 4yz$$

into the form

$$a(X^2 + Y^2) + bZ^2 .$$

Verify that

$$\mathbf{H'AH} = \text{diag } (a, a, b),$$

where $\mathbf{A}$ is the symmetric matrix of the form Q, and use this relation to derive the inverse of $\mathbf{A}$.

10. Obtain the general solutions of the following systems.

(a) $\ddot{y}_1 = 20y_1 + 5y_2,$ (b) $\ddot{y}_1 = 9y_1 - 2y_2,$

    $\ddot{y}_2 = 19y_1 + 6y_2 .$ $\ddot{y}_2 = 5y_1 + 2y_2 .$

CHAPTER 3

Introduction to Ordinary Differential Equations

3.1. INTRODUCTION

In this and the next three chapters several methods are presented for solving the more common types of differential equation. Many problems which arise in science and engineering may be described mathematically by a differential equation, often by many such equations. In its most general form a differential equation is a relation of equality between a mixture of an unknown function, the variable on which it depends and at least one of its derivatives with respect to that variable; the solution of such an equation is a statement, not containing derivatives, of the unknown function in terms of the independent variable. The task of extracting the required function from such a mixture can be a stimulating exercise, and the simplicity of many methods for achieving this can make their application satisfying.

A solution cannot always be obtained as a functional relation between the variables, though nearly always the hope is that this may be possible. A method which yields this type of solution is said to be analytic, as is the solution itself, in the sense that the stages of solution analyse the equations exactly by concepts of calculus and algebra.

In practice equations often arise which cannot be solved by analytic methods and then approximate numerical solutions may be sought. A numerical solution is a set of corresponding values for the variables over a specified range. Although numerical solutions lack generality and the elegance of analytic solutions they are hardly less useful, for ultimately the solution of any practical problem is stated as a set of numbers.

The advent of high-speed computers during the past two decades has meant that very many problems, expressible as differential equations, can be solved numerically with little effort, whereas previously their solution could not have been considered or at best perhaps could have been achieved analytically under limiting and simplifying restrictions.

The numerical way of solving a differential equation under certain conditions may convert the problem into one of linear algebra so, although this idea is not developed here, the material of previous chapters may be needed in what appears to be an entirely separate field of study.

The main purpose of this chapter is to present some principles which should aid the understanding of what is implied by a solution to a differential equation and then to treat first-order equations briefly. This is done by illustrating a solution graphically and by obtaining an approximate numerical solution before the analytic methods are explained.

3.2. DEFINITIONS

If y is a function of x then, in terms of these variables, a *differential equation* is a relation between the dependent variable y and the independent variable x which includes at least one derivative of y with respect to x. Either x or y, possibly both, may be absent from the equation.

Examples are

$$x^2 y'' - xy' + y = 0, \tag{2.1}$$

$$y' = 2, \tag{2.2}$$

$$yy'' = (y')^2 + yy', \tag{2.3}$$

$$F(x, y, y') = 0, \tag{2.4}$$

where y', y'' are abbreviated forms of dy/dx, d^2y/dx^2 respectively.

This notation is extended to y''', $y^{iv}, \ldots$ and $y^{(n)}$ denotes $d^n y/dx^n$. When time t is the independent variable it is customary for a dot to denote d/dt, as in the pair of equations below.

When dependent variables x and y are functions of a single independent variable t, the most general form of differential equation relates x, y, t and some of the derivatives of x and of y with respect to t. For example

$$5\ddot{x} + \dot{y} + 2x = 4\cos t, \tag{2.5}$$

$$3\dot{x} + y = 8t\cos t. \tag{2.6}$$

When there is but one independent variable, as in the previous examples, the derivatives in a related differential equation are all total and the equation is said to be *ordinary*; such equations are the main concern of these four chapters.

A *partial* differential equation is related to a function of two or more variables, such as $z = F(x, y)$, and therefore contains partial derivatives. An example is Laplace's equation

$$\frac{\partial^2 z}{\partial x^2} + \frac{\partial^2 z}{\partial y^2} = 0. \tag{2.7}$$

The *order* of a differential equation is the order of the highest derivative present; thus the even-numbered equations above are of first order and the rest are of second order. The *degree* of a differential equation is the degree of the highest power of the highest ordered derivative present; thus equation (2.3) is of degree one because the degree of y'' is one. This form of the definition is slightly simpler than a more general form that is sometimes given, but it is sufficient for the equations to be considered here.

3.3. SOLUTION OF A DIFFERENTIAL EQUATION

The general ordinary differential equation of order n with one dependent variable y is of the form

$$F(x, y, y', y'', \ldots, y^{(n)}) = 0. \tag{3.1}$$

If the substitution in (3.1) of a function $y = f(x)$ and its derivatives produces an identity in x, then $f(x)$ is said to be a *solution*, or an *integral*, of the differential equation and to define an *integral curve* in the x–y plane.

Example 3.1.

Show that $y = xe^x$ is a solution of the equation

$$y'' + y' - 2y = 3e^x.$$

Upon substitution of y, y' and y'' the left-hand side becomes $(x+2)e^x + (x+1)e^x - 2xe^x$, which simplifies to $3e^x$. Thus the differential equation is reduced to the identity $3e^x \equiv 3e^x$, showing that the given function is a solution.

The process of solving a differential equation is called integration and usually it requires a special method suitable for the type of equation in question. Integration in the limited sense of the integral calculus is often called *quadrature*. When the solution of a differential equation is reduced to direct integration the equation is said to be solved by quadratures.

In this short account it is assumed that the functions $F(x)$ and $f(x)$ are acceptable in the mathematical sense and therefore possess any necessary properties, such as continuity and differentiability, over a required range.

Problems 3.3

1. Classify the following differential equations by order and by degree.

(a) $x^4 y''' - y^2 (y'')^3 = 2.$ (d) $yy' + (y'')^3 = \sin x.$

(b) $y' = y(1 + x^2).$ (e) $(yy')^2 - e^x y^4 = 2x.$

(c) $\dfrac{\partial^2 y}{\partial t^2} - 3 \dfrac{\partial x}{\partial t} = 0.$

2. Verify that the following differential equations are satisfied by the corresponding functions.

(a) $xy' = y(2 + x \cot x); \quad y = x^2 \sin x.$

(b) $y' = y(1 - \tan x); \quad y = e^x \cos x.$

(c) $x^2 (y'' - y') = 2(x+1)y; \quad y = x^2 e^x.$

(d) $xyy'' = y'(y - xy'); \quad y = \sqrt{(1 + x^2)}.$

(e) $x^2 (y''' + y'' - y' - y) = x - 1 - x^2 \{1 + (x+1) \log x\};$
$y = x(e^{-x} + \log x).$

3. Integrate equation (2.2) by quadratures.

4. Show that the equation

$$\{1 + (y')^2\} y^2 = a^2$$

is satisfied by

$$(x - C)^2 + y^2 = a^2,$$

which represents circles of constant radius a centred on the x-axis at $x = C$, where C can take any value. Show that the equation is satisfied also by the envelopes $y = \pm a$ to the circles.

3.4. DIRECTION FIELDS AND ISOCLINES

The solution of a first-order differential equation is now illustrated graphically. Suppose for simplicity of explanation that the equation can be expressed in the form

$$\frac{dy}{dx} = g(x, y). \tag{4.1}$$

A solution of (4.1) is a function $y = f(x)$, which defines a curve in the x–y plane, such that $y' \equiv g\{x, f(x)\}$. Although the function y is not known, the differential equation (4.1) enables the slope dy/dx of the curve $y = f(x)$ to be evaluated at any point (x, y) and is said to define a *direction field*. A direction field may be shown by constructing at each of a set of points (x, y) a short line, called a *line element*, with slope $y' = g(x, y)$. An *isocline* is a curve in the direction field at every point of which the line elements have the same slope, so that $y' = g(x, y)$ is constant. A set of isoclines is a set of curves $g(x, y) = k$, one for each value of the parameter k. A direction field is most easily constructed by first drawing a set of isoclines, as shown in the following example.

Example 4.1.

Construct the direction field of the equation

$$\frac{dy}{dx} = \frac{x}{2} - 1 \tag{4.2}$$

and form the integral curve which passes through the origin.

In this simple example, the right-hand side of the equation being a function of x only, the isoclines are the set of straight lines $x/2 - 1 = k$, shown in Fig. 3.1 for seven values of k. Along

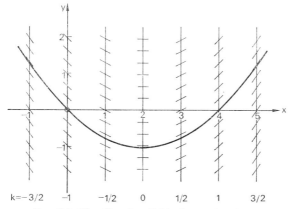

FIG. 3.1. The direction field of equation (4.2)

each isocline several line elements are drawn with slopes equal to the respective value of k. As a line element shows the direction of the integral curve which passes through a point it may be considered as approximating to a short arc of the curve. By joining a sequence of elements, the direction of each leading to the next, an integral curve is formed; the curve shown in Fig. 3.1 is the one which passes through the origin.

Equation (4.2) is solvable by direct integration, that is by quadratures, to give

$$y = \tfrac{1}{4}x^2 - x + C, \qquad (4.3)$$

where C is an arbitrary constant. This solution defines a set of parabolas, one for each value of C, and the direction field gives the impression of such a set over the limited range of the figure; the curve through the origin is for $C = 0$.

Problems 3.4

1. By means of isoclines sketch the direction field of the equation $y' = x^2 + y^2$ and trace the integral curve which passes through the origin.

2. Sketch the direction field of the equation $y' + y = x^2$ and trace the integral curve which passes through $(0, 2)$. Compare this approximate curve with that defined by the solution $y = x^2 - 2x + 2$ to the differential equation.

3. Sketch the direction field of the equation $y' + y = 2x$ and trace the integral curve which passes through the origin; it has the equation $y = 2(x - 1 + e^{-x})$. What is special about the isocline $k = 2$?

4. Sketch the direction field of the equation $y' + y = e^x$ and trace the integral curve which passes through $(0, 1)$. This curve should have the shape of the catenary $y = \cosh x$.

3.5. NUMERICAL SOLUTION

An approximate solution of equation (4.2) is now illustrated by an elementary numerical method due to Euler. The method is based on the assumption that the slope dy/dx at any given point on an integral curve is approximately equal to $\delta y/\delta x$, where δx, δy are increments in x, y taken from the same point. By starting the process at a given point the solution is derived as a sequence of points which is continued until the required range is covered. The increment in the independent variable, called the step length, is usually kept at the same value and, to ensure a fairly good approximate solution, this should be as small as is practically possible.

Example 5.1.

Derive an approximate solution of equation (4.2) which passes through the origin, over the range $0 \leqslant x \leqslant 3$. Since $dy/dx = (x/2) - 1$, the assumption that $dy/dx = \delta y/\delta x$ gives $\delta y = y' \, \delta x = \left(\frac{1}{2}x - 1\right)\delta x$. To limit the amount of numerical work a fairly large step length $\delta x = \frac{1}{2}$ is taken here. The results are shown in Table 1 where the set of six points (x, y)

TABLE 1. Numerical Solution of Equation (4.2)

x	y	δx	y'	$\delta y = y' \, \delta x$	$x + \delta x$	$y + \delta y$
0	0	$\frac{1}{2}$	-1	$-\frac{1}{2}$	$\frac{1}{2}$	$-\frac{1}{2}$
$\frac{1}{2}$	$-\frac{1}{2}$	$\frac{1}{2}$	$-\frac{3}{4}$	$-\frac{3}{8}$	1	$-\frac{7}{8}$
1	$-\frac{7}{8}$	$\frac{1}{2}$	$-\frac{1}{2}$	$-\frac{1}{4}$	$\frac{3}{2}$	$-\frac{9}{8}$
$\frac{3}{2}$	$-\frac{9}{8}$	$\frac{1}{2}$	$-\frac{1}{4}$	$-\frac{1}{8}$	2	$-\frac{5}{4}$
2	$-\frac{5}{4}$	$\frac{1}{2}$	0	0	$\frac{5}{2}$	$-\frac{5}{4}$
$\frac{5}{2}$	$-\frac{5}{4}$	$\frac{1}{2}$	$\frac{1}{4}$	$\frac{1}{8}$	3	$-\frac{9}{8}$

forming the solution appears in the first pair of columns. The initial point is the origin and from this a step $\delta x = \frac{1}{2}$ produces an increment $\delta y = -\frac{1}{2}$, so the next point in the sequence is $\left(\frac{1}{2}, -\frac{1}{2}\right)$ shown in the final pair of columns. The process is continued at this point and is repeated until the range is covered in six steps. The points of the numerical solution are plotted in Fig. 3.2 and joined by a smooth curve, which can be compared with the correct integral curve $y = \frac{1}{4}x^2 - x$.

The accuracy of the above numerical solution is poor and this is to be expected from the use of such an unrefined method with a large step length. However, the example serves to show the principles upon which some of the more elegant and highly accurate methods are based and therefore gives some insight into the processes used by a computer.

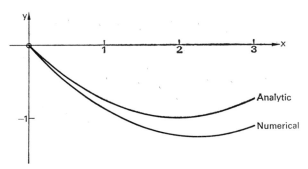

FIG. 3.2. Numerical solution of equation (4.2)

Problems 3.5

1. Use Euler's method and a step of length $\frac{1}{4}$ to obtain a numerical solution of equation (4.2), satisfied at the origin, over the range $0 \leqslant x \leqslant 2$. Sketch the integral curve and compare the solution with that in Example 5.1.

2. Use Euler's method and a step of length $\frac{1}{4}$ to obtain a numerical solution of the equation

$$\frac{dy}{dx} = x + y,$$

satisfied at the origin, over the range $0 \leqslant x \leqslant 2$. Sketch the integral curve and compare it with that of the analytic solution $y = e^x - x - 1$.

Also obtain a numerical solution for the same range and step by Euler's improved method, as now described, and sketch the integral curve. Let $P_n = (x_n, y_n)$, $P_{n+1} = (x_n + \delta x, y_{n+1})$, $n = 0, 1, 2, \ldots, 7$, be any two adjacent points in the numerical solution. The value of y' at

$P_0 = (0, 0)$ is 0 by the differential equation, so $\delta y = 0$, $P_1 = \left(\frac{1}{4}, 0\right)$ and the value of y' at P_1 is $\frac{1}{4}$, as before. The value of y' at P_1 is now modified to be the mean of its (correct) value at P_0 and the value just computed at P_1; the mean is $\frac{1}{8}$. From this is computed a modified change in y between P_0 and P_1, $\delta y = \frac{1}{32}$, whence a further value of y' at P_1 is $\frac{9}{32}$. Again the mean of this value and the value at P_0 is taken, so that $y' = \frac{9}{64}$ and $\delta y = \frac{9}{256}$. This iterative process is continued until changes in the computed values of δy between P_0 and P_1 are small enough to be insignificant. The process is then applied to the next interval, between P_1 and P_2, keeping the value of y' just computed at P_1 fixed whilst the value of y' at P_2 is successively modified. Each interval between P_n and P_{n+1} of the complete range is treated in this way.

3.6. GENERAL, PARTICULAR AND SINGULAR SOLUTIONS

The *general* solution of the nth order differential equation (3.1) is a set of functions or curves

$$G(x, y, C_1, C_2, \ldots, C_n) = 0, \tag{6.1}$$

each member of which is a solution of the differential equation. $C_1, C_2, \ldots, C_n$ are arbitrary constants, or parameters, and (6.1) is said to define an n-parameter family of curves. Conversely, an n-parameter family of curves is the general solution to some differential equation of order n.

The features of a general solution have been shown by solution (4.3) to equation (4.2). As the equation is of first order its general solution contains one arbitrary constant, the one-parameter family of curves consisting of parabolas. Conversely, a single differentiation of solution (4.3) eliminates C and the corresponding differential equation (4.2) is formed.

Example 6.1.

Obtain the differential equation whose general solution is the set of all circles centred at the origin.

The family of circles is $x^2 + y^2 = C^2$, where the parameter C is the radius. Differentiation eliminates C to give the differential equation $x+yy' = 0$, or $y' = -x/y$. The isoclines $y' = k$ are the set of straight lines $y = -(1/k)x$ through the origin, a result readily seen from the geometry of the circles.

In general the elimination of the n parameters from an n-parameter family requires n differentiations to provide sufficient equations for the elimination to be made, the result of which is a differential equation of order n containing no arbitrary constants.

Example 6.2.

Obtain the differential equation whose general solution is the set of all circles in the x–y plane.

The general equation to a circle is

$$x^2 + y^2 + C_1 x + C_2 y + C_3 = 0, \tag{i}$$

which contains three parameters.

$$\frac{d}{dx} \text{ (i):} \qquad 2x + 2yy' + C_1 + C_2 y' = 0. \tag{ii}$$

$$\frac{d}{dx} \text{ (ii):} \quad 2 + 2(y')^2 + 2yy'' + C_2 y'' = 0. \tag{iii}$$

$$\frac{d}{dx} \text{ (iii):} \qquad 6y'y'' + 2yy''' + C_2 y''' = 0. \tag{iv}$$

The elimination of C_2 between (iii) and (iv) gives the third-order differential equation

$$\{1 + (y')^2\}\, y''' = 3y'(y'')^2.$$

A *particular* solution is derived from the general solution by assigning particular values to the n arbitrary constants. Thus the particular solution of (4.2) for which $C = 0$ is the integral curve $y = \frac{1}{4}x^2 - x$. In practice a particular solution is derived by imposing conditions on the general solution which enable appropriate values of the parameters to be found.

In sections 4 and 5 a solution to a first-order equation was derived by specifying a point through which the required integral curve should pass; such a solution is a particular solution and in general it is unique, in the sense that no other integral curve passes through the same point. The solution of a second-order equation is not quite so easy to illustrate. Since the general solution contains two arbitrary constants, two conditions are needed for their values to be found. The conditions may be that an integral curve should pass through a given point and have there a given slope. By starting with such a pair of *initial* conditions an approximate numerical solution may be developed in a way similar, though rather more complicated, to that illustrated for a first order equation. Another pair of conditions, known as *boundary* conditions, requires an integral curve to pass through a pair of specified points.

Example 6.3.

The general solution of equation (2.1) is $y = (A \log x + B)x$. Derive the particular solution which satisfies the conditions $y = 0$ and $y' = 1$ when $x = 1$. (The notation A, B, C, ... for arbitrary constants is preferred in specific examples to the more general notation C_1, C_2, C_3, ...) $y(1) = 0$ gives $B = 0$, whence $y' = A(1 + \log x)$. $y'(1) = 1$ gives $A = 1$, so the particular solution is $y = x \log x$.

A general solution is so described because of the generality of the n parameters. For most applications the general solution of a given equation contains all solutions of interest, the particular solutions, though it may not contain all solutions of the differential equation. A *singular* solution is defined as one which is not derivable from the general solution; usually it represents an envelope to the family of curves defined by the general solution, as in number **4** of Problems 3.3.

Problems 3.6

1. Obtain the differential equation which has a general solution representing the family of straight lines in the x–y plane, excepting those parallel to the y axis.

2. Obtain the differential equation in Example 6.2 by considering, in relation to a circle, the expression

$$\frac{y''}{\sqrt{\{1+(y')^2\}^3}}$$

for the curvature at a point (x, y) on a curve $y = f(x)$.

3. Derive the ordinary differential equations which have the following general solutions, where A and B are arbitrary constants.

(a) $y = Ae^x + Be^{-x}$.

(d) $4(y+A) = (x+A)^2$.

(b) $y = Ae^x - 1/A$.

(e) $y = (Ax+B)e^{2x}$.

(c) $y = A \sin x + B \cos x$.

4. From each of the general solutions given in **3** derive the particular solution which is satisfied at the origin and, when the related differential equation is of second order, is such that $y' = 1$ at the origin.

5. How can the function $y = x - 1$ be described in relation to the differential equation whose general solution is given in 3(d)?

3.7. EQUATIONS OF THE FIRST ORDER AND FIRST DEGREE

There are several types of first-order equation which can be solved by routine methods using techniques of algebra and calculus with which the student is likely to be familiar. An equation of first order and first degree may be expressed in either of the forms

$$P(x, y) + Q(x, y)\, y' = 0,$$
$$P(x, y)\, dx + Q(x, y)\, dy = 0, \qquad (7.1)$$

where P and Q do not include y'.

EQUATIONS WHICH HAVE SEPARABLE VARIABLES

If P and Q are functions of either variable only, say $P = R(x)$ and $Q = S(y)$, then the general solution of (7.1) is obtainable by direct integration, that is by quadratures; it is

$$\int R(x)\,dx + \int S(y)\,dy = C,$$

where C is an arbitrary constant.

Example 7.1.

Obtain the general solution of $2yy' + \cos x = 3x^2$.
In the form of (7.1) the equation is

$$(\cos x - 3x^2)\,dx + 2y\,dy = 0,$$

whence by quadratures the general solution is

$$\sin x - x^3 + y^2 = C.$$

When possible a general solution is usually expressed in a form such as

$$y^2 = C + x^3 - \sin x,$$

where y or a simple function of y is the subject of the equation

Example 7.2.

Obtain the general solution of $\cot y + xy' = 0$.
In the form of (7.1) the equation is

$$dx/x + \tan y\,dy = 0,$$

whence by quadratures

$$\log x + \log \sec y = -\log C.$$

Here the arbitrary constant is expressed as $-\log C$, rather than C, to facilitate the simplification of the solution for, upon taking antilogs, the general solution then assumes the neat

form
$$\cos y = Cx.$$

A slightly more general case occurs when $P(x, y)$ and $Q(x, y)$ are products of a function of x only and a function of y only. Suppose $P = R(x) T(y)$ and $Q = S(y) U(x)$, then (7.1) has the form
$$R(x) T(y) dx + S(y) U(x) dy = 0$$

and division by $U(x) T(y)$ gives
$$\frac{R(x)}{U(x)} dx + \frac{S(y)}{T(y)} dy = 0, \tag{7.2}$$

an expression in which the two variables x, y are separated. Again, the general solution is obtained by quadratures.

Example 7.3.

Obtain the general solution of
$$x(y^2 - 1) dx + 2y(x^2 - 3x + 2) dy = 0.$$

In the form of (7.2) the equation is
$$\frac{x \, dx}{(x-1)(x-2)} + \frac{2y \, dy}{(y^2 - 1)} = 0,$$

whence by partial fractions and then by quadratures
$$-\log (x-1) + 2 \log (x-2) + \log (y^2 - 1) = \log C$$

which simplifies to
$$y^2 = 1 + \frac{C(x-1)}{(x-2)^2}.$$

Problems 3.7a

Obtain the general solutions of the following differential equations.

1. $xy' - 2 \tan y = 0.$

2. $y' + 1 = \tan x.$

3. $(1 + x^2)y' + xy = 0.$

4. $(xy'+y) \log x + 2y = 0.$

5. $xy' + 2y(1 + x \cot x) = 0.$

6. $e^{x+y}y' = 2x.$

EQUATIONS WITH HOMOGENEOUS COEFFICIENTS

A function $F(x, y)$ is said to be homogeneous of degree n in x and y if

$$F(kx, ky) = k^n F(x, y).$$

For example, the function

$$F(x, y) = \frac{3y^4}{x} \sin (y/x) - \frac{x^5}{x^2 - y^2}$$

is homogeneous of degree 3 in x and y because $F(kx, ky) = k^3 F(x, y)$. The two following properties of homogeneous functions are needed.

If $P(x, y)$ and $Q(x, y)$ are both homogeneous and of the same degree, then the function P/Q is homogeneous of degree zero.

If $F(x, y)$ is homogeneous of degree zero in x and y, then $F(x, y)$ is a function of y/x only.

Suppose the coefficients P, Q of dx, dy in (7.1) are homogeneous functions of the same degree. By the properties of such functions equation (7.1) may be expressed in the form

$$F(y/x) + \frac{dy}{dx} = 0,$$

which suggests that the ratio y/x should be treated as a single variable. Let $v = y/x$; then

$$\frac{dy}{dx} = v + x \frac{dv}{dx}$$

and (7.1) becomes

$$x \frac{dv}{dx} + v + F(v) = 0$$

or, upon separation of the variables,

$$\frac{dv}{v+F(v)}+\frac{dx}{x}=0.$$

Thus an equation of the form (7.1) which has homogeneous coefficients of the same degree is solvable by quadratures through the substitution $y = vx$, or $x = vy$.

Example 7.4.

Obtain the general solution of the equation

$$(y^2+xy+x^2)+x^2y' = 0.$$

The coefficients are each homogeneous of degree 2. Let $y = vx$, then

$$x^2(v^2+v+1)+x^2\left(v+x\frac{dv}{dx}\right) = 0$$

which simplifies to

$$\frac{dv}{(v+1)^2}+\frac{dx}{x}=0.$$

By quadratures,

$$\frac{-1}{(v+1)}+\log x =-\log C$$

and, since $v = y/x$, the general solution is

$$Cx = \exp\left(\frac{x}{x+y}\right).$$

Problems 3.7b

Obtain the general solutions of the following differential equations.

1. $(y+ax)y' = (ay-x)$.

2. $x^2y' = 3x(2x+3y)+2y^2$.

3. $xyy' = (4x+y)^2$.

4. $xy' = y + x \sin^2 (y/x)$.

5. $x(x+y)^2 y' = y(x^2 + 3xy + y^2)$.

6. $\left(1 + \dfrac{x}{y}\right) dy = \left(1 - \dfrac{y}{x}\right) dx$.

EQUATIONS WITH LINEAR COEFFICIENTS

If the coefficients P, Q in (7.1) are linear in x and y then the equation has the form

$$(a_1 x + b_1 y + c_1)\, dx + (a_2 x + b_2 y + c_2)\, dy = 0. \qquad (7.3)$$

In the special case of $c_1 = c_2 = 0$, the coefficients P, Q are both homogeneous of degree 1 and the equation may be solved by the methods already considered. This suggests that, when c_1 and c_2 are not both zero, P and Q should be transformed into equivalent linear functions which have zero constant terms. The linear equations

$$\left.\begin{array}{r} a_1 x + b_1 y + c_1 = 0 \\ a_2 x + b_2 y + c_2 = 0 \end{array}\right\} \qquad (7.4)$$

represent a pair of lines which intersect or are parallel. The case of coincidence is of no interest; also cases where both a_1 and b_1 or a_2 and b_2 are zero are not relevant to this analysis. Suppose that the lines intersect at (h, k), then the transformation

$$x = X + h, \quad y = Y + k$$

relates (h, k) to the origin of the X–Y coordinate system and (7.3) becomes

$$(a_1 X + b_1 Y)\, dX + (a_2 X + b_2 Y)\, dY = 0,$$

which has the form suggested.

Example 7.5.

Obtain the general solution of the equation

$$(x + y - 3)\, dx + (x - y + 1)\, dy = 0.$$

The transformed equation is

$$(X+Y)\,dX+(X-Y)\,dY = 0,$$

where $X = x-1$ and $Y = y-2$ as $(h, k) = (1, 2)$. Let $Y = vX$, then

$$(1+v)+(1-v)\left(v+X\frac{dv}{dX}\right) = 0$$

which, upon separation of the variables, becomes

$$\frac{(v-1)\,dv}{(v^2-2v-1)}+\frac{dX}{X} = 0.$$

By quadratures the general solution is

$$\tfrac{1}{2}\log(v^2-2v-1)+\log X = \tfrac{1}{2}\log C', \quad \text{say,}$$

or
$$(v^2-2v-1)\,X^2 = C'$$

which, upon substitution for v, X and Y, becomes

$$y^2-x^2-2xy-2y+6x = C,$$

where $C = C'+1$.

The situation is simpler when the lines represented by (7.4) are parallel for then (a_1x+b_1y) is a multiple of (a_2x+b_2y) and the substitution $v = a_1x+b_1y$ reduces both P and Q to functions of one variable.

Example 7.6.

Obtain the general solution of the equation

$$(2x-3y+1)+(2x-3y+3)\,y' = 0$$

and the particular solution satisfied by $x = 1$, $y = -1$.
Let $v = 2x-3y$, then $y' = \left(2-\dfrac{dv}{dx}\right)\bigg/3$ and

$$3(v+1)+(v+3)\left(2-\frac{dv}{dx}\right) = 0$$

which, upon separation of the variables, becomes

$$dx - \frac{(v+3)\,dv}{(5v+9)} = 0$$

or, in partial fractions,

$$dx - \frac{dv}{5} - \frac{6\,dv}{5(5v+9)} = 0.$$

By quadratures the general solution is

$$25x - 5v - 6 \log C(5v+9) = 0$$

which, upon substitution for v, becomes

$$5(x+y) = 2 \log C(10x - 15y + 9).$$

The required particular solution follows when $C = \frac{1}{34}$.

Problems 3.7c

Obtain the general solutions of the following differential equations.

1. $(x-2)y' + x - y + 1 = 0$.
2. $(x-2y)y' = 4y + x - 6$.
3. $(3x-6y-2)y' = x - 2y + 1$.
4. $(3y-x+4)y' + y - 3x = 0$.
5. $(y+x+2)^2\, y' = 2(y+1)^2$.

EXACT EQUATIONS

Suppose that the general solution of a differential equation has the form

$$\varphi(x, y) = C,$$

then the total differential $d\varphi$ is given by

$$d\varphi = \frac{\partial \varphi}{\partial x}\,dx + \frac{\partial \varphi}{\partial y}\,dy = 0.$$

This differential equation has the form of (7.1) provided P and Q can be identified with $\partial\varphi/\partial x$ and $\partial\varphi/\partial y$ respectively, that is if

$$P = \frac{\partial\varphi}{\partial x} \quad \text{and} \quad Q = \frac{\partial\varphi}{\partial y}.$$

The elimination of φ between this pair of relations, see section 4 of Chapter 7, shows that

$$\frac{\partial P}{\partial y} = \frac{\partial Q}{\partial x}. \tag{7.5}$$

When equation (7.1) is such that this condition is satisfied it is said to be *exact* for it is then in a form which can be integrated immediately, without any of the manipulations seen to be necessary for the other types of equation considered so far.

Example 7.7.

Obtain the general solution of

$$P\,dx + Q\,dy \equiv (2xy - e^x \sin y)\,dx + (x^2 - e^x \cos y)\,dy = 0.$$

Now $\partial P/\partial y = 2x - e^x \cos y = \partial Q/\partial x$, so the equation is exact. Although the integration of the equation in this example is simple, and may be achieved by inspection, the general routine is now followed. Since $P = \partial\varphi/\partial x = 2xy - e^x \sin y$, φ may be derived from a partial integration with respect to x but the result is liable to include an arbitrary function of y, say $f(y)$. Likewise another expression for φ follows from $Q = \partial\varphi/\partial y = x^2 - e^x \cos y$ by a partial integration with respect to y which will generally include an arbitrary function of x, say $g(x)$. These two integrations give

$$\varphi = \int (2xy - e^x \sin y)\,dx = x^2 y - e^x \sin y + f(y)$$

and

$$\varphi = \int (x^2 - e^x \cos y)\,dy = x^2 y - e^x \sin y + g(x).$$

For these two forms of φ to be equivalent the arbitrary functions f and g must be chosen so that $f(y) = g(x) = C'$, where C' is an arbitrary constant. In practice C' may be omitted because an arbitrary constant C arises from the integral of the right-hand side of the differential equation and the general solution is

$$x^2y - e^x \sin y = C.$$

Example 7.8.

Obtain the general solution of

$$(3x^2 - 2y^2)\, dx + (\sinh y - 4xy)\, dy = 0.$$

The equation is exact, because $\partial P/\partial y = -4y = \partial Q/\partial x$, and the general procedure gives

$$\varphi = \int (3x^2 - 2y^2)\, dx = x^3 - 2xy^2 + f(y)$$

and

$$\varphi = \int (\sinh y - 4xy)\, dy = \cosh y - 2xy^2 + g(x).$$

These two forms of φ are equivalent if $f(y) = \cosh y$ and $g(x) = x^3$, so the general solution is

$$x^3 - 2xy^2 + \cosh y = C.$$

Problems 3.7d

Obtain the general solutions of the following differential equations.

1. $(1 - \tan x)\, dx + dy = 0.$

2. $\dfrac{y^2}{x} + 2yy' \log x = 0.$

3. $(x \sin y - \sin x)y' = \cos y + y \cos x.$

4. $2(x + \cos 2x - y^3 e^{2x}) + 3(y^2 - \sin 3y - y^2 e^{2x})y' = 0.$

5. $\dfrac{2(x+3)}{(y+4)} = \dfrac{(x+3)^2}{(y+4)^2}\, y'.$

INTEGRATING FACTORS

The differential equation which has $\varphi(x, y) = y/x = C$ as its general solution is $d\varphi = (-y/x^2)\,dx + (1/x)\,dy = 0$ and this, by its very mode of derivation, is exact. If the equation is multiplied by x^2 it then has the form $-y\,dx + x\,dy = 0$ which is not exact. Conversely if this equation is multiplied by $1/x^2$ it then becomes exact and the factor $1/x^2$ is called an integrating factor, as its use produces an equation which can be integrated immediately. In general an integrating factor is a function $\mu(x, y)$, not necessarily of both x and y, such that the equation

$$P(x, y)\,\mu(x, y)\,dx + Q(x, y)\,\mu(x, y)\,dy = 0 \qquad (7.6)$$

is exact. It can be shown that an equation of the form (7.1) always possesses an integrating factor, though it is not always an easy matter to find one. A little thought will show that the methods given for enabling variables to be separated really rely on the introduction of an integrating factor. There the resulting equations, which are of the form $X(x)\,dx + Y(y)\,dy = 0$ prior to integration, are exact and satisfy the simple condition $\partial X/\partial y = \partial Y/\partial x = 0$. Since equation (7.6) is exact the coefficients of dx and dy must satisfy a relation of the form (7.5), namely

$$\frac{\partial}{\partial y}(P\mu) = \frac{\partial}{\partial x}(Q\mu),$$

which leads to

$$P\,\frac{\partial \mu}{\partial y} - Q\,\frac{\partial \mu}{\partial x} + \mu\left(\frac{\partial P}{\partial y} - \frac{\partial Q}{\partial x}\right) = 0. \qquad (7.7)$$

This partial differential equation is generally much harder to solve for μ than is the original ordinary differential equation for y. In practice the search for an integrating factor is often based on a wise choice from experience and a knowledge of some standard forms. This aspect of the subject is not pursued here, but there is a simplification worth noting which arises when a given equation has an integrating factor which is a function of one variable only. Suppose that $\mu = \mu(x)$, then (7.7) becomes

$$\frac{d\mu}{dx}\bigg/\mu = \left\{\frac{\partial P}{\partial y} - \frac{\partial Q}{\partial x}\right\}\bigg/Q, \qquad (7.8)$$

which implies that the right-hand side is a function of x only. Conversely if the right-hand side is a function of x only then there exists an integrating factor which is a function of x only, a test which is easy to apply. A similar argument applies for μ to be a function of y only.

Example 7.9.

Obtain the general solution of

$$(1+xy)\,dx + x(x+y)\,dy = 0.$$

Now

$$\left\{\frac{\partial P}{\partial y} - \frac{\partial Q}{\partial x}\right\}\bigg/Q = (x-2x-y)/x(x+y) = -1/x,$$

so the equation has an integrating factor μ which is a function of x only. By (7.8) $d\mu/\mu = -dx/x$, whence $\mu = 1/x$. Note that the inclusion of an arbitrary constant with an integrating factor is not necessary for, in forming (7.6), it would produce a multiple of $P\,dx + Q\,dy$, which is zero. In the form of (7.6) the equation is

$$\left(\frac{1}{x}+y\right)dx + (x+y)\,dy = 0,$$

which is exact because $\dfrac{\partial}{\partial y}(P\mu) = 1 = \dfrac{\partial}{\partial x}(Q\mu)$. The general solution is therefore of the form $\varphi(x, y) = C$, where

$$\varphi = \int \left(\frac{1}{x}+y\right) dx = \log x + xy + f(y)$$

or $\qquad \varphi = \int (x+y)\, dy = xy + \tfrac{1}{2}y^2 + g(x).$

Clearly $\qquad f(y) = \tfrac{1}{2}y^2, \quad g(x) = \log x \quad$ and
$$\varphi(x, y) = \log x + xy + \tfrac{1}{2}y^2 = C.$$

Problems 3.7e

Each of the functions $\log x$, $e^{x\,\sin x}$, $(1+x^2)^{-1/2}$, xe^{xy}, $x\sin^2 x$ is an integrating factor suitable for one of the differential equations below. For each equation select, or derive, an appropriate integrating factor and hence obtain its general solution.

1. $(xy+2)\, dx + x^2\, dy = 0.$
2. $\{(x^2+y^2)\,(\sin x + x\cos x) + 2x\}\, dx + 2y\, dy = 0.$
3. $xy\, dx + (1+x^2)\, dy = 0.$
4. $y(2+\log x)\, dx + x\log x\, dy = 0.$
5. $2y(1+x\cot x)\, dx + x\, dy = 0.$

THE LINEAR EQUATION

A linear differential equation of the first order may always be expressed in the general form

$$a_1(x)\,\frac{dy}{dx} + a_0(x)y = f(x),$$

where the notation is derived from that to be used in Chapter 4. Such an equation is linear in y and dy/dx. The general form is inconvenient for solution so it is divided by $a_1(x)$ to give the standard form

$$\frac{dy}{dx} + p(x)\,y = q(x), \tag{7.9}$$

the notation being conventional. In the form of (7.1) this equation is

$$\{p(x)\,y - q(x)\}\,dx + dy = 0, \qquad (7.10)$$

where $P = py - q$ and $Q = 1$. Consideration of (7.8) shows that an integrating factor $\mu(x)$ exists such that

$$d\mu/\mu = p\,dx$$

whence $$\mu(x) = \exp\left\{\int p(x)\,dx\right\},$$

whereby (7.10) may be converted into the exact equation

$$\mu(x)\,\{p(x)\,y - q(x)\}\,dx + \mu(x)\,dy = 0 \qquad (7.11)$$

which may be solved by the usual method.

Example 7.10.

Obtain the general solution of

$$xy' + (1-x)\,y = x(2-x).$$

In standard form the equation is

$$y' + \frac{(1-x)}{x}\,y = 2 - x,$$

so $$p(x) = \frac{1}{x} - 1, \quad \int p(x)\,dx = \log x - x$$

and $$\mu = \exp\,(\log x - x) = x e^{-x}.$$

The exact equation corresponding to (7.11) is

$$x e^{-x}\left\{\frac{(1-x)}{x}\,y - (2-x)\right\}\,dx + x e^{-x}\,dy = 0,$$

whence

$$\varphi = \int \{(1-x)\,e^{-x}y - (2x - x^2)\,e^{-x}\}\,dx = x e^{-x}y - x^2 e^{-x} + f(y)$$

and

$$\varphi = \int x e^{-x}\,dy = x e^{-x}y + g(x),$$

so that $f(y) = 0$, $g(x) = -x^2 e^{-x}$ and the general solution is

$$xe^{-x}y - x^2 e^{-x} = C \quad \text{or} \quad y = x + \frac{C}{x} e^x.$$

Although this is the usual way of solving an exact equation it is not the best way when applied to the special case of a linear equation. An approach which enables the nature of the integrals to be conceived more easily is to multiply the standard form (7.9) by the integrating factor μ to produce the equation

$$\mu \frac{dy}{dx} + p\mu y = \mu q, \tag{7.12}$$

the left-hand side of which is the derivative $\dfrac{d}{dx}(\mu y)$ since $d\mu/dx = p\mu$. Thus by quadratures the equation is reduced to

$$\mu y = \int \mu q \, dx + C$$

and the general solution follows upon solution of the integral.

Example 7.11.

Obtain the general solution of

$$\cos^4 xy' - (\sin 2x \cos^2 x) y - 1 = 0.$$

In standard form and after some simplification the equation is

$$y' - 2y \tan x = \sec^4 x.$$

Now $\int p \, dx = -2 \int \tan x \, dx = \log \cos^2 x$ so $\mu = \cos^2 x$ and in the form of (7.12) the equation is

$$\cos^2 xy' - 2y \sin x \cos x = \sec^2 x$$

or

$$\frac{d}{dx}(y \cos^2 x) = \sec^2 x.$$

Thus $\qquad y \cos^2 x = \int \sec^2 x \, dx + C = \tan x + C$

whence $\qquad\quad y = (C + \tan x) \sec^2 x.$

Problems 3.7f

Obtain the general solutions of the following differential equations.

1. $y' \sin 2x + 2y = 2(x \cos x)^2$.

2. $xy' + (3x+1)y = (e^{-2x}+1)e^{-2x}$.

3. $(x^2+1)y' + xy = \sqrt{(x^2+1)}$.

4. $2(x+2)(x+1)y' + 3(2x+3)y - 4(x^2+3x+2)^{-1/2} = 0$.

5. $xy' - 2y = x(1 - \log x)$.

6. $x(xy'+y) + 2 = 0$.

THE BERNOULLI EQUATION

This is an equation which may be expressed in the standard form

$$y' + p(x)y = q(x)y^n, \tag{7.13}$$

where p and q are functions of x only and n is any real number. If $n = 1$ the variables are separable so cases of interest occur when $n \neq 1$. Division of (7.13) by y^n gives

$$y^{-n}y' + py^{1-n} = q \tag{7.14}$$

and, since

$$\frac{d}{dx}(y^{1-n}) = (1-n)y^{-n}y',$$

the substitution $v = y^{1-n}$ is suggested. This change in variable leads to the linear equation

$$\frac{dv}{dx} + (1-n)pv = (1-n)q.$$

Example 7.12.

Obtain the general solution of

$$x^2y' - xy + y^{3/2} = 0.$$

In the form of (7.14) the equation is

$$y^{-3/2}y' - \frac{1}{x}\,y^{-1/2} = -\frac{1}{x^2}$$

which, with $v = y^{1-3/2} = y^{-1/2}$ and $v' = -\frac{1}{2}y^{-3/2}y'$, becomes the linear equation

$$v' + \frac{1}{2x}\,v = \frac{1}{2x^2}\,.$$

This has an integrating factor $x^{1/2}$, so

$$x^{1/2}v = \frac{1}{2}\int \frac{dx}{x^{3/2}} + C = -x^{-1/2} + C$$

and $\qquad\qquad v = y^{-1/2} = (Cx^{1/2}-1)/x,$

whence the general solution is

$$y = \frac{x^2}{(C\sqrt{x}-1)^2}\,.$$

Problems 3.7g

Obtain the general solutions of the following differential equations.

1. $2(x+yy') = x^2+y^2$.

2. $2(x+yy') + (x^2+y^2)(\sin x + x \cos x) = 0$.

THE RICCATI EQUATION

The most general type of Riccati equation has the form

$$y' + p(x)y = q(x)y^2 + r(x), \tag{7.15}$$

where p, q, r are functions of x only and $q(x) \not\equiv 0$. If $q(x) \equiv 0$ the equation is simply linear and if $r(x) \equiv 0$ it is of Bernoulli type. Although there is no general method for obtaining the general solution of a Riccati equation, there is a procedure which relies on the knowledge of a particular solution, $u(x)$

say, and this may be apparent upon inspection of the equation. The method is to introduce a new dependent variable v through the relation $y = u + 1/v$ which changes (7.15), after some manipulation, into the linear form

$$v' + (2qu - p)v + q = 0. \tag{7.16}$$

Example 7.13.

Obtain the general solution of

$$y' + (2x^2 + 1)y = xy^2 + (x^3 + x + 1).$$

Upon inspection the equation is seen to be satisfied by $y = x$, so let u be this particular solution and substitute for p, q and u in (7.16) to obtain

$$v' - v = -x.$$

This equation has an integrating factor e^{-x} so

$$v e^{-x} = -\int x e^{-x} \, dx + C = (x+1)e^{-x} + C,$$

whence

$$v = C e^x + x + 1$$

and, since $y = x + 1/v$, the general solution of the original equation is

$$y = x + 1/(Ce^x + x + 1).$$

This section has described most of the standard methods available for the solution of a differential equation of first order and first degree. In seeking the solution of such an equation the first step is an attempt to classify it as a type for which a method of solution is known. However, it may be that a given equation can be classified in more than one way and therefore be solved by more than one method. Many of the equations given so far in the sets of problems within this section occur more than once, though usually in thinly disguised forms to suggest a solution by the method being studied. If this has not been noticed already then it may be of interest to look back and find the similarities, perhaps by comparing solutions first.

Problems 3.7h

Classify by type the following differential equations and obtain their general solutions.

1. $(xy+1)y'+y^2 = 0.$ **2.** $(x^2-1)y'+xy = 1.$

3. $x^2(y+1)^2 (y'-1)+2x^2y = y^2+1.$

4. $e^{-xy} \cos (y/x) \{y \tan (y/x)+x^2y-x\}y' =$
$\quad e^{-xy} \cos (y/x) \{y \tan (y/x)-x^2y+x\}y/x.$

5. $y^3(y'-1) = x(x^2+3y^2y').$

6. $xy'+2x^3(y-x)^2+y = 2x$; a particular solution is $y = x.$

7. $(16y-7x+9)y' = y+8x+9.$ **8.** $(1-x-2y)y' = (x+y)y.$

9. $3x(2y-xy') = 2y^{5/2}.$ **10.** $y'+1+(x+y) \tan x = 0.$

CHAPTER 4

Ordinary Linear Differential Equations of the Second Order

4.1. DEFINITION OF A LINEAR EQUATION

A differential equation of order n is said to be linear when it can be expressed in the form

$$a_n(x)\,\frac{d^n y}{dx^n} + a_{n-1}(x)\,\frac{d^{n-1}y}{dx^{n-1}} + \ldots + a_2(x)\,\frac{d^2 y}{dx^2}$$

$$+ a_1(x)\,\frac{dy}{dx} + a_0(x)\,y = f(x), \qquad (1.1)$$

where $f(x)$ and the variable coefficients $a_r(x)$, $r = 0, 1, 2, \ldots, n$, of y and its derivatives are functions of x. The equation is linear in y and its derivatives. The most general linear equation of the second order has the form

$$a_2(x)\,\frac{d^2 y}{dx^2} + a_1(x)\,\frac{dy}{dx} + a_0(x)y = f(x). \qquad (1.2)$$

This type of equation is the subject of this chapter, though in most cases it will be simplified by treating the $a_r(x)$ as constants. The theory and methods of solution will be given mainly for the simplified version of equation (1.2), but they are applicable also to equations of higher order.

In equations (1.1) and (1.2) the term $f(x)$ is not of the same kind as the terms on the left-hand sides, each of which contains y or a derivative of y, so the equations are said to be inhomogeneous. When $f(x) \equiv 0$ the equations are said to be homogeneous as they consist of terms all having a similar form. The homogeneous equation

$$a_2(x)\,\frac{d^2y}{dx^2} + a_1(x)\,\frac{dy}{dx} + a_0(x)\,y = 0 \qquad (1.3)$$

is called a *reduced equation* in the sense that it is derived from the *complete equation* (1.2) by reducing $f(x)$ to zero. Many methods for solving the complete equation depend on knowing the general solution of the reduced equation, so it is important to study ways of finding this solution first.

4.2. SOLUTION OF THE REDUCED EQUATION

Equation (1.3) has the following important properties:

(a) If $y(x)$ is a solution of (1.3) then $A\,y(x)$ is also a solution.
(b) If $y_1(x)$ and $y_2(x)$ are solutions of (1.3) then their sum is also a solution.
(c) If $y_1(x)$ and $y_2(x)$ are linearly independent solutions of (1.3) then the general solution of (1.3) is $A\,y_1(x) + B\,y_2(x)$, where A and B are arbitrary constants.

Statements (a) and (b) should be verified by substitution in (1.3).

The form of equation (1.3), in which the $a_r(x)$ are variable, is too general to be of interest at this stage; methods of solution are given in Chapter 5. Many of the differential equations which describe physical behaviour, such as currents in circuits and electron ballistics, have coefficients a_r which are all constants; in such cases (1.3) may be expressed as

$$a_2\,\frac{d^2y}{dx^2} + a_1\,\frac{dy}{dx} + a_0 y = 0. \qquad (2.1)$$

This homogeneous equation may be considered as a reduced equation derived from a complete equation with constant coefficients. The methods for solving these simplified equations may be developed in a pleasing and systematic way, enhanced by the knowledge that there are many applications for which the solutions are of practical interest.

The substitution in (2.1) of

$$y = e^{mx}, \quad y' = me^{mx} = my \quad \text{and} \quad y'' = m^2 e^{mx} = m^2 y$$

gives the algebraic equation

$$(a_2 m^2 + a_1 m + a_0)y = 0,$$

which is satisfied when m is a root of the quadratic equation

$$a_2 m^2 + a_1 m + a_0 = 0, \tag{2.2}$$

called the *auxiliary equation* (A.E.). The solution of this equation consists, in general, of two distinct values of m, m_1 and m_2, for each of which the assumed function $y = e^{mx}$ is a solution of (2.1). These independent solutions may be expressed as

$$y_1(x) = e^{m_1 x} \quad \text{and} \quad y_2(x) = e^{m_2 x}.$$

Property (c) above shows that the general solution of (2.1) has the form

$$y = Ae^{m_1 x} + Be^{m_2 x}. \tag{2.3}$$

Example 2.1.

Obtain the general solution of the equation $y'' - y' - 6y = 0$ and derive the particular solution which satisfies the conditions $y = 1$ and $y' = 2$ when $x = 0$.

The A.E. is $m^2 - m - 6, = (m+2)(m-3), = 0$ so $m_1 = -2$, $m_2 = 3$ (or vice versa) and the general solution is

$$y = Ae^{-2x} + Be^{3x}.$$

The condition $y(0) = 1$ gives $1 = A + B$. (i)
Now $y' = -2Ae^{-2x} + 3Be^{3x}$, so the condition $y'(0) = 2$ gives $2 = -2A + 3B$. (ii) Equations (i) and (ii) have the solution

10*

$A = \frac{1}{5}$, $B = \frac{4}{5}$ so the required particular solution of the differential equation is

$$5y = e^{-2x} + 4e^{3x}.$$

When the roots of the auxiliary equation are $\pm m$, solution (2.3) becomes $y = A'e^{mx} + B'e^{-mx}$. By using the relations $e^{\pm mx} = \cosh mx \pm \sinh mx$ this solution may, if preferred, be put in the form

$$y = A \cosh mx + B \sinh mx, \tag{2.4}$$

where $A = A' + B'$ and $B = A' - B'$. Solution (2.3) is rewritten with A' and B', both arbitrary constants, merely to preserve A and B for the final result (2.4); this is also done for the derivation of result (2.5) below.

Problems 4.2a

For each of the differential equations below obtain the general solution and the particular solution that satisfies the given conditions.

1. $y'' + y' - 6y = 0$; $y(0) = 1$, $y'(0) = -3$.
2. $y'' - 5y' + 4y = 0$; $y(0) = 0$, $y'(0) = 3$.
3. $6y'' + y' - 2y = 0$; $y \to 0$ as $x \to \infty$, $y'(0) = -2$.
4. $y'' - 4y = 0$; $y(0) = 2$, $y'(0) = 0$.
5. $y'' + 3y' = 0$; $y(0) = 2$, $y'(0) = 3$.

When m_1 and m_2 are complex numbers the general solution of (2.1) is still given by (2.3), but this form of solution is then not so easy to analyse; more convenient forms are now derived. The complex roots will be denoted by $\lambda \pm i\mu$, where λ and μ are real, so that $m_1 = \lambda + i\mu$ and $m_2 = \lambda - i\mu$, say, are conjugates of each other. The general solution of (2.1) is

$$y = A'e^{m_1 x} + B'e^{m_2 x} = A'e^{(\lambda + i\mu)x} + B'e^{(\lambda - i\mu)x}$$
$$= e^{\lambda x}(A'e^{i\mu x} + B'e^{-i\mu x}).$$

By using the relations $e^{\pm i\mu x} = \cos \mu x \pm i \sin \mu x$, this solution takes the form

$$y = e^{\lambda x}\{(A'+B') \cos \mu x + i(A'-B') \sin \mu x\}$$

or

$$y = e^{\lambda x}(A \cos \mu x + B \sin \mu x), \tag{2.5}$$

in which A is put for $(A'+B')$ and B for $i(A'-B')$. If $\lambda = 0$ the roots are imaginary, $\pm i\mu$, and the solution is

$$y = A \cos \mu x + B \sin \mu x. \tag{2.6}$$

The presence in (2.5) of two oscillating functions, the sine and cosine, sometimes makes the analysis of a solution rather difficult. To eliminate the sine, for instance, and thereby clarify the meaning of the solution introduce the ratio $B/A = \tan \alpha$, shown in Fig. 4.1, and this gives

$$A = \sqrt{(A^2+B^2)} \cos \alpha \quad \text{and} \quad B = \sqrt{(A^2+B^2)} \sin \alpha.$$

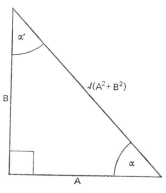

Fig. 4.1.

Substitution for A and B in (2.5) gives

$$y = \sqrt{(A^2+B^2)} \, e^{\lambda x}(\cos \alpha \cos \mu x + \sin \alpha \sin \mu x)$$

or

$$y = Ce^{\lambda x} \cos (\mu x - \alpha), \tag{2.7}$$

where now C, equal to $+\sqrt{(A^2+B^2)}$, and α, equal to $\tan^{-1} B/A$, are the two arbitrary constants. The elimination of the cosine may be achieved in a similar way by using the ratio $A/B = \tan \alpha'$ and this gives

$$y = C e^{\lambda x} \sin (\mu x+\alpha'), \qquad (2.8)$$

the proof being left as an exercise. When the solution of (2.1) entails complex values of m the required form, (2.5), (2.6), (2.7) or (2.8), is quoted at the start, there being no need to follow the above analysis each time. Solution (2.5) is the form most commonly used.

Example 2.2.

Obtain the general solution of the equation $y''-4y'+13y=0$ and derive the particular solution for which $y(0) = 3$ and $y'(0) = 18$.

The A.E. is $m^2-4m+13 = 0$, $m = 2\pm i3$ and by (2.5) the general solution is

$$y = e^{2x}(A \cos 3x+B \sin 3x).$$

The required particular solution is

$$y = e^{2x}(3 \cos 3x+4 \sin 3x).$$

In forms (2.7) and (2.8) $C = 5$, $\alpha = \tan^{-1} \frac{4}{3}$ and $\alpha' = \tan^{-1} \frac{3}{4}$. The particular solution may be derived directly from form (2.7) or form (2.8) of the general solution. By (2.7)

$$y = Ce^{2x} \cos (3x-\alpha)$$

whence $\quad y' = Ce^{2x} \{2 \cos (3x-\alpha)-3 \sin (3x-\alpha)\}.$

The condition $y(0) = 3$ gives $C = 3 \sec \alpha$, where $\sec \alpha$ is positive since C is defined to be positive, and $y'(0) = 18$ gives $\tan \alpha = \frac{4}{3}$. As $\sec^2 \alpha = 1+\tan^2 \alpha = \frac{25}{9}$, $\sec \alpha = \frac{5}{3}$ and $C = 5$, so the particular solution is

$$y = 5 e^{2x} \cos \left(3x-\tan^{-1} \tfrac{4}{3}\right).$$

The use of form (2.8) in this way is left as an exercise. In general it is easier to start with form (2.5) and to derive from the values of A and B whichever final form is required for the particular solution.

Problems 4.2b

For each of the differential equations below obtain the general solution, in the form of (2.6), and the particular solution that satisfies the given conditions.

1. $y'' + 4y' + 13y = 0$; $y(0) = 4$, $y'(0) = 7$.
2. $y'' - 2y' + 2y = 0$; $y(0) = y'(0) = 2$.
3. $y'' + 3y' + 3y = 0$; $y(0) = 0$, $y'(0) = 3$.
4. $y'' + 6y' + 13y = 0$; $y(0) = 6$, $y'(0) = -2$.
5. $y'' + 16y = 0$; $y(\pi/2) = y'(\pi/2) = 1$.

The remaining case to consider is when the auxiliary equation has only one distinct root m, giving the solution $y = Ae^{mx}$; this is not the general solution since it does not contain two arbitrary constants. Suppose the general solution has the form $y = ve^{mx}$, where v is a function of x to be found, then

$$y' = (v' + mv)e^{mx} \quad \text{and} \quad y'' = (v'' + 2mv' + m^2v)e^{mx}.$$

Substitution of y, y' and y'' in (2.1) gives

$$\{a_2(v'' + 2mv' + m^2v) + a_1(v' + mv) + a_0v\}\, e^{mx} = 0,$$

which implies that

$$a_2v'' + (2a_2m + a_1)v' + (a_2m^2 + a_1m + a_0)v = 0.$$

By (2.2) the coefficient of v is zero and as $m = -a_1/2a_2$, the condition for equal roots, the coefficient of v' is also zero. This leaves the condition $v'' = 0$ which by quadratures gives v as $Ax + B$. The general solution therefore is

$$y = (Ax + B)e^{mx}, \tag{2.9}$$

which is quoted when the auxiliary equation is seen to have equal roots.

Example 2.3.

Obtain the solution of the equation $y'' + 8y' + 16y = 0$ which satisfies the conditions $y = 2$ and $y' = 0$ when $x = 0$. The A.E. is $m^2 + 8m + 16 = 0$, $m = -4$ and the general solution is $y = (Ax + B)e^{-4x}$. The required solution is

$$y = 2(4x + 1)e^{-4x}.$$

Problems 4.2c

For each of the differential equations below obtain the general solution and the particular solution that satisfies the given conditions.

1. $y'' + 6y' + 9y = 0$; $y(0) = y'(0) = 1$.
2. $4y'' + 12y' + 9y = 0$; $y(0) = 3$, $y'(0) = -\frac{1}{2}$.
3. $y'' - 2y' + y = 0$; $y(0) = 2$, $y'(0) = 4$.
4. $y'' - 10y' + 25y = 0$; $y(0) = y'(0) = 1$.
5. $L\ddot{y} + R\dot{y} + y/C = 0$, $CR^2 = 4L$, where C, R, L are constants;
   $y(0) = 0$, $\dot{y}(0) = 1$.

The important results of this section will soon become familiar through practice; for convenience they are listed in the summary below.

Roots of the auxiliary equation	*Nature of solution to equation (2.1)*	
Real and distinct, m_1, m_2	$A e^{m_1 x} + B e^{m_2 x}$	(2.3)
Real and distinct,		
$m_1 = -m_2 = m$	$A \cosh mx + B \sinh mx$	(2.4)
Complex, $\lambda \pm i\mu$	$e^{\lambda x}(A \cos \mu x + B \sin \mu x)$	(2.5)
	or $C e^{\lambda x} \cos (\mu x - \alpha)$	(2.7)
	or $C e^{\lambda x} \sin (\mu x + \alpha')$	(2.8)
Imaginary, $\pm i\mu$	$A \cos \mu x + B \sin \mu x$	(2.6)
Equal, m	$(Ax + B) e^{mx}$	(2.9)

The methods of solution given in this section may be extended to homogeneous equations of higher order. Problems 4.2d use examples of such equations and, should any doubts arise, the solutions given should indicate the nature of the extensions.

Problems 4.2d

Obtain the general solutions of the following differential equations.

1. $y''' - 3y'' + 2y' = 0.$ 2. $y''' - 6y'' + 9y' = 0.$

3. $y''' - 3y'' + 3y' - y = 0.$ 4. $y''' + 2y'' + 5y' - 26y = 0.$

5. $y^{iv} - 6y''' + 9y'' = 0.$

4.3. THE COMPLETE EQUATION AND THE NATURE OF ITS SOLUTION

The basis of methods for solving the complete equation (1.2) is contained in the following theorem:

If $y_1(x)$ and $y_2(x)$ are linearly independent solutions of the reduced equation (1.3) and $y_p(x)$ is a particular solution of the complete equation (1.2) then

$$y = Ay_1(x) + By_2(x) + y_p(x) \tag{3.1}$$

is the general solution of the complete equation.

That (3.1) is a solution may be verified by substitution in (1.2) which becomes, when like functions are arranged together,

$$A(a_2y_1'' + a_1y_1' + a_0y_1) + B(a_2y_2'' + a_1y_2' + a_0y_2) \\ + (a_2y_p'' + a_1y_p' + a_0y_p) = f(x).$$

Since $y_1(x)$ and $y_2(x)$ are solutions of the reduced equation (1.3) the coefficients of A and B vanish and, as y_p is a solution of the complete equation (1.2), the remaining expression on the left equals $f(x)$.

Solution (3.1) is often expressed in the concise form

$$y = y_c + y_p,$$

where y_c, equal to $Ay_1(x) + By_2(x)$, is called the *complementary function* (C.F.). The term y_p is referred to as a *particular integral* (P.I.) of the complete equation. For reasons given in the previous section, the complete equation is taken to have the form

$$a_2 \frac{d^2y}{dx^2} + a_1 \frac{dy}{dx} + a_0 y = f(x), \tag{3.2}$$

where the a_r are constant.

Example 3.1.

Obtain the general solution of the equation

$$y'' - 5y' + 4y = 4x - 5. \tag{3.3}$$

The reduced equation $y'' - 5y' + 4y = 0$ is solved first. The A.E. is $m^2 - 5m + 4 = 0$, $m = 1$ or 4 so the C.F. is $y_c = Ae^x + Be^{4x}$. As no method has yet been given for finding a particular integral, it will suffice to verify and accept that $y_p = x$ is a solution of the complete equation. Thus the general solution is

$$y = y_c + y_p = Ae^x + Be^{4x} + x.$$

It is important to realise that y_p is *any* particular solution of the complete equation and therefore it can take more than one form. For instance it should be verified that $y_p = e^x + x$ and $y_p = 2e^{4x} + x$ are also particular solutions of (3.3). Note, however, that the terms e^x and $2e^{4x}$ can be combined with the complementary function; the important term is x.

4.4. DERIVATION OF PARTICULAR INTEGRALS BY OPERATOR METHODS

Section 4.2 has shown that the derivation of complementary functions is a well-defined process which produces results quickly. The derivation of particular integrals is not such an easy process and there are several methods available, some of which are direct whereas others progress by trial and adjustment. In this chapter operator methods only are described. These are direct and they produce the most commonly required results quickly but, as their mathematical justification is often not clear to the user, their application tends to rely on the correct use of rules. The rules are only stated here but their use is well illustrated and, where possible, an indication of their bases is given.

The derivative of y with respect to x, denoted by dy/dx, may be considered symbolically as the result of d/dx operating on y. For convenience of notation d/dx is replaced by D, called "operator D", and Dy is another way of denoting the derivative of y with respect to a given independent variable, in this case x. Although Dy has more of an algebraic character than the calculus form dy/dx, a feature to be exploited, it is stressed that the two forms have the same meaning. The nth derivative of y is denoted by $D^n y$ in operator form, D^n standing for d^n/dx^n. An important feature of operator D is that it operates only on whatever is written to its right, in the product sense. This is evident when its meaning is respected for clearly $y(d/dx)$ would not be written instead of dy/dx. Provided the meaning of D is understood, the symbol D may be used according to most rules of algebra, and it is this property which forms the basis of the operator methods.

For example, if u, v are functions of x and a, b are constants then the following statements should be found correct:

$$Dau = aDu, \quad D(au+bv) = aDu+bDv;$$
$$Du+au = (D+a)u = (a+D)u;$$
$$D^2y+2aDy+a^2y = (D^2+2aD+a^2)y = (D+a)^2y;$$
$$D^2y+(a+b)Dy+aby = \{D^2+(a+b)D+ab\}y$$
$$= (D+a)(D+b)y.$$

Integration of $y(x)$ is the inverse process of differentiation. Since the derivative is denoted by Dy, where in the algebraic sense D "multiplies" y, the integral is denoted by $(1/D)y$ and the operation $D^{-n}y$ is interpreted as the nth successive integral of y, arbitrary constants being omitted.

To simplify the algebra, without loss in generality, equation (3.2) is taken with a_2 as unity,

$$\frac{d^2y}{dx^2} + a_1\frac{dy}{dx} + a_0y = f(x). \tag{4.1}$$

In operator form (4.1) is

$$D^2y + a_1Dy + a_0y = (D^2 + a_1D + a_0)y = f(x).$$

If the sum of operations on y, in this case $D^2 + a_1D + a_0$, is denoted by $F(D)$ then (4.1) becomes

$$F(D)\,y = f(x).$$

The basis of the operator method is that this simple expression may be rearranged as

$$y = \frac{1}{F(D)}f(x),$$

where y is a required particular integral of (4.1), provided the effect of the inverse operator function $1/[F(D)]$ on $f(x)$ is interpreted correctly. The interpretation will depend on the nature of $f(x)$ and, by taking some common forms of $f(x)$, a set of rules for this is now presented.

$f(x)$ of the form e^{ax}

The truth of the statements

$$D^n e^{ax} = a^n e^{ax} \quad \text{and} \quad F(D)e^{ax} = F(a)e^{ax}$$

s easily proved. Based on these the inverse operator function s used according to the rule

$$\frac{1}{F(D)}\,e^{ax} = \frac{1}{F(a)}\,e^{ax}, \tag{4.2}$$

provided $F(D)$ does not contain a factor $(D-a)$ as this would make $F(a)$ vanish.

Examples 4.1.

Obtain particular integrals of the following differential equations.

(a) $y'' + y' - 2y = e^{-3x}$.

$$y_p = \frac{1}{F(D)} e^{-3x} = \frac{1}{(D^2 + D - 2)} e^{-3x}$$

$$= \frac{e^{-3x}}{(-3)^2 - 3 - 2} = \frac{e^{-3x}}{4}.$$

(b) $y'' + y' - 6y = 4e^x$.

$$y_p = \frac{4}{(D^2 + D - 6)} e^x = -e^x.$$

(c) $y'' + 3y' + 2y = 6e^x + 12e^{2x}$.

$$y_p = 6 \frac{1}{(D^2 + 3D + 2)} e^x$$

$$+ 12 \frac{1}{(D^2 + 3D + 2)} e^{2x} = e^x + e^{2x}.$$

Note that constant factors of $f(x)$ may be placed above $F(D)$ or before $1/[F(D)]$. It is often preferable to use $F(D)$ in factorised form and this may be known from the derivation of the complementary function, since the auxiliary equation is $F(m) = 0$.

Example 4.2.

Obtain the general solution of $y'' + 4y' + 3y = e^{-2x}$.

A.E. $F(m) = m^2 + 4m + 3 = (m+3)(m+1) = 0$;

$$m = -3, -1.$$

C.F. $y_c = Ae^{-3x} + Be^{-x}$.

P.I. $y_p = \dfrac{1}{F(D)}\, e^{-2x} = \dfrac{1}{(D+3)(D+1)}\, e^{-2x} = -e^{-2x}.$

G.S. $y = y_c + y_p = Ae^{-3x} + Be^{-x} - e^{-2x}.$

Rule (4.2) fails when $F(D)$ contains $(D-a)$ as a factor. If $(D-a)$ occurs r times in $F(D)$ then $F(D)$ has the form $g(D)(D-a)^r$, where $g(D)$ is the remaining factor, which for some equations may be just unity, when $(D-a)^r$ has been extracted from $F(D)$. The modified form of rule (4.2) for this case of failure is

$$\frac{1}{F(D)}\, e^{ax} = \frac{1}{g(D)(D-a)^r}\, e^{ax} = \frac{e^{ax}x^r}{g(a)\,r!}. \qquad (4.3)$$

Example 4.3.

Obtain particular integrals of the following equations:

(a) $y'' + y' - 2y = 6e^x.$

$$y_p = \frac{6}{(D+2)(D-1)}\, e^x = 6\frac{e^x}{3}x = 2xe^x.$$

(b) $y'' + 4y' + 4y = 4e^{-2x}.$

$$y_p = \frac{4}{1(D+2)^2}\, e^{-2x} = 4\frac{e^{-2x}x^2}{1\,2!} = 2x^2 e^{-2x}.$$

(c) $2y'' + y' - y = 3e^{x/2}.$

$$y_p = \frac{3}{(2D-1)(D+1)}\, e^{x/2}$$

$$= \frac{3}{2}\,\frac{1}{(D-\tfrac{1}{2})(D+1)}\, e^{x/2} = \frac{3}{2}\,\frac{e^{x/2}}{(3/2)}\, x$$

$$= xe^{x/2}.$$

Note, as in (c), that the coefficient of D in the vanishing factor, $(2D-1)$, must be made unity in order to apply rule (4.3) literally and correctly.

Problems 4.4a

Obtain a particular integral for each of the differential equations below.

1. $y'' + 5y' + 6y = 2e^{-x}$.

2. $y'' + y' + y = e^{-x}$.

3. $5y'' - 3y' + 2y = 4e^x$.

4. $3y'' - 5y' - 2y = 7e^{2x}$.

5. $y'' - 4y' + 4y = e^{3x} + 2e^{2x}$.

6. $3y'' - 7y' + 2y = e^{2x} + e^{x/3}$.

7. $y'' - 9y = 6 \cosh 3x$.

8. $y''' - 4y' = 8e^{2x}$.

9. $y''' - 6y'' + 11y' - 6y = 2e^x + e^{2x} + 2e^{3x}$.

10. $y''' + 6y'' + 12y' + 8y = 6e^{-2x}$.

$f(x)$ of the form $\cos ax$ or $\sin ax$

Each time a sine or cosine is differentiated twice the original function is restored but the sign is changed and a constant factor is introduced. In operator form

$$D^2 \cos ax = -(a^2) \cos ax \quad \text{and} \quad D^2 \sin ax = -(a^2) \sin ax,$$

so the operation D^2 is equivalent to multiplication by $-(a^2)$. In general

$$D^{2n} \cos ax = \{-(a^2)\}^n \cos ax,$$

likewise for $\sin ax$. The notation $F(D^2)$ here refers to terms in D^2 and to terms which include D^2 as a factor within $F(D)$. For example, if $F(D) = 3D^3 + D^2 + 5D + 6$ then $F(D^2)$ refers to the term D^2 and to the factor D^2 in $3D^3$; the operation $F(D^2) \cos 2x$ produces $(-12D - 4 + 5D + 6) \cos 2x$, which in turn gives $(2 - 7D) \cos 2x = 2 \cos 2x + 14 \sin 2x$.   Similar remarks apply to the inverse operator function $1/[F(D^2)]$, but the total effect of $1/[F(D)]$ on the operand cannot be completed until any single D that may be present is converted into D^2 by algebraic means. The inverse operation $\{1/[F(D)]\} \cos ax$ is made according to the rule

$$\frac{1}{F(D^2)} \cos ax = \frac{1}{F\{-(a^2)\}} \cos ax \qquad (4.4)$$

and likewise for sin ax.

Example 4.4.

Obtain a particular integral of $y'' + 3y' + 2y = \cos 2x$.

$y_p = \dfrac{1}{F(D)} \cos 2x = \dfrac{1}{(D^2 + 3D + 2)} \cos 2x$ which, by (4.4),

$= \dfrac{1}{(-4 + 3D + 2)} \cos 2x = \dfrac{1}{(3D - 2)} \cos 2x$. The single D is

converted into D^2 giving $y_p = \dfrac{(3D + 2)}{(9D^2 - 4)} \cos 2x$ which, again

by (4.4), $= \dfrac{(3D + 2)}{(-36 - 4)} \cos 2x$. The numerator is applied as

a normal operation to give finally

$$y_p = \tfrac{1}{20}(3 \sin 2x - \cos 2x).$$

In most examples of this type the essential operations are with powers of D^2, so it is preferable to use $F(D)$ in unfactorised form.

Rule (4.4) fails when $F(D)$ contains a factor $(D^2 + a^2)$. In such a case the best method is to represent cos ax by $\mathcal{Re}\ e^{iax}$ and sin ax by $\mathcal{Im}\ e^{iax}$, express $(D^2 + a^2)$ in complex factor form and use rule (4.3).

Example 4.5.

Obtain a particular integral of $y'' + 9y = \sin 3x$.

$$y_p = \frac{1}{D^2 + 9} \sin 3x = \mathcal{Im} \frac{1}{(D + 3i)(D - 3i)} e^{i3x} = \mathcal{Im} \frac{e^{i3x}}{6i} x$$

$$= \mathcal{Im} -i\frac{x}{6}(\cos 3x + i \sin 3x) = -\frac{x}{6} \cos 3x.$$

When $f(x)$ has the form $\cosh ax$ or $\sinh ax$ a particular integral may be found by expressing the function in exponential form and by using rules (4.2) and (4.3). Number 7 of Problems 4.4a may be solved in this way. The above treatment of the trigonometric functions is easily modified to apply to the hyperbolic functions. Since

$$D^2 \cosh ax = a^2 \cosh ax \quad \text{and} \quad D^2 \sinh ax = a^2 \sinh ax,$$

there being no change in sign, rule (4.4) becomes

$$\frac{1}{F(D^2)} \cosh ax = \frac{1}{F(a^2)} \cosh ax, \qquad (4.5)$$

similarly for $\sinh ax$. This rule fails when $F(D)$ contains a factor $(D^2 - a^2)$, but this is easily overcome by a change to exponential form and use of (4.3).

Problems 4.4b

Obtain a particular integral for each of the differential equations below.

1. $y'' - 3y' + 2y = \sin x.$ 2. $y'' + 2y' + 5y = \cos x.$
3. $y'' + 6y' + 9y = \cos 3x.$ 4. $y'' + y' + y = \cos x + \sin x.$
5. $4y'' - 2y' + y = 4 \sin x \cos x.$ 6. $y'' + 4y = 4 \cos 2x.$
7. $y^{iv} + 3y'' + 2y = 6 \sin 2x.$ 8. $y^{iv} + 3y'' - 4y = 20 \cos 2x.$
9. $y'' - 3y' + 3y = \cosh 3x.$ 10. $y'' - y = 2 \sinh x.$

$f(x)$ *of the form* x^n.

The result of $\{1/[F(D)]\} x^n$, n being a positive integer, is derived by expanding $1/[F(D)]$ as a series of terms in ascending powers of D up to and including the term in D^n. It is important that the powers of D should ascend so that the effect on x^n of all operators of higher order than n is merely to produce zeros.

Examples 4.6.

Obtain particular integrals of the following equations.

(a) $y'' - y' = x^2$.

$$y_p = \frac{1}{D(D-1)} x^2 = -\frac{1}{D} (1-D)^{-1} x^2$$

$$= -\frac{1}{D} (1 + D + D^2 + D^3 + \ldots) x^2$$

$$= -\left(\frac{1}{D} + 1 + D + D^2 + \ldots\right) x^2 = -\left(\frac{x^3}{3} + x^2 + 2x + 2\right).$$

To derive this solution each term of the series for $(1-D)^{-1}$ is multiplied by $1/D$ before the operations on x^2 are performed. An equivalent approach is to let the series operate on x^2 first and then to operate on the result by $1/D$, as follows.

$$y_p = -\frac{1}{D} (x^2 + 2x + 2) = -\left(\frac{x^3}{3} + x^2 + 2x\right).$$

Although this result differs from the previous one, the term -2 being absent, it is still a particular integral for, in this example, any constant is part of the complementary function.

(b) $y'' + 3y' + 2y = x$.

$$y_p = \frac{1}{(D+1)(D+2)} x.$$

$\dfrac{1}{F(D)}$ may be expressed either as $(1+D)^{-1}(2+D)^{-1}$ or by partial fractions as $(1+D)^{-1} - (2+D)^{-1}$. Thus

$$y_p = (1+D)^{-1} (2+D)^{-1} x = (1-D+ \ldots) \left(\frac{1}{2} - \frac{D}{4} + \ldots\right) x$$

$$= \left(\frac{1}{2} - \frac{3D}{4} + \ldots\right) x = \frac{x}{2} - \frac{3}{4},$$

or

$$y_p = \{(1+D)^{-1}-(2+D)^{-1}\}\,x$$

$$= \left\{(1-D+\ldots)-\left(\frac{1}{2}-\frac{D}{4}+\ldots\right)\right\}x = \left(\frac{1}{2}-\frac{3D}{4}+\ldots\right)x$$

$$= \frac{x}{2}-\frac{3}{4}, \quad \text{as before.}$$

(c) $y''+3y'+y = x^2$.

$$y_p = \{1+D(3+D)\}^{-1}\,x^2$$
$$= \{1-D(3+D)+D^2(3+D)^2+\ldots\}\,x^2$$
$$= (1-3D+8D^2+\ldots)\,x^2 = x^2-6x+16.$$

Note again that when $1/[F(D)]$ involves the use of $1/D$, as an integrator, it is not necessary to add an arbitrary constant when finding a particular integral.

Problems 4.4c

Obtain a particular integral for each of the differential equations below.

1. $y''+y = x^2$.
2. $y''-6y'+9y = 27x$.
3. $y''-4y' = x^3$.
4. $y''+5y'+6y = x^2$.
5. $y'''-3y''+3y'-y = x^4$.

$f(x)$ of the form $v(x)e^{ax}$

Here $f(x)$ is a product of two functions of x, $v(x)$ and e^{ax}; $v(x)$ is often of the form $\cos bx$, $\sin bx$ or x^n. Consider

$$Dv(x)\,e^{ax}, = e^{ax}(Dv+av) = e^{ax}(D+a)\,v,$$

and

$$D^2v(x)\,e^{ax}, = e^{ax}(D^2v+2aDv+a^2v) = e^{ax}(D+a)^2\,v.$$

These forms may be generalised as

$$D^nv(x)\,e^{ax} = e^{ax}(D+a)^n\,v,$$

whence it follows that

$$F(D)\, v(x)\, e^{ax} = e^{ax} F(D+a)\, v(x).$$

The corresponding inverse operation is interpreted according to the rule

$$\frac{1}{F(D)}\, e^{ax} v(x) = e^{ax} \frac{1}{F(D+a)}\, v(x). \qquad (4.6)$$

The operation is performed in two stages. Initially e^{ax} is moved to the left of the operator function and every D in $F(D)$ is replaced by $(D+a)$; this completes the effect of $1/[F(D)]$ on e^{ax}. Finally $1/[F(D+a)]$ operates on $v(x)$ according to any of the rules already given. A relation such as (4.6) is called a shift theorem.

Examples 4.7.

   (*a*) Obtain particular integrals of the following equations.

   (i) $y'' - 3y' + 2y = 2\, e^{2x} \cos x.$

$$y_p = \frac{2}{(D-2)(D-1)}\, e^{2x} \cos x = 2\, e^{2x} \frac{1}{D(D+1)} \cos x$$

$$= 2\, e^{2x} \frac{1}{(D-1)} \cos x = 2\, e^{2x} \frac{(D+1)}{(D^2-1)} \cos x$$

$$= e^{2x} (\sin x - \cos x).$$

   (ii) $y'' - 2y' - 3y = x\, e^{2x}.$

$$y_p = \frac{1}{(D-3)(D+1)}\, x e^{2x} = e^{2x} \frac{1}{(D-1)(D+3)}\, x$$

$$= -e^{2x}(1+D+ \,\dots)\left(\frac{1}{3} - \frac{D}{9} + \,\dots\right) x$$

$$= -\frac{e^{2x}}{9}\, (3x+2).$$

(iii) $y'' + 6y' + 18y = 6 e^{-3x} \sin 3x$.

$$y_p = \frac{6}{(D^2 + 6D + 18)} e^{-3x} \sin 3x$$

$$= 6 e^{-3x} \frac{1}{(D-3)^2 + 6(D-3) + 18} \sin 3x$$

$$= 6 e^{-3x} \frac{1}{(D^2 + 9)} \sin 3x = -x e^{-3x} \cos 3x,$$

by the result of Example 4.5.

(iv) $y'' + 2y' + 3y = x \cos x$.

$$y_p = \frac{1}{(D^2 + 2D + 3)} x \cos x = \mathcal{Re} \frac{1}{(D^2 + 2D + 3)} x e^{ix}$$

$$= \mathcal{Re} \, e^{ix} \frac{1}{(D+i)^2 + 2(D+i) + 3} x$$

$$= \mathcal{Re} \frac{e^{ix}}{2(1+i)} \frac{1}{\{1 + D + D^2/2(1+i)\}} x$$

$$= \mathcal{Re} \, (1-i) \frac{e^{ix}}{4} \{1 - (D + \ldots) + \ldots\} x$$

$$= \mathcal{Re} \, (1-i) \frac{e^{ix}}{4} (x-1) = \frac{(x-1)}{4} (\cos x + \sin x).$$

(b) Obtain the integral of $e^{ax} \sin bx$.

$$\int e^{ax} \sin bx \, dx = \frac{1}{D} e^{ax} \sin bx = e^{ax} \frac{1}{(D+a)} \sin bx$$

$$= e^{ax} \frac{(D-a)}{(D^2 - a^2)} \sin bx$$

$$= \frac{e^{ax}}{(a^2 + b^2)} (a \sin bx - b \cos bx).$$

The function $e^{ax} \cos bx$ may be treated similarly.

The shift theorem (4.6) may be used when rule (4.2) fails. To illustrate this the equation in Example 4.3(a) is taken again.

$$y'' + y' - 2y = 6 e^x. \qquad y_p = \frac{6}{(D+2)(D-1)} e^x 1,$$

where $v(x)$ is simply unity. By (4.6)

$$y_p = 6 e^x \frac{1}{D(D+3)} 1 = 6 e^x \frac{1}{D} \left(\frac{1}{3} - \frac{D}{9} + \dots \right) 1 = 6 e^x \frac{x}{3}$$

$$\text{or} \quad 6 e^x \left(\frac{x}{3} - \frac{1}{9} \right),$$

depending respectively on whether the integration $1/D$ is performed after or before the terms in the series operate on unity. The extra term is part of the complementary function. This lengthy application of the shift theorem is worth noting but clearly it is inferior to the methods already given for dealing with cases of failure. However, these methods really only state the final result of using the shift theorem, eliminate the working and suppress any terms which can be combined with the complementary function.

Problems 4.4d

Obtain a particular integral for each of the differential equations below.

1. $y'' + y' + y = 13 e^x \sin x.$ **2.** $y'' + y' + y = 3x e^x.$

3. $y'' - 8y' + 41y = 10 e^{4x} \sin 5x.$ **4.** $y'' + 2y' + 37y = 12 e^{-x} \cos 6x.$

5. $y'' - 2y' + y = x^2 e^x.$ **6.** $y'' + 4y' + 4y = x^2 e^{-3x}.$

7. $y'' + 2y' + 4y = 8x \sin 2x.$ **8.** $y'' + y' + y = x e^x \cos x.$

9. $y'' + 2y' + y = x e^{2x} \sin 3x.$ **10.** $y'' - 8y' + 17y = x^2 e^{4x} \cos x.$

$f(x)$ equal to a constant

When $f(x)$ is constant, k say, a particular integral is obtained by equating k to the term in y or, if this is absent, to the term containing the least order derivative of y and then by solving the resulting equation for y.

Examples 4.8.

Obtain particular integrals of the following equations:

(a) $y'' + 2y' + 3y = 6.$ $3y_p = 6,$ $y_p = 2.$

(b) $y'' - 2y' = 4.$ $-2y'_p = 4,$ $y_p = -2x.$

(c) $y''' + y'' = 2.$ $y''_p = 2,$ $y_p = x^2.$

These results also follow from rules (4.2) and (4.3) with $a = 0$.

Problems 4.4e

For each of the differential equations below obtain the general solution and the particular solution that satisfies the given conditions.

1. $y'' - 3y' + 2y = 2(e^x - 1)^2;$ $y(0) = y'(0) = 0.$

2. $y'' + 5y' + 6y = 12 \cos^2 x;$ $y(0) = 1, y'(0) = 0.$

3. $y'' + 3y' - 4y = e^x(x + 1)^2;$ $y(0) = 0, y'(0) = 1/375.$

4. $y'' - 2y' + 2y = 20(1 + \sin x)^2;$ $y(0) = y'(0) = 0.$

5. $y'' + 8y' + 16y = 4e^{-4x} (\sin x + \cos x)^2;$ $y(0) = y'(0) = 0.$

4.5. GENERAL EXPRESSIONS FOR A PARTICULAR INTEGRAL

A general formula is now derived which gives a particular integral of a second-order equation in terms of $f(x)$ and the roots m_1, m_2 of the auxiliary equation. Equation (4.1) is

$$(D^2 + a_1 D + a_0)y = f(x)$$

and this may be expressed as

$$F(D)y = (D - m_1)(D - m_2)y = f(x).$$

It follows that a particular integral may be put in the form

$$y_p = \frac{1}{F(D)} f(x) = \frac{1}{(m_1 - m_2)} \left\{ \frac{1}{(D - m_1)} - \frac{1}{(D - m_2)} \right\} f(x)$$

$$= \frac{1}{(m_1 - m_2)} \left[\frac{1}{(D - m_1)} e^{m_1 x} \{ e^{-m_1 x} f(x) \} \right.$$

$$\left. - \frac{1}{(D - m_2)} e^{m_2 x} \{ e^{-m_2 x} f(x) \} \right]$$

which, by (4.6), becomes

$$y_p = \frac{1}{(m_1 - m_2)} \left\{ e^{m_1 x} \int e^{-m_1 x} f(x) \, dx - e^{m_2 x} \int e^{-m_2 x} f(x) \, dx \right\}.$$

$$(5.1)$$

When the auxiliary equation has equal roots m the analysis is similar in principle. In this case

$$F(D)y = (D - m)^2 y = f(x)$$

and

$$y_p = \frac{1}{(D - m)^2} f(x) = \frac{1}{(D - m)^2} e^{mx} \{ e^{-mx} f(x) \}$$

which, again by (4.6), becomes

$$y_p = e^{mx} \int \left\{ \int e^{-mx} f(x) \, dx \right\} dx. \qquad (5.2)$$

If arbitrary constants are included with the integrals then (5.1) and (5.2) give the general solution of the equation.

Formulae similar to those derived may be obtained for equations of higher order. The value of such formulae depends on the ease with which the integrations can be performed and, clearly, in most cases they will not be preferred to the operator methods.

4.6. APPLICATIONS TO ELECTRICAL CIRCUITS

The methods given in previous sections are now used to solve several differential equations which apply to simple circuits. Each circuit consists of an electromotive force $E(t)$ volts applied to a set of components in series. The components, which have constant characteristics, are an inductance of L henrys, a resistance of R ohms and a capacitance of C farads; each set is a combination of some or all of these. For each circuit the e.m.f. is first taken as constant and then as alternating. The instantaneous current i ampères in the circuit and instantaneous charge q coulombs on the capacitor occur as dependent variables with respect to time, t seconds. The current is the rate of change of charge, that is $i = dq/dt$. The instantaneous drop in voltage across each of the L, R, C components is $L(di/dt)$, Ri, q/C volts respectively. The differential equation for a circuit is formulated by equating the sum of the voltage drops across each component to the applied e.m.f. To simplify the analysis the e.m.f. is applied to each circuit at time $t = 0$ and the initial charge $q(0)$ on the capacitor is taken to be zero. In this section $\sqrt{-1}$ is denoted by j.

INDUCTANCE AND RESISTANCE

The differential equation is

$$L \frac{di}{dt} + Ri = E(t).$$

(i) Constant e.m.f. $E(t) = E$.

A.E. $\qquad Lm + R = 0, \quad m = -R/L.$

C.F. $\qquad i_c = Ae^{-Rt/L}.$

P.I. $\qquad i_p = E/R.$

G.S. $\qquad i = Ae^{-Rt/L} + \dfrac{E}{R}.$

The condition $i(0) = 0$ gives $A = -E/R$, so the expression for the current at time t is

$$i = \frac{E}{R}(1 - e^{-Rt/L}),$$

showing that as t increases the value of the current approaches E/R.

(ii) Alternating e.m.f. $E(t) = E_m \sin \omega t$.

The C.F. is given in (i).

P.I. $i_p = \dfrac{E_m}{(LD+R)} \sin \omega t = E_m \dfrac{(LD-R)}{(L^2D^2-R^2)} \sin \omega t$

$\qquad = \dfrac{E_m}{(R^2+\omega^2L^2)}(R \sin \omega t - \omega L \cos \omega t).$

The method for deriving form (2.8) gives

$$i_p = \frac{E_m}{\sqrt{(R^2+\omega^2L^2)}} \sin (\omega t - \alpha), \quad \text{where} \quad \tan \alpha = \frac{\omega L}{R}.$$

G.S. $i = Ae^{-Rt/L} + \dfrac{E_m}{\sqrt{(R^2+\omega^2L^2)}} \sin (\omega t - \alpha).$

The condition $i(0) = 0$ gives $A = \dfrac{E_m \sin \alpha}{\sqrt{(R^2+\omega^2L^2)}}$, whence

$$i = \frac{E_m}{\sqrt{(R^2+\omega^2L^2)}} \{\sin (\omega t - \alpha) + e^{-Rt/L} \sin \alpha\}.$$

This expression for the current is composed of two terms. Provided L/R is not very large the value of the exponential term tends quickly to zero as t increases and its contribution to the current is said to be "transient". The term which remains when the transient is negligible gives the "steady state" value of the current; it is

$$i = \frac{E_m}{\sqrt{(R^2+\omega^2L^2)}} \sin (\omega t - \alpha),$$

where $\sqrt{(R^2+\omega^2L^2)}$ is called the impedance. The variations in current lag behind those in the applied voltage by the phase angle α, called the phase lag.

RESISTANCE AND CAPACITANCE

The differential equation is

$$Ri+\frac{q}{C} = E(t), \quad \text{where} \quad i = \frac{dq}{dt}.$$

It is convenient to solve the differential equation for q,

$$R\frac{dq}{dt}+\frac{q}{C} = E(t),$$

and to derive i from the solution q.

(i) Constant e.m.f. $E(t) = E$.

A.E. $\qquad Rm+\dfrac{1}{C} = 0, \quad m = -1/CR.$

C.F. $\qquad q_c = Ae^{-t/CR}.$

P.I. $\qquad q_p = CE.$

G.S. $\qquad q = Ae^{-t/CR}+CE.$

The condition $q(0) = 0$ gives $A = -CE$ so the expression for the charge at time t is

$$q = CE(1-e^{-t/CR}),$$

whence that for the current is

$$i = \frac{E}{R}\,e^{-t/CR}.$$

The current flows until the charge on the capacitor is CE.

(ii) Alternating e.m.f. $E(t) = E_m \sin \omega t$.

The C. F. is given in (i).

P.I. $\quad q_p = \dfrac{CE_m}{(CRD+1)} \sin \omega t = CE_m \dfrac{(CRD-1)}{(C^2R^2D^2-1)} \sin \omega t$

$\qquad = \dfrac{CE_m}{(\omega^2C^2R^2+1)} (\sin \omega t - \omega CR \cos \omega t)$

$\qquad = \dfrac{-CE_m}{\sqrt{(\omega^2C^2R^2+1)}} \cos (\omega t + \alpha),$

where $\qquad\qquad\qquad \tan \alpha = \dfrac{1}{\omega CR}.$

G.S. $\quad q = Ae^{-t/CR} - \dfrac{CE_m}{\sqrt{(\omega^2C^2R^2+1)}} \cos (\omega t + \alpha).$

The condition $q(0) = 0$ gives $A = \dfrac{CE_m \cos \alpha}{\sqrt{(\omega^2C^2R^2+1)}}$. The C.F. is transient and the charge in the steady state is given by q_p. Since $i = dq/dt$, the steady-state current is

$$i = \dfrac{\omega CE_m}{\sqrt{(\omega^2C^2R^2+1)}} \sin (\omega t + \alpha).$$

The impedance is $\sqrt{(R^2+1/\omega^2C^2)}$ and there is a phase lead α.

INDUCTANCE, RESISTANCE AND CAPACITANCE

The differential equation is

$$L\frac{di}{dt} + Ri + \frac{q}{C} = E(t),$$

which will be solved in the form

$$L\frac{d^2i}{dt^2} + R\frac{di}{dt} + \frac{i}{C} = \frac{dE(t)}{dt}.$$

(i) Constant e.m.f. $E(t) = E$.

A.E.
$$LCm^2 + CRm + 1 = 0,$$
$$m = \{-CR \pm \sqrt{(C^2R^2 - 4LC)}\}/2LC.$$

The nature of the solution depends on the roots m_1, m_2 of the A.E., as shown in section 4.2, and these depend on the term $C^2R^2 - 4LC$. When $C^2R^2 - 4LC > 0$, that is $CR^2 > 4L$, the roots are real, distinct and always negative. The solution

$$i = Ae^{m_1 t} + Be^{m_2 t}$$

shows that the current decreases exponentially to zero. When $CR^2 < 4L$ the roots have the form $-(R/2L) \pm jp$, where p^2 stands for the positive quantity $(4LC - C^2R^2)/4L^2C^2$. The solution

$$i = e^{-Rt/2L}(A \cos pt + B \sin pt)$$

shows that the current oscillates but is attenuated by the presence of the resistance. When $CR^2 = 4L$ the roots are equal and the solution

$$i = (At + B)e^{-Rt/2L}$$

is the critical state between the previous two types of variation in current.

(ii) Alternating e.m.f. $E(t) = E_m \sin \omega t$.

The C.F. is given in (i) and is always transient. The steady-state current is

$$i = \frac{\omega E_m}{(LD^2 + RD + 1/C)} \cos \omega t = \frac{\omega E_m}{(RD + 1/C - \omega^2 L)} \cos \omega t$$

$$= \frac{\omega E_m}{\{\omega^2 R^2 + (1/C - \omega^2 L)^2\}} \{\omega R \sin \omega t + (1/C - \omega^2 L) \cos \omega t\}$$

$$= \frac{\omega E_m}{\sqrt{\{\omega^2 R^2 + (1/C - \omega^2 L)^2\}}} \sin (\omega t + \alpha),$$

$$\text{where} \quad \tan \alpha = (1/\omega C - \omega L)/R.$$

The impedance is $\sqrt{\{R^2+(1/\omega C-\omega L)^2\}}$ and the phase difference is α. When the frequency ω of the applied e.m.f. is such that $\omega^2 = 1/LC$ the impedance is least and the amplitude of the current has its greatest value E_m/R. In this condition of resonance the phase angle α is zero and the current is in phase with the e.m.f. When $1/\omega C < \omega L$ α is negative and the current lags behind the e.m.f. so that the net reactance $(1/\omega C-\omega L)$ is inductive. When $1/\omega C > \omega L$ α is positive and the current leads the e.m.f., the net reactance being capacitive.

Problems 4.6

1. A capacitor C is charged through a resistance R by means of a constant e.m.f. E, the circuit being closed at $t = 0$ when the capacitor is uncharged. For the values $C = 5 \times 10^{-4}$ F, $R = 2000$ ohm, $E = 4000$ V calculate the current at time $t = 0$ and determine the time taken for the current to become half its initial value.

2. An e.m.f. $E \sin \omega t$ is applied at time $t = 0$ to a capacitor C and an inductance L in series, the initial charge and current being zero. Show that the current at time t is given by

$$i = \frac{\omega EC}{(1-LC\omega^2)}\left(\cos \omega t - \cos \frac{t}{\sqrt{LC}}\right),$$

provided $LC\omega^2 \neq 1$.

3. An e.m.f. $E \sin \omega t$ is applied at time $t = 0$ to an L, R, C series circuit, the initial charge and current being zero. Show that the smallest value of R needed to prevent natural, unforced, oscillations in the current is $2\sqrt{L/C}$. Taking this value of R and the condition for resonance derive an expression for the current at time t.

4. A charged capacitor C, initially at voltage V, discharges through a resistance R and inductance L in series. Derive an expression, not containing L, for the charge and an expression, not containing C, for the current at time t under the condition that the variation in charge is just non-oscillatory.

5. A constant e.m.f. E is applied at time $t = 0$ to an L, R, C series circuit, the initial charge and current being zero. For the values $L = 1$ H, $R = 600$ ohm, $C = 10^{-5}$ F and $E = 100$ V, determine the time taken for the current to attain its greatest value.

4.7. SIMULTANEOUS DIFFERENTIAL EQUATIONS

A set of simultaneous differential equations involves two or more dependent variables and a single independent variable which is usually t, representing time in applications. The number of equations should equal the number of dependent variables. The solution of such equations is a set of relations expressing each dependent variable in terms of the independent variable only.

Usually the first stage in the solution of simultaneous differential equations is similar in principle to the solution of simultaneous algebraic equations, where all but one of the variables are eliminated by suitable combinations of the equations. The result is a differential equation in one dependent variable only and it is solved by any suitable method. Solutions for the other dependent variables should be obtained in the easiest ways and these will depend upon the nature of the equations. In general the best approach is to use the solution obtained first as a basis for substitution in the given equations rather than to repeat the elimination process for another variable. Different methods of solution are liable to produce varying numbers of seemingly arbitrary constants in the general solution, that is the set of solutions. There cannot be more independent arbitrary constants in the general solution than the sum of the orders of the differential equations. There may be less and the actual number is the degree of the operator $D \equiv d/dt$ in the determinant $\Delta \neq 0$ of the coefficients of the dependent variables. If more than this number of arbitrary constants are obtained, then the extra ones must be dependent on the others and should be eliminated. These remarks are now illustrated by examples.

Example 7.1.

Obtain the general solution of the equations

$$5\ddot{x}+\dot{y}+2x = 4\cos t, \quad 3\dot{x}+y = 8t\cos t.$$

These are equations (2.5) and (2.6) of Chapter 3; they are arranged so that the dependent variables are in columns on the left and the functions of t are on the right.

$$(5D^2+2)x+Dy = 4\cos t, \tag{a}$$

$$3Dx+y = 8t\cos t. \tag{b}$$

The order of (a) is 2 and of (b) is 1; thus there cannot be more than three independent arbitrary constants in the general solution. In fact there are less because the degree of D in

$$\Delta = \begin{vmatrix} (5D^2+2) & D \\ 3D & 1 \end{vmatrix}$$

is 2 and this is the actual number.

The elimination of y from (a) is achieved by forming $(a)-D(b)$ which gives

$$(D^2+1)x = -2\cos t+4t\sin t.$$

The solution of this equation may be derived by methods given in previous sections and is

$$x = A\cos t+B\sin t-t^2\cos t. \tag{c}$$

The solution for y is obtained most easily by the substitution of Dx, obtainable from (c), in (b) giving

$$y = 3(A\sin t-B\cos t)+14t\cos t-3t^2\sin t. \tag{d}$$

The general solution of the equations consists of (c) and (d); it contains the correct number of arbitrary constants.

Example 7.2.

Obtain the general solution of the equations

$$2\dot{x} + \dot{y} - 2x - 2y = 4e^t, \quad \dot{x} + \dot{y} + 4x + 2y = 5e^{-t}.$$

In the required form the equations are

$$2(D-1)x + (D-2)y = 4e^t, \tag{a}$$

$$(D+4)x + (D+2)y = 5e^{-t}. \tag{b}$$

The elimination of x by the operations $2(D-1)$(b)$-(D+4)$(a) leads to

$$(D^2+4)y = -20(e^t + e^{-t}) = -40 \cosh t$$

which has the solution

$$y = A \cos 2t + B \sin 2t - 8 \cosh t. \tag{c}$$

The easiest way to obtain x is by the elimination of $\dot{x}$; thus (a)-2(b) gives

$$10x = -\dot{y} - 6y + 6 \cosh t - 14 \sinh t,$$

into which the substitution of y and $\dot{y}$ from (c) provides the solution

$$5x = -(3A+B) \cos 2t + (A-3B) \sin 2t + 27 \cosh t - 3 \sinh t. \tag{d}$$

Suppose, however, that this method is not chosen and that y is eliminated from (a) by the operations $(D+2)$(a)$-(D-2)$(b) giving

$$(D^2+4)x = 12e^t + 15e^{-t} = 27 \cosh t - 3 \sinh t$$

which has the solution

$$x = E \cos 2t + F \sin 2t + 3(9 \cosh t - \sinh t)/5. \tag{e}$$

The general solution of (a) and (b) has two independent arbitrary constants only, as the degree of D in Δ is 2. This means that two of the constants A, B, E, F in (c) and (e) must be eliminated, E and F say. The elimination can be achieved by substitution of solutions (c) and (e) in (a), or in (b), and, since this gives an identity, a comparison of coefficients leads to the relations $5E = -(3A+B)$ and $5F = A-3B$ in accordance with the result (d) of the other method.

Example 7.3.

The constant e.m.f. E is applied at $t = 0$ to the network shown in Fig. 4.2 when no energy is stored in the inductances. Obtain an expression for the current i at time t.

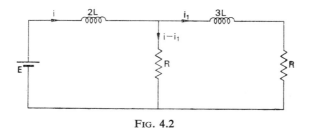

FIG. 4.2

The differential equations for each part of the network are

$$2L\frac{di}{dt}+R(i-i_1) = E \quad \text{and} \quad R(i-i_1) = 3L\frac{di_1}{dt}+Ri_1,$$

or

$(2LD+R)i-Ri_1 = E$ (a)   and $Ri-(3LD+2R)i_1 = 0.$ (b)

The elimination of i_1 by the operations $(3LD+2R)$(a)$-R$(b) leaves

$$(6L^2D^2+7RLD+R^2)i = 2ER$$

which has the general solution

$$i = \frac{2E}{R} + Ae^{-Rt/L}+Be^{-Rt/6L}.$$

Substitution of i and di/dt in (a) gives

$$i_1 = \frac{E}{R} - Ae^{-Rt/L} + \frac{2}{3} Be^{-Rt/6L}.$$

The conditions $i = i_1 = 0$ when $t = 0$ yield $A = -E/5R$, $B = -9E/5R$ and the required expression is

$$i = \frac{E}{5R} (10 - e^{-Rt/L} - 9e^{-Rt/6L}).$$

Problems 4.7

Obtain the solutions of the following systems which satisfy the conditions stated.

1. $\dot{x} + y = 3 \cos 2t$,
   $\dot{y} - x = 3 \sin 2t$; $x = 1, y = 4$ when $t = \frac{1}{2}\pi$.

2. $\dot{x} - 2x + \dot{y} = 4t - 1$,
   $\dot{y} - y - x = 1 + 3t - t^2$; $x(0) = \dot{y}(0) = 1$.

3. $\ddot{y} - y + 5\dot{x} = t$,
   $-\ddot{x} + 4x + 2\dot{y} = 2$; $x(0) = \dot{x}(0) = y(0) = \dot{y}(0) = 0$.

4. $5\ddot{y} - \dot{x} + 2y = 4 \cos t$,
      $3\dot{y} - x = 8t \cos t$; $y = 1, \dot{y} = 0$ when $t = 0$.

5. $t\dot{x} - 2\dot{y} - 2y = 5 - 4e^t$,
   $t\dot{x} + ty = te^t + t^3 + 1$; $x(1) = y(0) = 1$.

4.8. THE EULER LINEAR EQUATION

This type of equation has the general form (1.1) in which the variable coefficients $a_r(x)$ are constant multiples of x^r only. Equations of this type are also called *homogeneous* linear equations because of the similarity in form of the terms on the left; the name does not imply that the function $f(x)$ on the right is necessarily zero. Such equations may be transformed into equations with constant coefficients by the substitution

12*

$x = e^t$, t becoming the independent variable. The solution of the transformed equation is converted into the solution of the original equation through the relation $t = \log x$.

Consider the general Euler equation of the second order

$$ax^2 \frac{d^2y}{dx^2} + bx \frac{dy}{dx} + cy = f(x), \tag{8.1}$$

where a, b, c are constants. Let $x = e^t$; then $dx/dt = e^t = x$ and, since

$$\frac{dy}{dx} = \frac{dy}{dt}\frac{dt}{dx}, \quad \frac{dy}{dx} = \frac{dy}{dt}\frac{1}{x}$$

or

$$x\frac{dy}{dx} = \frac{dy}{dt}. \tag{8.2}$$

Now

$$\frac{d^2y}{dx^2} = \frac{d}{dx}\left(\frac{dy}{dx}\right) = \frac{d}{dt}\left(\frac{1}{x}\frac{dy}{dt}\right)\frac{dt}{dx}$$

$$= \left\{\frac{1}{x}\frac{d^2y}{dt^2} - \left(\frac{1}{x^2}\frac{dx}{dt}\right)\frac{dy}{dt}\right\}\frac{dt}{dx} = \frac{1}{x^2}\left(\frac{d^2y}{dt^2} - \frac{dy}{dt}\right)$$

so that

$$x^2\frac{d^2y}{dx^2} = \frac{d^2y}{dt^2} - \frac{dy}{dt}. \tag{8.3}$$

Thus equation (8.1) is transformed into

$$a(\ddot{y} - \dot{y}) + b\dot{y} + cy = f(e^t)$$

or

$$a\ddot{y} + (b-a)\dot{y} + cy = f(e^t)$$

which, being an equation with constant coefficients, can be solved by standard methods.

Results (8.2) and (8.3) of the transformation may be expressed in the forms

$$x\,dy/dx = Dy,$$
$$x^2\,d^2y/dx^2 = D(D-1)y,$$

where $D \equiv d/dt$, and their generalisation is the simple sequence

$$x^3 \, d^3y/dx^3 = D(D-1)(D-2)y,$$

. . . .

$$x^n \, d^ny/dx^n = D(D-1)(D-2)(D-3) \ldots (D-n+1)y.$$

Example 8.1.

Obtain the general solution of the equation

$$x^2y'' - xy' + y = x^2.$$

The transformed equation is

$$\ddot{y} - 2\dot{y} + y = e^{2t}$$

having the general solution

$$y = (At+B)e^t + e^{2t},$$

which in terms of x is

$$y = (A \log x + B)x + x^2.$$

Problems 4.8

Obtain the general solutions of the following differential equations.

1. $x^2y'' - 3xy' + 4y = 2x^2$.　　　　**2.** $x^2y'' - 4xy' + 6y = x^3$.

3. $2x^2y'' + 7xy' + 2y = 1/x$.　　　**4.** $x^3y''' + 2x^2y'' - xy' + y = 1$.

5. $xy'' + y' = x^2$.　　　　　　　　**6.** $9x^2y'' + 2y = \sqrt[3]{x^2}$.

7. $x^2y'' - xy' + 2y = x \log x$.　　**8.** $x^2y'' - xy' + 2y = 2x \cos(\log x)$.

9. $x^2y'' - 5xy' + 10y = 10(\log x^5)^2$.

10. $(x-1)^2y'' + 3(x-1)y' - 3y = x^2$;　　let $u = x-1$.

CHAPTER 5

Solution in Power Series of Differential Equations

5.1. INTRODUCTION

This chapter is concerned almost entirely with the solution of differential equations which have the form (4.1.3), that is

$$a_2(x)y'' + a_1(x)y' + a_0(x)y = 0, \qquad (1.1)$$

where at least one of the coefficients a_r is a function of the independent variable. The methods of solution are applicable also to linear equations of any order and to inhomogeneous equations, such as (4.1.2). The general solution of (1.1) has the form $Ay_1(x) + By_2(x)$, where y_1 and y_2 are linearly independent solutions. These solutions are derived as convergent infinite power series which generally do not represent known functions, though sometimes it is possible to recognise a solution as the expansion of an elementary function such as the sine or exponential. Some solutions contain a finite number of terms only.

Certain series solutions occur so frequently that they are taken to define functions whose properties are worthy of detailed study and whose values, therefore, may become tabulated over ranges likely to be of practical use. The names of Bessel and Legendre, to quote but two, are associated with such functions and a simple treatment of Bessel functions is given at the end of this chapter.

5.2. OUTLINE OF THE METHOD

Before the detailed methods of solution are presented some theoretical discussion is advisable. However, at this stage the method is applied to a simple equation so that the purpose of the theory will be appreciated more readily.

Example 2.1.

Obtain the general solution of

$$y'' + y = 0. \tag{2.1}$$

Although this equation has constant coefficients and is solved more easily by the direct methods already given, its solution provides an adequate example of the series method. Suppose the equation is satisfied by an infinite series of the form

$$y = c_0 + c_1 x + c_2 x^2 + c_3 x^3 + \ldots + c_n x^n + \ldots, \tag{2.2}$$

then $\quad y' = c_1 + 2c_2 x + 3c_3 x^2 + 4c_4 x^3 + \ldots$

and $\quad y'' = 2c_2 + 6c_3 x + 12c_4 x^2 + 20c_5 x^3 + \ldots,$

where the c_n are undetermined constants. Substitute for y and y'' in (2.1) and combine terms which relate to the same powers of x to obtain

$$(c_0 + 2c_2) + (c_1 + 6c_3)x + (c_2 + 12c_4)x^2 + (c_3 + 20c_5)x^3 + \ldots \equiv 0.$$

This identity is satisfied if all the coefficients of x^n ($n = 0$, 1, 2, ...) are zero; this set of conditions gives

$$c_2 = -c_0/2, \quad c_4 = -c_2/12 = c_0/24,$$
$$c_6 = -c_4/30 = -c_0/720, \ldots$$

and $\quad c_3 = -c_1/6, \quad c_5 = -c_3/20 = c_1/120, \ldots.$

Thus two of the constants, normally taken to be c_0 and c_1, remain arbitrary and the supposed solution is

$$y = c_0\left(1 - \frac{x^2}{2!} + \frac{x^4}{4!} - \frac{x^6}{6!} + \ldots\right) + c_1\left(x - \frac{x^3}{3!} + \frac{x^5}{5!} - \ldots\right),$$

$= c_0 y_1 + c_1 y_2$, say. This is the general solution because each of y_1 and y_2 satisfies (2.1) and they are linearly independent. The series y_1 could be taken as the *definition* of what is already a familiar function, namely $\cos x$, and y_2 could be taken to define $\sin x$. When a series solution can be identified with a known function it is said to be expressed in *closed form*, but for most series this cannot be done.

Problems 5.2

Obtain the general solutions, in powers of x, of the following differential equations by the series method. All the series can be expressed in closed form, but in some cases the form is unlikely to become apparent until a comparison is made with the solution obtainable by direct methods.

1. $y'' - y = 0$.   **2.** $y'' - 4y' + 4y = 0$.

3. $y' + y = x$.   **4.** $(x^2 + 1)y' - (x - 1)^2 y = 0$.

5. $y' \cos x + y \sin x = \cos^2 x$.

(The circular functions should be expanded as power series before the method is applied.)

5.3. POWER SERIES

Methods of solution by series require certain operations to be performed on power series, such as differentiation, addition and possibly multiplication. The relevant properties of power series are now stated without proofs; some remarks would need to be qualified in a complete account.

A power series is defined as an infinite series of the form

$$\sum_{n=0}^{\infty} c_n(x - x_0)^n = c_0 + c_1(x - x_0) + c_2(x - x_0)^2 + \ldots, \quad (3.1)$$

where x, the *centre* x_0 and the constants c_n are taken to be real here. It is essential to know the range of values for x over which the series is convergent. Clearly if $x = x_0$ the series is convergent and its value is c_0; in some cases x_0 may be the only value of x for which the series is convergent. When the series is convergent for values of x other than x_0, these values form an interval, the convergence interval, with x_0 as the midpoint. When this interval is finite the series is convergent for all x within the interval and is divergent for values of x outside the interval. If the series is convergent for *all* values of x then the interval is infinite. The *radius of convergence* R of a power series is the distance from the centre to each end point of the convergence interval, as shown in Fig. 5.1.

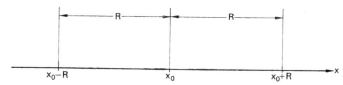

FIG. 5.1. Radius of convergence

Thus a series is convergent when $|x-x_0| < R$ and is divergent when $|x-x_0| > R$. Convergence at either end point of R depends on the particular series. It may be possible to evaluate R from the formula

$$\frac{1}{R} = \lim_{n \to \infty} \left| \frac{c_{n+1}}{c_n} \right|. \tag{3.2}$$

Example 3.1.

For the series

$$\sum_{n=0}^{\infty} \frac{x^n}{n!} = 1 + x + \frac{x^2}{2!} + \ \dots$$

$$c_n = \frac{1}{n!} \quad \text{and by (3.2)} \quad \frac{1}{R} = \lim_{n \to \infty} \frac{1/(n+1)!}{1/n!}$$

$$= \lim_{n \to \infty} \frac{1}{(n+1)} = 0.$$

Thus R is infinite and the series converges for all values of x. This is the series for the function e^x and it represents this function for all values of x.

Example 3.2.

For the series

$$\sum_{n=0}^{\infty} (-1)^n (n+1)x^n = 1-2x+3x^2-4x^3+ \ldots$$

$$1/R = \lim_{n \to \infty} |-(n+2)/(n+1)| = 1.$$

This series is the binomial expansion of the function $(1+x)^{-2}$ but it represents the function only when $|x| < 1$.

In these examples the centre x_0 is zero and the series simply involve powers of x. Most of the examples and problems to be considered are of this form or can be converted to it by a change of variable.

The following properties of power series expanded about the same centre are used in the series solution of differential equations.

(a) A power series may be differentiated term by term; the convergence interval of the derived series equals that of the original.
(b) Two power series may be added term by term; the sum series converges within that interval of the component series which is the smaller.
(c) Two power series may be multiplied term by term; the product series converges within the smaller of the two convergence intervals.
(d) Two power series may be divided provided the denominator series does not vanish at the centre; the quotient series converges in some interval which contains the centre.

(e) If the sum of a convergent power series is identically zero throughout the convergence interval then each coefficient of the series is zero.

The vital information needed about a given differential equation is whether or not it has solutions which can be represented by power series. This is provided by the nature of some terms in the equation; the conditions which they must satisfy are based on the following concept.

DEFINITION. *A function of x which can be represented by a power series in powers of $(x-x_0)$ with non-zero radius of convergence is said to be* analytic, *or* regular, *at $x = x_0$.*

Before the following theorem can be applied it is essential to divide the differential equation by the coefficient of the highest order derivative; equation (4.1.2) then becomes

$$y'' + p(x)y' + q(x)y = r(x), \qquad (3.3)$$

say, and (1.1) takes a similar form with $r(x) \equiv 0$.

THEOREM 3.1. *If all the functions $p(x)$, $q(x)$ and $r(x)$ in (3.3) are analytic at $x = x_0$ then every solution of the equation is analytic at $x = x_0$.*

Values of x for which the conditions of the theorem are satisfied are called *ordinary points* of the differential equation; all others are called *singular points* and at such a point the equation is said to have a singularity.

The question arises of how to test whether or not a function, such as p or q, is analytic. The answer is implied by the next theorem which gives a relation between the coefficients of a power series and the derivatives, at the centre, of the function it may represent.

THEOREM 3.2. *If there exists a power series*

$$c_0 + c_1(x - x_0) + c_2(x - x_0)^2 + \ldots + c_n(x - x_0)^n + \ldots$$

which represents a function $f(x)$ within an interval $|x-x_0| < R$ then the coefficients are given by $c_n = f^{(n)}(x_0)/n!$.

Thus a power series representation of a function is the Taylor series expansion of that function and it is unique:

$$f(x) = f(x_0) + f'(x_0)(x-x_0) + \frac{f''(x_0)}{2!}(x-x_0)^2 + \ldots$$
$$+ \frac{f^{(n)}(x_0)}{n!}(x-x_0)^n + \ldots.$$

This shows that for a function to be analytic it must be single-valued and must possess derivatives of all orders at the point x_0. In nearly all cases a function $f(x)$ may fail to be analytic because x_0 is a pole, a term defined in section 6.7, of $f(x)$ or of one or more of its derivatives. The range over which a solution is valid is related to the nature of the equation as follows.

THEOREM 3.3. *If the expansions of p, q, and r in (3.3) are valid over the range $|x-x_0| < R$ then the expansion of every solution of (3.3) is valid also in this range.*

This means that if p, q and r are polynomials, which they frequently are, then every power series expansion of every solution of (3.3) is valid for *all* values of x. The theorem does not imply that a solution cannot exist with a range of convergence equal to or greater than R.

Example 3.3.

Consider Legendre's equation

$$(1-x^2)y'' - 2xy' + \nu(\nu+1)y = 0,$$

where ν is a constant, and express it in the form of (3.3):

$$y'' - \frac{2x}{(1-x^2)}y' + \frac{\nu(\nu+1)}{(1-x^2)}y = 0. \tag{3.4}$$

The general solution expanded about the ordinary point $x_0 = 0$ is $y = c_0 y_1 + c_1 y_2$ where

$$y_1 = 1 - \frac{\nu(\nu+1)}{2!} x^2 + \frac{\nu(\nu-2)(\nu+1)(\nu+3)}{4!} x^4 - \cdots$$

and

$$y_2 = x - \frac{(\nu-1)(\nu+2)}{3!} x^3 + \frac{(\nu-1)(\nu-3)(\nu+2)(\nu+4)}{5!} x^5 - \cdots$$

These two series are convergent for $|x| < 1$, which is also the convergence interval of $p(x)$ and of $q(x)$ in (3.4), thereby illustrating Theorem 3.3. In many applications ν is a positive integer n. If n is either zero or even then y_1 reduces to a polynomial of degree n; y_2 is reduced likewise if n is odd. These polynomial solutions, each multiplied by a properly defined constant, are called Legendre polynomials and are denoted by $P_n(x)$; they all satisfy $P_n(1) = 1$. They are the only solutions of the equation which are finite *both* at $x = 1$ and at $x = -1$. Their importance in applications is a consequence of this property because x is often the cosine of a real angle and the *complete* range $-1 \leqslant x \leqslant 1$ is significant.

Problems 5.3

Determine the real values of x that are not ordinary points of the following differential equations.

1. $(x^2 - 4)y'' + (x+2)y' + (x-2)y = 0$.
2. $xy'' + (e^x - 1)y' + y \sin x = 0$.
3. $y'' \cos x + y' \sin x + (x-1)y = 0$.
4. $y'' + \sqrt{x}y' + ye^x = 0$.
5. $y'' \sin x + xy' + xy \sec x = 0$.
6. $(x^2 + x)(x^3 - 1)y''' + (x^2 + x + 1)y'' - (x+1)y' - xy = 0$.

5.4. SUCCESSIVE DIFFERENTIATION

The main principles of the previous section can be shown by a slightly different approach. Express equation (1.1) in the form

$$y'' = -\frac{a_1(x)}{a_2(x)}\, y' - \frac{a_0(x)}{a_2(x)}\, y. \tag{4.1}$$

At an ordinary point x_0 of this equation the coefficients of y and y' are analytic and, therefore, the equation itself determines the value of y'' from given values of y and y' at such a point. Theorem 3.1 states that every solution y is analytic at x_0 and theorem 3.2 shows that the expansion of a solution about x_0 has the form

$$y(x) = y(x_0) + y'(x_0)\,(x - x_0) + \frac{y''(x_0)}{2!}\,(x - x_0)^2 + \ \dots$$

$$+ \frac{y^{(n)}(x_0)}{n!}\,(x - x_0)^n + \ \dots. \tag{4.2}$$

The values of the derivatives at x_0 in this series may be obtained by successive differentiations of (4.1) and substitution of x_0 for x. This method for obtaining a series solution is likely to be of most use when values of y and y' are given at $x = 0$, so that $x_0 = 0$, and the corresponding particular solution only is required; even so the form of the derivative is liable to become complicated as the order increases.

Example 4.1.

Expand the solution of $xy'' + xy' + 2y = 0$ in powers of $(x-1)$, given that $y = 0$ and $y' = 1$ when $x = 1$. Now $x = 1$ is an ordinary point of the equation and the initial conditions give $y''(1) = -1$. Successive differentiations of the equation give:

$$xy''' + (1+x)y'' + 3y' = 0, \quad \text{whence} \quad y'''(1) = -1,$$
$$xy^{\text{iv}} + (2+x)y''' + 4y'' = 0, \quad \text{whence} \quad y^{\text{iv}}(1) = 7,$$
$$xy^{\text{v}} + (3+x)y^{\text{iv}} + 5y''' = 0, \quad \text{whence} \quad y^{\text{v}}(1) = -23$$

and so on. Substitution in (4.2) gives the first few terms of t h required solution:

$$y = (x-1) - \frac{(x-1)^2}{2!} - \frac{(x-1)^3}{3!} + 7\frac{(x-1)^4}{4!} - 23\frac{(x-1)^5}{5!} + \ldots$$

Problems 5.4

By successive differentiation obtain particular solutions, in the form of series expanded about values of x_0 implied by the conditions given, of the following differential equations. Obtain three, or more, non-zero coefficients in addition to those implied directly by the conditions stated.

1. $y'' + e^x y' - xy = 0; \quad y = 1, \quad y' = 0 \quad \text{at} \quad x = 0.$

2. $(x^2 - 1)y'' + 4xy' + (x^2 + 1)y = 0; \quad y = 0, \quad y' = 1 \quad \text{at} \quad x = 0.$

Use Leibnitz's theorem to show that at $x = 0$

$$y_{n+2} = (n^2 + 3n + 1)y_n + n(n-1)y_{n-2}, \quad (n \geqslant 2)$$

where $y_n \equiv d^n y/dx^n$, and by means of this relation derive coefficients of x^n in the series solution.

3. $(x^2 + 1)y'' + xy' - 4y = 0; \quad y = 0, y' = 2 \quad \text{at} \quad x = 0.$

Show that at $x = 0$

$$y_{n+2} = (4 - n^2)y_n. \quad (n \geqslant 0).$$

4. $(x^2 - 1)y'' + 2xy' - 12y = 0.$

Show that this is Legendre's equation with $v = 3$. Obtain a finite series solution in powers of $(x-1)$, given that $y = 1$ at $x = 1$, showing that at $x = 1$

$$y_{n+1} = \frac{(n+4)(3-n)}{2(n+1)} y_n \quad (n \geqslant 0).$$

Express the solution as a polynomial in powers of x and deduce that $P_3(x) = \frac{1}{2}(5x^3 - 3x)$. (See Example 3.3.)

5. $y'' + (x-2)y = 0; \quad y = 1, y' = 0 \quad \text{at} \quad x = 2.$

5.5. ORDINARY POINTS

The method for solution in series outlined in section 2 is now presented in a more elegant form. When a solution is to be expanded, or developed, about an *ordinary point* x_0 it can be expressed in the form

$$y = c_0 + c_1(x - x_0) + c_2(x - x_0)^2 + \ldots + c_n(x - x_0)^n + \ldots$$

$$= \sum_0^\infty c_n(x - x_0)^n. \tag{5.1}$$

A solution is often expanded about $x_0 = 0$ and then (5.1) takes the simpler form of (2.2). If desired a change from $(x - x_0)$ to t, say, could be made and the series for t would have the simple form of (2.2). The student is advised to reconsider Example 4.1 and to change the independent variable from x to $t + 1$; the differential equation becomes $(1 + t)\ddot{y} + (1 + t)\dot{y} + 2y = 0$ subject to the conditions $y = 0$ and $\dot{y} = 1$ when $t = 0$. Since $t = 0$ is an ordinary point of the transformed equation the solution relative to that point is in powers of t and this can be an advantage when an expression for the general term of the series is required.

Example 5.1.

Expand in powers of x the general solution of

$$(x^2 - 1)y'' - xy' + y = 0.$$

The solution "in powers of x" implies that $x_0 = 0$, which is an ordinary point of the equation. The solution progresses in a neat form if the terms of the equation are arranged in blocks so that the terms in each block have equal powers of x after substitution of the series for y and its derivatives. In such a form the equation is

$$(x^2 y'' - xy' + y) - y'' = 0.$$

By (5.1) substitute $y = \sum\limits_{0}^{\infty} c_n x^n$, $y' = \sum\limits_{1}^{\infty} n c_n x^{n-1}$ and

$y'' = \sum\limits_{2}^{\infty} n(n-1) c_n x^{n-2}$ to obtain

$$\sum_{0}^{\infty} \{ n(n-1)c_n - nc_n + c_n \} x^n - \sum_{0}^{\infty} n(n-1)c_n x^{n-2} = 0.$$

Although the *initial* value of n for the summations is not strictly zero for all four terms, the factors n and $(n-1)$ ensure that it is effectively so. After simplification the above expression becomes

$$\sum_{0}^{\infty} (n-1)^2 c_n x^n - \sum_{0}^{\infty} n(n-1) c_n x^{n-2} = 0.$$

Now make both series refer to the same general power of x. This is largely a matter of choice, but it is often advisable to take the least power, $(n-2)$ in this case. To do this replace each n in the first series by $(n-2)$ and change the lower limit to 2; this gives

$$\sum_{2}^{\infty} (n-3)^2 c_{n-2} x^{n-2} - \sum_{0}^{\infty} n(n-1) c_n x^{n-2} = 0.$$

The constants c_n are determined by equating to zero the coefficients of each power of x. The contribution from the first series does not start until $n = 2$. The second series is simply y'' and, although it does not start effectively until $n = 2$, it may be used formally to show that both c_0 and c_1 are arbitrary:

When $n = 0$ the second series gives $0c_0 = 0$, thus c_0 is arbitrary.
When $n = 1$ the second series gives $0c_1 = 0$, thus c_1 is arbitrary.
When $n \geqslant 2$ both series give $(n-3)^2 c_{n-2} - n(n-1)c_n = 0$, which may be expressed as

$$c_n = \frac{(n-3)^2}{n(n-1)} c_{n-2}. \tag{5.2}$$

An expression such as (5.2), which relates the coefficients t
each other, is called a *recurrence relation*. By substitution c
values for n (5.2) gives $c_2 = c_0/2$, $c_3 = 0c_1$ whence $c_3 = 0$
$c_4 = c_2/12 = c_0/24$, $c_5 = c_3/5 = 0$ and $c_7 = c_9 = \ldots = 0$
$c_6 = 3c_4/10 = c_0/80$, and so on. Thus the general solution

$$y = c_0\left(1+\frac{x^2}{2}+\frac{x^4}{24}+\frac{x^6}{80}+ \ldots\right)+c_1x, \equiv y_1+y_2.$$

The recurrence relation may be used to derive an expressio
for the general term in a series, but this is liable to be rathe
complicated. The relation (5.2) shows that

$$c_{n-2} = \frac{(n-5)^2}{(n-2)(n-3)}\, c_{n-4}$$

and

$$c_{n-4} = \frac{(n-7)^2}{(n-4)(n-5)}\, c_{n-6}.$$

This sequence may be continued as far as c_0 and then th
general expression for c_n in terms of c_0 is

$$c_n = \frac{(n-3)^2}{n(n-1)}\cdot\frac{(n-5)^2}{(n-2)(n-3)}\cdot\frac{(n-7)^2}{(n-4)(n-5)} \, \cdots \, \frac{(3)^2}{(6)(5)}$$
$$\times \frac{(1)^2}{(4)(3)}\frac{(-1)^2}{(2)(1)}\, c_0,$$

$n \geqslant 2$ and even.

Since n is even, replace n by $2n$ so that n may take consecutiv
integral values; then the solution y_1 may be expressed as

$$y_1 = c_0 \sum_0^\infty \frac{(2n-3)^2\,(2n-5)^2\,\ldots\,(3)^2\,(1)^2\,(-1)^2}{(2n)!}\, x^{2n}.$$

Note that for a given value of n the first n terms only, countin
from the right, are taken in the numerator, whilst direc
substitution of n is made in the denominator.

A two-term recurrence relation such as (5.2) generates the coefficients c_n easily. Some equations yield three-term relations and they are more difficult to manage; for example see number **2** of Problems 5.2.

Example 5.2.

Expand the general solution of Airy's equation $y'' - xy = 0$ about the ordinary point $x = 0$.

The stages of the solution are similar to those explained in the previous example and are shown below with little comment.

$$\sum_0^\infty n(n-1)c_n x^{n-2} - \sum_0^\infty c_n x^{n+1} = 0,$$

$$\sum_0^\infty n(n-1)c_n x^{n-2} - \sum_3^\infty c_{n-3} x^{n-2} = 0.$$

When $n = 0$ the first series gives $0c_0 = 0$, thus c_0 is arbitrary.
When $n = 1$ the first series gives $0c_1 = 0$, thus c_1 is arbitrary.
When $n = 2$ the first series gives $2c_2 = 0$, thus $c_2 = 0$.
When $n \geqslant 3$ both series give $n(n-1)c_n - c_{n-3} = 0$, whence the recurrence relation

$$c_n = \frac{c_{n-3}}{n(n-1)}.$$

This gives $c_3 = c_0/3\cdot2$, $c_4 = c_1/4\cdot3$, $c_5 = c_2/5\cdot4 = 0$ whence $c_5 = c_8 = \ldots = 0$, $c_6 = c_3/6\cdot5 = c_0/6\cdot5\cdot3\cdot2$, $c_7 = c_4/7\cdot6 = c_1/7\cdot6\cdot4\cdot3$, and so on. The general solution is

$$y = c_0\left(1 + \frac{x^3}{3\cdot2} + \frac{x^6}{6.5.3.2} + \ldots\right)$$

$$+ c_1\left(x + \frac{x^4}{4\cdot3} + \frac{x^7}{7.6.4.3} + \ldots\right).$$

It is left as an exercise to show that this may be expressed in the general form

$$y = c_0 \sum_0^\infty \frac{(3n-2)(3n-5) \ldots (7)(4)(1)}{(3n)!} x^{3n}$$

$$+ c_1 \sum_0^\infty \frac{(3n-1)(3n-4) \ldots (8)(5)(2)}{(3n+1)!} x^{3n+1}.$$

These examples indicate the general approach for obtaining a solution by series about an ordinary point. The details of a solution depend on the particular equation and, in some cases, the elementary method of section 2 may be preferred for analysing the nature of the first few coefficients in the series.

Problems 5.5

Obtain the general solutions, as series in powers of x, of the following differential equations.

1. $2y'' + xy' - 4y = 0$.

2. $(x^2+1)y'' + 2xy' = 0$.

By direct integration of the equation deduce a series expansion of $\tan^{-1} x$.

3. $(x^2-2)y'' - 2xy' + 2y = 0$.   **4.** $(4x^2-1)y'' - 6xy' + 4y = 0$.

5. $(x^2+1)y'' - xy' + y = 0$.

Verify that in closed form one solution is $y = x \sinh^{-1} x - \sqrt{(1+x^2)}$.

6. $(x^2-1)y'' + 4xy' + 2y = 0$.   **7.** $(2x^2+1)y'' + 11xy' + 9y = 0$.

8. $(x-1)y'' - xy' + y = 0$.   **9.** $y'' = (x-3)y$.

10. $y''' + xy = 0$.

5.6. REGULAR SINGULAR POINTS

The reasoning about series solutions given so far has probably left the impression that expansions should be made about ordinary points only and that singular points should be avoided. There is a class of singular points about which series solutions can be obtained and of these the most important are termed regular singular points. Solutions developed about such points are likely to have several advantages over those related to ordinary points, though the techniques of derivation are often more complicated.

DEFINITION. *If, when the differential equation* $y'' + p(x)y' + q(x)y = 0$ *is expressed in the form*

$$(x - x_0)^2 y'' + (x - x_0)\bar{p}(x)y' + \bar{q}(x)y = 0, \qquad (6.1)$$

the functions $\bar{p}(x) = (x - x_0)\,p(x)$ *and* $\bar{q}(x) = (x - x_0)^2\,q(x)$ *are analytic at* $x = x_0$ *then* x_0 *is a regular singular point of the equation.*

THEOREM 6.1. *A differential equation of the form (6.1) in which* $\bar{p}$ *and* $\bar{q}$ *are analytic at* $x = x_0$ *has at least one solution of the form*

$$\begin{aligned} y &= (x - x_0)^\varrho \{c_0 + c_1(x - x_0) + c_2(x - x_0)^2 + \ \ldots\} \\ &= (x - x_0)^\varrho \sum_0^\infty c_n(x - x_0)^n, \end{aligned} \qquad (6.2)$$

where the index ϱ*, which need not be an integer, is chosen so that* c_0 *is always the coefficient of the lowest power* ϱ *of* $(x - x_0)$*; that is* $c_0 \neq 0$.

If the expansions of $\bar{p}$ and $\bar{q}$ are valid for $|x - x_0| < R$ then the series solution is valid over the same range at least, except perhaps at $x = x_0$.

Example 6.1.

Obtain the general solution of

$$2xy'' + (2x - 1)y' + 2y = 0$$

in series expanded about $x = 0$.

In the form of (6.1) the equation is $x^2 y'' + x\left(x - \frac{1}{2}\right)y' + xy = 0$, where $\bar{p}(x) = \left(x - \frac{1}{2}\right)$ and $\bar{q}(x) = x$ are analytic at $x = 0$, a regular singular point. Rearrange the equation in the convenient form

$$(2xy'' - y') + 2(xy' + y) = 0$$

and substitute $y = \sum_0^\infty c_n x^{\varrho+n}, \quad y' = \sum_0^\infty (\varrho+n)c_n x^{\varrho+n-1}$

and $y'' = \sum_0^\infty (\varrho+n)(\varrho+n-1)c_n x^{\varrho+n-2}$

to obtain

$$\sum_0^\infty \{2(\varrho+n)(\varrho+n-1) - (\varrho+n)\} c_n x^{\varrho+n-1}$$

$$+ 2\sum_0^\infty \{(\varrho+n)+1\} c_n x^{\varrho+n} = 0,$$

which simplifies to

$$\sum_0^\infty (\varrho+n)(\varrho+n-\tfrac{3}{2})c_n x^{\varrho+n-1} + \sum_0^\infty (\varrho+n+1)c_n x^{\varrho+n} = 0.$$

Replace each n in the second series by $(n-1)$ and change the lower limit to 1; this gives

$$\sum_0^\infty (\varrho+n)(\varrho+n-\tfrac{3}{2})c_n x^{\varrho+n-1} + \sum_1^\infty (\varrho+n)c_{n-1} x^{\varrho+n-1} = 0. \quad (6.3)$$

When $n = 0$ the first series gives the *indicial equation* for the index ϱ,

$$\varrho(\varrho - \tfrac{3}{2})c_0 = 0,$$

showing that $\varrho = 0$ or $\tfrac{3}{2}$ since by definition $c_0 \neq 0$.

Taking $\varrho = \tfrac{3}{2}$ in (6.3) gives the recurrence relation

$$c_n = -\frac{1}{n} c_{n-1} \qquad (n \geqslant 1),$$

whence $c_1 = -c_0, c_2 = -c_1/2 = c_0/2!, c_3 = -c_2/3 = -c_0/3!, \ldots$. Thus one solution is

$$y_1 = c_0 x^{3/2}\left(1 - x + \frac{x^2}{2!} - \frac{x^3}{3!} + \ldots\right), = c_0 x^{3/2} e^{-x}$$

in closed form.

Taking $\varrho = 0$ in (6.3) gives for a different set of coefficients c_n, say b_n, the recurrence relation

$$b_n = \frac{-2}{(2n-3)} b_{n-1} \qquad (n \geqslant 1),$$

whence $b_1 = 2b_0$, $b_2 = -2b_1 = -4b_0$, $b_3 = -2b_2/3 = 8b_0/3$, $b_4 = -2b_3/5 = -16b_0/15$. Thus another solution, independent of y_1, is

$$y_2 = b_0\left(1 + 2x - 4x^2 + \tfrac{8}{3}x^3 - \tfrac{16}{15}x^4 + \ \ldots\right).$$

The general solution is $y = y_1 + y_2$, which is valid for all x because both $\bar{p}$ and $\bar{q}$ are analytic for all x.

Example 6.2.

Obtain the general solution of

$$x(x+1)y'' + (x+5)y' - 4y = 0$$

in series expanded about $x = 0$, a regular singular point.

As the main steps of the solution are similar to those in the example above they are followed with little comment.

$$(x^2 y'' + xy' - 4y) + (xy'' + 5y') = 0.$$

$$\sum_0^\infty \{(\varrho+n)(\varrho+n-1) + (\varrho+n) - 4\}c_n x^{\varrho+n}$$

$$+ \sum_0^\infty \{(\varrho+n)(\varrho+n-1) + 5(\varrho+n)\}c_n x^{\varrho+n-1} = 0,$$

$$\sum_0^\infty (\varrho+n+2)(\varrho+n-2)c_n x^{\varrho+n} + \sum_0^\infty (\varrho+n)(\varrho+n+4)c_n x^{\varrho+n-1} = 0,$$

$$\sum_1^\infty (\varrho+n+1)(\varrho+n-3)c_{n-1} x^{\varrho+n-1}$$

$$+ \sum_0^\infty (\varrho+n)(\varrho+n+4)c_n x^{\varrho+n-1} = 0. \qquad (6.4)$$

When $n = 0$ the second series gives the indicial equation $\varrho(\varrho+4)c_0 = 0$ and $\varrho = 0$ or -4. With $\varrho = 0$ (6.4) gives the recurrence relation

$$c_n = \frac{-(n+1)(n-3)}{n(n+4)} c_{n-1} \qquad (n \geqslant 1),$$

whence $c_1 = 4c_0/5, \ c_2 = c_0/5, \ c_3 = 0, \ c_4 = c_5 = \ldots = 0$ and

$$y_1 = c_0\left(1+\frac{4x}{5}+\frac{x^2}{5}\right).$$

With $\varrho = -4$ (6.4) gives the recurrence relation

$$n(n-4)b_n = -(n-3)(n-7)b_{n-1} \qquad (n \geqslant 1).$$

Due to the factor $(n-4)$ it is advisable not to use a relation with b_n expressed explicitly until $n > 4$. The relation shows that $b_1 = 4b_0, \ b_2 = 5b_1/4 = 5b_0, \ b_3 = 0, \ b_4$ is arbitrary, $b_5 = 4b_4/5, \ b_6 = b_5/4 = b_4/5, \ b_7 = b_8 = \ldots = 0$, whence the solution

$$y = b_0 x^{-4}(1+4x+5x^2)+b_4\left(1+\frac{4x}{5}+\frac{x^2}{5}\right), \equiv y_2+y_1,$$

which is the general solution, valid for all $x \neq 0$, because y_1 and y_2 are independent. In this rather special case the second solution contains the first.

If the indicial equation has equal roots then the process outlined above yields one solution only; difficulties may also arise when the roots differ by an integer, though the solution to Example 6.2 was straightforward. Some treatments of this subject which do not consider the nature of the point about which a series solution is to be developed, and even some that do, quote (6.2) as the only series to use. Expansions about ordinary points can certainly be derived from (6.2) but the algebra is then heavier than if (5.1) were used. It is worth treating Examples 5.1 and 5.2 in this way to appreciate the difference.

There are several ways of dealing with the awkward cases when the roots of the indicial equation differ by an integer and when they are equal. None of these methods is particularly easy and the one chosen is likely to depend on how many terms of the series solution are sufficient for the accuracy required. To complete this section one method is illustrated, due to G. Frobenius, and in the following section another method is evolved from an enquiry into the independence of solutions.

THEOREM 6.2. *When the indicial equation has equal roots ϱ, two linearly independent solutions to (6.1) expanded about $x = 0$ are always of the form*

$$y_1 = x^\varrho \sum_0^\infty c_n x^n,$$

which is the familiar series, and

$$y_2 = y_1 \log x + x^\varrho \sum_1^\infty b_n x^n. \tag{6.5}$$

The Frobenius method is derived by regarding the differential equation, with an indicial equation having equal roots ϱ, as the limiting form of an equation whose indicial equation has slightly unequal roots, ϱ and $\varrho + \varepsilon$. Two solutions of this equation are

$$y_1 = x^\varrho \sum_0^\infty c_n x^n \quad \text{and} \quad y_2 = x^{\varrho + \varepsilon} \sum_0^\infty c_n' x^n,$$

where the coefficients c_n, c_n' are functions of ϱ, $\varrho + \varepsilon$ respectively. As $\varepsilon \to 0$ the two distinct solutions merge to the single solution y_1. However, the difference $y_2 - y_1$ is also a solution and although this tends to zero as $\varepsilon \to 0$ the solution $(y_2 - y_1)/\varepsilon$ tends to $\partial y_1/\partial \varrho$. Thus it seems reasonable to suppose that a second solution is

$$\frac{\partial y_1}{\partial \varrho} = \left\{ x^\varrho \sum_0^\infty c_n x^n \right\} \log x + x^\varrho \sum_0^\infty \frac{dc_n}{d\varrho} x^n, \qquad (6.6)$$

where the coefficients $dc_n/d\varrho$ are the b_n of (6.5) for the value of ϱ and, as c_0 is independent of ϱ, the second term has the same form as that in (6.5) where n starts at 1.

Example 6.3.

Obtain the general solution of

$$x^2 y'' + x(x-1)y' - (x-1)y = 0$$

in series expanded about the regular singular point $x = 0$. The initial steps of the solution are as in previous examples.

$$(x^2 y'' - xy' + y) + (x^2 y' - xy) = 0,$$

$$\sum_0^\infty \{(\varrho+n)(\varrho+n-1) - (\varrho+n) + 1\}c_n x^{\varrho+n}$$

$$+ \sum_0^\infty \{(\varrho+n) - 1\}c_n x^{\varrho+n+1} = 0,$$

$$\sum_0^\infty (\varrho+n-1)^2 c_n x^{\varrho+n} + \sum_0^\infty (\varrho+n-1)c_n x^{\varrho+n+1} = 0,$$

$$\sum_0^\infty (\varrho+n-1)^2 c_n x^{\varrho+n} + \sum_1^\infty (\varrho+n-2)c_{n-1} x^{\varrho+n} = 0.$$

When $n = 0$ the first series gives the indicial equation $(\varrho-1)^2 = 0$ and $\varrho = 1$ twice. The recurrence relation for this only value of ϱ is

$$c_n = \frac{-(n-1)}{n^2} c_{n-1} \qquad (n \geqslant 1),$$

whence $c_1 = c_2 = \ldots = 0$ and a solution of the equation is

$$y_1 = c_0 x.$$

In order to derive a second solution y_2 it is necessary to express y_1 as a function of ϱ; the value $\varrho = 1$ is not inserted until the final stage of the derivation. In terms of ϱ the recurrence relation is

$$c_n = \frac{-(\varrho+n-2)}{(\varrho+n-1)^2}\, c_{n-1} \qquad (n \geqslant 1)$$

and this enables c_n to be expressed relative to c_0 as

$$c_n = \frac{-(\varrho+n-2)}{(\varrho+n-1)^2} \cdot \frac{-(\varrho+n-3)}{(\varrho+n-2)^2} \cdots \frac{-(\varrho)}{(\varrho+1)^2} \cdot \frac{-(\varrho-1)}{\varrho^2}\, c_0$$

by successive reduction of the type used in Example 5.1. The expression for c_n simplifies to

$$c_n = \frac{(-1)^n(\varrho-1)}{\varrho(\varrho+1)\ldots(\varrho+n-2)(\varrho+n-1)^2}\, c_0, \equiv (-1)^n f(\varrho)c_0, \text{ say.}$$

It remains to differentiate c_n, that is effectively $f(\varrho)$, with respect to ϱ and this is achieved most easily by taking logarithms first. Thus

$$\log f(\varrho) = \log(\varrho-1) - \log\varrho - \log(\varrho+1) - \ldots$$
$$- \log(\varrho+n-2) - 2\log(\varrho+n-1)$$

and

$$\frac{df}{d\varrho} = f(\varrho)\left\{ \frac{1}{(\varrho-1)} - \frac{1}{\varrho} - \frac{1}{(\varrho+1)} - \ldots \right.$$
$$\left. - \frac{1}{(\varrho+n-2)} - \frac{2}{(\varrho+n-1)} \right\}.$$

When $\varrho = 1$

$$\frac{dc_n}{d\varrho} = (-1)^n f'(1)c_0 = \frac{(-1)^n}{1 \cdot 2 \ldots (n-1)n^2}\, c_0$$
$$= \frac{(-1)^n}{nn!}\, c_0 = b_n \qquad (n \geqslant 1)$$

and substitution in (6.5) or (6.6) gives a second solution

$$y_2 = c_0 x \log x + c_0 x \sum_1^\infty \frac{(-1)^n}{nn!} x^n.$$

Since y_1 and y_2 are each solutions of the differential equation, and as c_0 is arbitrary, each solution may be multiplied by a different arbitrary constant. In practice it is usual to set $c_0 = 1$ at an early stage of the solution and later to introduce arbitrary constants A, B for the general solution. For this example the general solution is

$$y = y_1 + y_2 = Ax + Bx \log x + Bx \sum_1^\infty \frac{(-1)^n}{nn!} x^n$$

$$= (A + B \log x)x + B \sum_1^\infty \frac{(-1)^n}{nn!} x^{1+n}.$$

When the roots of the indicial equation differ by an integer, the general solution may or may not include a solution which involves $\log x$. The method of Frobenius may be used with certain refinements to deal with the awkward case when a logarithmic term occurs.

Problems 5.6

Obtain the general solutions, as series in powers of x, of the following differential equations. Try to identify some of the series as known functions.

1. $2xy'' + (1 - 2x)y' - y = 0.$ 2. $xy'' - (x - 4)y' - 2y = 0.$

3. $x^2y'' + x(x + 1)y' - y = 0.$ 4. $x(x-1)y'' + 2(2x-1)y' + 2y = 0.$

5. $4xy'' + 2y' + y = 0.$ 6. $x(x-1)y'' + (x-3)y' - 4y = 0.$

7. $4x^2y'' + 4xy' + (x^2 - 1)y = 0.$

8. $2x(2x-1)y'' - (4x^2 + 1)y' + (2x+1)y = 0.$

5.7. RELATION BETWEEN SOLUTIONS

In previous sections the assertion has often been made that two given solutions of a second-order differential equation are linearly independent. This property of the solutions is often evident upon inspection but it is as well to have a more reliable test which can be applied to all cases. Let the differential equation be expressed in the form

$$y'' + p(x)y' + q(x)y = 0 \qquad (7.1)$$

and have the general solution

$$y = Ay_1 + By_2,$$

where y_1, y_2 are linearly independent and A, B are arbitrary constants.

Two functions $y_1(x)$, $y_2(x)$ are linearly dependent when one can be expressed as a constant multiple of the other, that is when

$$Ay_1 + By_2 = 0 \qquad (7.2)$$

for all x where A, B are non-zero constants; otherwise the functions are linearly independent. Differentiate (7.2) to obtain

$$Ay_1' + By_2' = 0. \qquad (7.3)$$

The elimination of either constant between (7.2) and (7.3) leads to the necessary condition for linear dependence

$$W = \begin{vmatrix} y_1 & y_2 \\ y_1' & y_2' \end{vmatrix} \equiv 0, \qquad (7.4)$$

where the determinant W is called the Wronskian of the two functions y_1, y_2. If y_1, y_2 are solutions of a homogeneous linear differential equation then the stronger condition $W = 0$ is *sufficient* for their linear dependence. Conversely, and more to the point, two solutions are linearly independent if

$$W \neq 0, \tag{7.5}$$

for it can be shown that W either vanishes identically or does not vanish at all. Note, however, that W may vanish at a singular point of the equation without implying that the solutions are dependent. To illustrate this point consider the solutions to Example 6.1 at $x = 0$.

By means of the Wronskian a relation can be derived between the two solutions y_1, y_2. Substitution of these solutions in (7.1) gives

$$y_1'' + py_1' + qy_1 = 0 \quad \text{and} \quad y_2'' + py_2' + qy_2 = 0.$$

Multiply these equations by y_2, y_1 respectively and subtract to obtain

$$y_2 y_1'' - y_1 y_2'' + p(y_2 y_1' - y_1 y_2') = 0$$

which is equivalent to

$$\frac{dW}{dx} + pW = 0.$$

This first-order equation has the solution

$$W = C \exp\left(-\int p \, dx\right), \tag{7.6}$$

where C is an arbitrary constant. Now

$$W = y_1 y_2' - y_2 y_1' = y_1^2 \frac{d}{dx}(y_2/y_1)$$

which upon integration gives

$$y_2 = y_1 \int \frac{W}{y_1^2} \, dx, \tag{7.7}$$

the arbitrary constant being omitted as its presence would add only a multiple of y_1 to y_2. The required relation is obtained by substituting (7.6) in (7.7) and is

$$y_2 = Cy_1 \int \frac{1}{y_1^2} \left\{ \exp\left(-\int p \, dx \right) \right\} dx. \qquad (7.8)$$

The satisfaction of being able to state the form of a second solution y_2 in terms of a known solution y_1 is justified only if the integrals are reasonably easy to obtain.

Example 7.1.

Obtain a second solution of the equation

$$x^2 y'' + x(x-1)y' - (x-1)y = 0$$

given that one solution is $y_1 = x$.

Now
$$\int p \, dx = \int \frac{(x-1)}{x} \, dx = x - \log x$$

and
$$\exp\left(-\int p \, dx \right) = xe^{-x}.$$

By (7.8) a second solution is

$$y_2 = Cx \int \frac{e^{-x}}{x} \, dx = Cx \int \left\{ \frac{1}{x} - 1 + \frac{x}{2!} - \frac{x^2}{3!} + \frac{x^3}{4!} - \dots \right\} dx$$

$$= Cx \left\{ \log x - x + \frac{x^2}{2 \cdot 2!} - \frac{x^3}{3 \cdot 3!} + \frac{x^4}{4 \cdot 4!} - \dots \right\}$$

$$= Cx \log x + C \sum_1^\infty \frac{(-1)^n}{nn!} x^{1+n}.$$

This solution was obtained by the method of Frobenius in Example 6.3.

Example 7.2.

Obtain independent solutions of the equation

$$x(x-1)y'' + 2(x-1)y' - 2y = 0$$

valid in the region of the regular singular point $x = 0$.

The first solution will be found by the series method of Examples 6.1 and 6.2.

$$(x^2y'' + 2xy' - 2y) - (xy'' + 2y') = 0,$$

$$\sum_0^\infty \{(\varrho+n)(\varrho+n-1) + 2(\varrho+n) - 2\} c_n x^{\varrho+n}$$

$$-\sum_0^\infty \{(\varrho+n)(\varrho+n-1) + 2(\varrho+n)\} c_n x^{\varrho+n-1} = 0,$$

$$\sum_0^\infty (\varrho+n-1)(\varrho+n+2)c_n x^{\varrho+n} - \sum_0^\infty (\varrho+n)(\varrho+n+1)c_n x^{\varrho+n-1} = 0,$$

$$\sum_1^\infty (\varrho+n-2)(\varrho+n+1)c_{n-1}x^{\varrho+n-1}$$

$$-\sum_0^\infty (\varrho+n)(\varrho+n+1)c_n x^{\varrho+n-1} = 0.$$

$$(7.9)$$

When $n = 0$ the second series gives the indicial equation $\varrho(\varrho+1)c_0 = 0$ and the roots $\varrho = 0$, -1 differ by an integer. With $\varrho = 0$ (7.9) gives the recurrence relation

$$c_n = \frac{(n-2)}{n} c_{n-1} \qquad (n \geqslant 1),$$

whence $c_1 = -c_0$, $c_2 = 0$ and $c_3 = c_4 = \ldots = 0$. One solution is therefore $y_1 = c_0(1-x)$. With $\varrho = -1$ (7.9) gives the recurrence relation

$$(n-1)b_n = (n-3)b_{n-1} \qquad (n \geqslant 1)$$

whence $0b_1 = -2b_0$. This implies that $b_0 = 0$ which violates the definition of b_0 and shows how roots which differ by an integer can upset the usual procedure. A second solution is now sought by means of (7.8) and the first solution is used in the form $y_1 = 1-x$ by setting $c_0 = 1$.

Now $$\int p \, dx = \int \frac{2}{x} \, dx = 2 \log x$$

and $$\exp\left(-\int p \, dx\right) = 1/x^2.$$

By (7.8) with $C = 1$ $$y_2 = (1-x) \int \frac{dx}{x^2(1-x)^2}$$

$$= (1-x) \int \left\{\frac{1}{x^2} + \frac{2}{x} + \frac{1}{(1-x)^2} + \frac{2}{(1-x)}\right\} dx$$

$$= (1-x) \left\{-\frac{1}{x} + 2 \log x + \frac{1}{(1-x)} - 2 \log (1-x)\right\}$$

$$= 2 - \frac{1}{x} + (1-x) \log \left(\frac{x}{1-x}\right)^2.$$

The result (7.8) may be derived by changing the dependent variable y of equation (7.1) to $y = vy_1$, where v is the new dependent variable and y_1 is a known solution of the equation. The substitution of y, y' and y'' in (7.1) gives after some rearrangement

$$y_1v'' + (2y_1' + py_1)v' + (y_1'' + py_1' + qy_1)v = 0.$$

Since y_1 satisfies (7.1), the coefficient of v is zero and the equation reduces to

$$\frac{v''}{v'} + \frac{2y_1'}{y_1} + p = 0$$

which upon integration becomes

$$\log v' + 2 \log y_1 = -\int p \, dx$$

or $\qquad v' = \dfrac{1}{y_1^2} \exp\left(-\int p\,dx\right),$

whence $\qquad v = \displaystyle\int \dfrac{1}{y_1^2} \left\{\exp\left(-\int p\,dx\right)\right\} dx$

and $\qquad y_2 = y_1 v = y_1 \displaystyle\int \dfrac{1}{y_1^2} \left\{\exp\left(-\int p\,dx\right)\right\} dx.$

If the two arbitrary constants of integration had been included then the result would have been the general solution of equation (7.1).

Problems 5.7

Obtain the general solutions of the following differential equations.

1. $x(x-1)y'' + 3xy' + y = 0.$ $\qquad$ **2.** $x(x-1)y'' + (4x-1)y' + 2y = 0.$

3. $xy'' - (2x-1)y' + (x-1)y = 0.$

4. $x(x+1)^2 y'' - (x^2-1)y' + (x-1)y = 0.$

5. $x^2(x+1)y'' + x(2x-1)y' + y = 0.$

6. $xy'' + (x+1)y' + 2y = 0.$ (To obtain a second solution of this equation use the technique shown for $Z_0(x)$ in section 5.10.)

5.8. THE GAMMA FUNCTION

This function is introduced here because it is needed in the next two sections. The gamma function is defined by

$$\Gamma(\nu) = \int\limits_0^\infty e^{-x} x^{\nu-1}\, dx \qquad (\nu > 0), \tag{8.1}$$

where ν is a parameter independent of x.

When $\qquad \nu = 1, \quad \Gamma(1) = \displaystyle\int\limits_0^\infty e^{-x}\, dx = 1. \tag{8.2}$

Replace ν in (8.1) by $\nu+1$ and integrate by parts, then

$$\Gamma(\nu+1) = \int\limits_0^\infty e^{-x} x^\nu\, dx = \left[-e^{-x}x^\nu\right]_0^\infty + \nu \int\limits_0^\infty e^{-x} x^{\nu-1}\, dx$$

whence $\qquad \Gamma(\nu+1) = \nu\,\Gamma(\nu). \tag{8.3}$

When ν is a positive integer n repeated application of (8.3) gives

$$\Gamma(n+1) = n\,\Gamma(n) = n(n-1)\,\Gamma(n-1) = \ldots$$
$$= n(n-1)(n-2) \ldots 3.2.1\,\Gamma(1)$$

which, together with (8.2), shows that

$$\Gamma(n+1) = n!. \tag{8.4}$$

This result shows why the gamma function is also called the general factorial function. There are tables of $\Gamma(\nu)$ for values of ν over short ranges, usually from 1 to 2. Such tables are sufficient because (8.3) enables the gamma function of any positive number to be expressed in terms of a tabulated function. For example

$$\Gamma\left(\tfrac{5}{2}\right) = \left(\tfrac{3}{2}\right)\Gamma\left(\tfrac{3}{2}\right) = \left(\tfrac{3}{2}\right)\left(\tfrac{1}{2}\right)\Gamma\left(\tfrac{1}{2}\right) = \left(\tfrac{3}{4}\right)(1\cdot7724) = 1\cdot3293.$$

An important relation is

$$\Gamma(\nu)\,\Gamma(1-\nu) = \pi \operatorname{cosec}(\nu\pi) \qquad (\nu \neq 0, \pm1, \pm2, \ldots).$$

Thus when $\nu = \tfrac{1}{2}$

$$\left\{\Gamma\left(\tfrac{1}{2}\right)\right\}^2 = \pi \operatorname{cosec}(\pi/2)$$
$$\text{or} \qquad \Gamma\left(\tfrac{1}{2}\right) = \sqrt{\pi}.$$

When ν takes non-integral negative values the gamma function is *defined* by (8.3), which would be used in the form $\Gamma(\nu) = \dfrac{1}{\nu}\Gamma(\nu+1)$. For example

$$\Gamma\left(-\tfrac{5}{2}\right) = \left(-\tfrac{2}{5}\right)\Gamma\left(-\tfrac{3}{2}\right) = \left(-\tfrac{2}{5}\right)\left(-\tfrac{2}{3}\right)\Gamma\left(-\tfrac{1}{2}\right)$$
$$= \left(-\tfrac{2}{5}\right)\left(-\tfrac{2}{3}\right)(-2)\,\Gamma\left(\tfrac{1}{2}\right) = -\tfrac{8}{15}\sqrt{\pi}.$$

Note that $\Gamma(\nu) \to \pm\infty$ as ν tends to zero or to a negative integer.

5.9. BESSEL'S DIFFERENTIAL EQUATION

The equation

$$x^2y'' + xy' + (x^2 - v^2)y = 0, \tag{9.1}$$

v being a constant, is known as Bessel's equation and its solutions are called Bessel functions. When v takes zero or integral values only it is a commendable and fairly standard practice to replace v by n; because of this m will be used here as the variable of summation for the series solutions. Bessel's equation has a regular singular point at the origin, so a solution of the form $y = \sum\limits_0^\infty c_m x^{\varrho+m}$ is assumed and the usual procedure followed. Rearrange (9.1) in the convenient form

$$(x^2y'' + xy' - v^2y) + x^2y = 0$$

and substitute for y, y' and y'' to obtain

$$\sum_0^\infty \{(\varrho+m)(\varrho+m-1) + (\varrho+m) - v^2\} c_m x^{\varrho+m} + \sum_0^\infty c_m x^{\varrho+m+2} = 0$$

which simplifies to

$$\sum_0^\infty \{(\varrho+m)^2 - v^2\} c_m x^{\varrho+m} + \sum_0^\infty c_m x^{\varrho+m+2} = 0.$$

Replace m by $m-2$ in the second series to obtain

$$\sum_0^\infty \{(\varrho+m)^2 - v^2\} c_m x^{\varrho+m} + \sum_2^\infty c_{m-2} x^{\varrho+m} = 0. \tag{9.2}$$

When $m = 0$ the first series gives the indicial equation $\varrho^2 - v^2 = 0$ and $\varrho = \pm v$.

With $\varrho = v$ the first series of (9.2) gives, for $m = 1$, $(2v+1)c_1 = 0$ which implies that $c_1 = 0$ since generally $(2v+1) \neq 0$. The recurrence relation is

$$c_m = \frac{-c_{m-2}}{2m(v+m/2)} \qquad (m \geqslant 2),$$

whence $c_2 = -c_0/2^2(v+1), \quad c_3 = c_5 = \ldots = 0,$

$c_4 = -c_2/2^22(v+2) = c_0/2^42(v+1)(v+2),$

$c_6 = -c_4/2^23(v+3)$

$\quad = -c_0/2^63!(v+1)(v+2)(v+3), \ldots.$

Thus one solution of Bessel's equation is

$$y_1 = c_0x^v \left\{ 1 - \frac{x^2}{2^21!(v+1)} + \frac{x^4}{2^42!(v+1)(v+2)} \right.$$

$$\left. - \frac{x^6}{2^63!(v+1)(v+2)(v+3)} + \cdots \right\}.$$

The standard form of this solution, denoted by $J_v(x)$, is obtained by including the factor $1/2^v\Gamma(v+1)$ and is, with $c_0 = 1$,

$$J_v(x) = \frac{(x/2)^v}{\Gamma(v+1)} \left\{ 1 - \frac{(x/2)^2}{1!(v+1)} + \frac{(x/2)^4}{2!(v+1)(v+2)} - \cdots \right\}$$

or

$$J_v(x) = \sum_0^\infty \frac{(-1)^m}{m!\Gamma(v+m+1)} \left(\frac{x}{2}\right)^{v+2m}. \tag{9.3}$$

With $\varrho = -v$ the analysis for a set of coefficients b_m is similar in form to that for the c_m and a second solution of Bessel's equation is

$$J_{-v}(x) = \sum_0^\infty \frac{(-1)^m}{m!\Gamma(-v+m+1)} \left(\frac{x}{2}\right)^{-v+2m}. \tag{9.4}$$

The solutions $J_v(x)$ and $J_{-v}(x)$ are called Bessel functions of the first kind of order v and, provided v is neither zero nor an integer, the general solution of Bessel's equation is

$$y = AJ_v(x) + BJ_{-v}(x),$$

since $J_v(x)$ and $J_{-v}(x)$ are linearly independent. If v is restricted to zero or integral values n then (9.3) is expressed as

$$J_n(x) = \sum_0^\infty \frac{(-1)^m}{m!\Gamma(n+m+1)} \left(\frac{x}{2}\right)^{n+2m}. \tag{9.5}$$

A similar expression for $J_{-n}(x)$ follows from (9.4). When ν is an integer the roots $\varrho = \pm n$ differ by an integer and the solutions $J_n(x)$, $J_{-n}(x)$ are not linearly independent. It can be shown by algebraic manipulation of the series for $J_{-n}(x)$ that the dependence is

$$J_{-n}(x) = (-1)^n J_n(x). \tag{9.6}$$

Clearly, if $\nu = 0$ the two solutions are identical.

For the cases when $\nu = n$ a second solution may be obtained by the method of Frobenius or by the result (7.8) which, although less general in form, will be used here. In Bessel's equation $p = 1/x$ and (7.8) shows that a second solution is

$$Z_n(x) = J_n(x) \int \frac{dx}{x\{J_n(x)\}^2}, \tag{9.7}$$

where $Z_n(x)$ is called a Bessel function of the second kind of order n. Thus when ν is zero or an integer the general solution of Bessel's equation is

$$y = AJ_n(x) + BZ_n(x). \tag{9.8}$$

It is more usual to take $Z_n(x)$ in a linear combination with $J_n(x)$ to form a second solution $Y_n(x)$, another Bessel function of the second kind. Thus

$$Y_n(x) = a\{Z_n(x) + bJ_n(x)\},$$

where a and b are constants. When $n = 0$ this solution has the form

$$Y_0(x) = \frac{2}{\pi} \{Z_0(x) + (\gamma - \log 2)J_0(x)\},$$

where Euler's constant γ, which has the value $0{\cdot}5772\ldots$, is defined as $\lim_{s\to\infty}(1+\frac{1}{2}+\frac{1}{3}+\ \ldots\ +\frac{1}{s}-\log s)$. The functions $Y_n(x)$ are well tabulated and another form of the general solution to Bessel's equation is

$$y = AJ_n(x)+BY_n(x). \tag{9.9}$$

To ensure a uniformity between the formal expressions and numerical tabulation, a form of the second solution is defined which is valid for all values of the order, both integral and non-integral. It is given by

$$Y_\nu(x) = \{J_\nu(x)\cos\nu\pi-J_{-\nu}(x)\}\ \text{cosec}\ \nu\pi$$

and

$$Y_n(x) = \lim_{\nu\to n} Y_\nu(x).$$

Although even this simple treatment of the solutions to Bessel's equation may appear complicated, it should be realised that the solutions, the Bessel functions, are widely tabulated. If a differential equation is recognised as or can be transformed into Bessel's equation then its solutions are likely to be available.

Example 9.1.

By means of the substitution $x = \log t$ obtain the general solution of the equation $y''+(e^{2x}-1)y = 0$. Derive the particular solution that satisfies the conditions $y(0) = 1$ and $y \to 0$ as $x \to -\infty$.

Now $\dfrac{dy}{dx} = \dfrac{dy}{dt}\dfrac{dt}{dx} = \dfrac{dy}{dt}\,e^x$, since $t = e^x$, and

$$\frac{d^2y}{dx^2} = \frac{d^2y}{dt^2}\frac{dt}{dx}e^x+\frac{dy}{dt}e^x = e^{2x}\frac{d^2y}{dt^2}+e^x\frac{dy}{dt}.$$

Thus the equation becomes $t^2\ddot{y}+t\dot{y}+(t^2-1)y=0$, which is Bessel's equation of order 1. By (9.9) a form of the general solution is

$$y = AJ_1(t)+BY_1(t) = AJ_1(e^x)+BY_1(e^x).$$

The given conditions, with tabulated values of J_1 and Y_1, show that

$$y(0) = 1 = AJ_1(1)+BY_1(1) = 0{\cdot}440A-0{\cdot}781B$$

and that, as $x \to -\infty$,

$$y \to 0 = AJ_1(0)+BY_1(0) = 0A - \infty B.$$

These are satisfied by $B = 0$ and $A = 2{\cdot}273$, so the required particular solution is

$$y = 2{\cdot}273J_1(e^x).$$

Problems 5.9

Obtain, in terms of Bessel functions, the general solutions of the following differential equations, using substitutions when given.

1. $y''+\dfrac{1}{x}y'+y = 0$.

2. $x^2y''+xy'+(a^2x^2-v^2)y = 0 \quad (a \neq 0)$.

(Obtain series solutions and compare them with those already derived for Bessels' equation. Alternatively make the substitution $u = ax$.)

3. $xy''+(1+2v)y'+xy = 0$; $\qquad y = ux^{-v}$.

4. $y''+\left(1+\dfrac{1}{4x^2}\right)y = 0$; $\qquad y = u\sqrt{x}$.

5. $xy''+y = 0$; $\qquad y = u\sqrt{x}, \quad t = 2\sqrt{x}$.

6. $3y''+xy = 0$; $\qquad y = u\sqrt{x}, \quad t = \dfrac{2\sqrt{x^3}}{3\sqrt{3}}$.

7. $x^2y''-2xy'+4(x^4-1)y = 0$; $\quad y = u\sqrt{x^3}, \quad t = x^2$.

8. Show that the substitution $y = u/\sqrt{x}$ transforms Bessel's equation (9.1) into

$$\frac{d^2u}{dx^2}+\left\{1-\frac{(4v^2-1)}{4x^2}\right\}u = 0.$$

Deduce the approximate form of the general solution to (9.1) when ν is small and x is large. State the values of ν for which this solution is exact for all x and express it in terms of Bessel functions.

5.10. BESSEL FUNCTIONS OF LOW ORDER

When $\nu = 0$ a solution of Bessel's equation is, from (9.5),

$$J_0(x) = \sum_0^\infty \frac{(-1)^m}{(m!)^2} \left(\frac{x}{2}\right)^{2m}, \tag{10.1}$$

since $\Gamma(m+1) = m!$. The first few terms in the expansion of $J_0(x)$ may be expressed in the form

$$J_0(x) = 1 - \frac{x^2}{2^2} + \frac{x^4}{2^2 4^2} - \frac{x^6}{2^2 4^2 6^2} + \frac{x^8}{2^2 4^2 6^2 8^2} - \cdots \tag{10.2}$$

and (9.7) shows that a second solution is given by

$$Z_0(x) = J_0(x) \int \frac{dx}{x\{J_0(x)\}^2}.$$

Although the integral is not a standard form, an approximation to Z_0 for small values of x may be derived from the expansion of $1/x\{J_0(x)\}^2$ in powers of x. Let the expansion include terms in x^3; then the integrand becomes

$$\begin{aligned}
\frac{1}{x\{J_0(x)\}^2} &= \frac{1}{x}\left\{1 - \frac{x^2}{2^2} + \frac{x^4}{2^2 4^2} - \cdots\right\}^{-2} \\
&= \frac{1}{x}\left\{1 - 2\left(-\frac{x^2}{2^2} + \frac{x^4}{2^2 4^2}\right)\right. \\
&\qquad \left. -2\left(\frac{3}{2}\right)\left(\frac{x^4}{2^4} + \cdots\right) + \cdots\right\} \\
&= \frac{1}{x} + \frac{x}{2} + \frac{5x^3}{32} + \cdots.
\end{aligned}$$

Thus
$$Z_0(x) = J_0(x) \int \left(\frac{1}{x} + \frac{x}{2} + \frac{5x^3}{32} + \ldots \right) dx$$

$$= J_0(x) \log x + \left(1 - \frac{x^2}{2^2} + \frac{x^4}{2^2 4^2} - \cdots \right)$$

$$\times \left(\frac{x^2}{4} + \frac{5x^4}{128} + \cdots \right)$$

$$= J_0(x) \log x + \frac{x^2}{4} - \frac{3x^4}{128} + \ldots.$$

The method of Frobenius may be used to show that the general form of this solution is

$$Z_0(x) = J_0(x) \log x - \sum_{1}^{\infty} \frac{(-1)^m}{(m!)^2} \left(\frac{x}{2} \right)^{2m} \phi(m)$$

where $\phi(m) = \sum_{s=1}^{m} 1/s$. Due to the term in $\log x$, $Z_0(x) \to -\infty$ as $x \to 0$ and this behaviour often makes the solution unsuitable for the physical conditions which the differential equation may describe; in such cases the constant B in the general solution (9.8) is taken to be zero.

When $\nu = 1$ (9.5) shows that

$$J_1(x) = \sum_{0}^{\infty} \frac{(-1)^m}{(m+1)(m!)^2} \left(\frac{x}{2} \right)^{1+2m},$$

since $\Gamma(m+2) = (m+1)m!$. The derivative of (10.1) with respect to x is

$$J_0'(x) = \sum_{1}^{\infty} \frac{m(-1)^m}{(m!)^2} \left(\frac{x}{2} \right)^{2m-1}.$$

In order to obtain a series which starts at $m = 0$ replace m by $(m+1)$, then

$$J_0'(x) = -\sum_{0}^{\infty} \frac{(m+1)(-1)^m}{\{(m+1)!\}^2} \left(\frac{x}{2} \right)^{2m+1}$$

and this series simplifies to that for $J_1(x)$, showing that

$$J_1(x) = -J_0'(x), \; = \frac{x}{2} - \frac{x^3}{2^2 4} + \frac{x^5}{2^3 4^2 6} - \cdots$$

from (10.2). This relation between $J_1(x)$ and $J_0(x)$ is similar to that between $\sin x$ and $\cos x$; the appropriate converse relation $J_1' = J_0$ is not true, though it can be a good approximation for

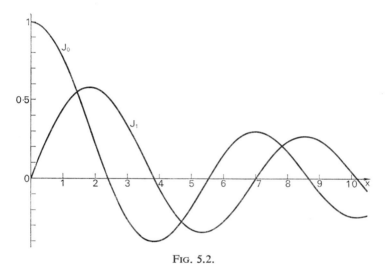

Fig. 5.2.

large values of x. Figure 5.2 shows that the graphs of J_0 and J_1 resemble damped cosine and sine functions, though the periods are not quite constant. Figure 5.3 shows that similar remarks apply to the relative properties of the graphs for Y_0 and Y_1, apart from their behaviour as $x \to 0$ when they both tend to $-\infty$. The positive values of x for which Bessel functions vanish are called the positive zeros of the functions; they are important in applications and are tabulated for the functions which are likely to occur most frequently.

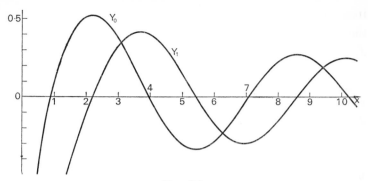

FIG. 5.3.

Consider, finally, the solutions to Bessel's equation when $\nu = \frac{1}{2}$. Although the roots $\varrho = \pm \frac{1}{2}$ of the indicial equation differ by an integer, the solutions $J_{1/2}$ and $J_{-1/2}$ are linearly independent; in fact the two Bessel functions J_ν and $J_{-\nu}$ are always independent when ν is an odd multiple of $\frac{1}{2}$. The two solutions for $\nu = \frac{1}{2}$ may be obtained by substitution in (9.3) and (9.4) but it is more interesting to return to the relation (9.2), replacing the c_m by b_m because $\varrho = -\nu$ is to be considered first. With $\varrho = -\frac{1}{2}$ the first series gives, for $m = 1$, $0 b_1 = 0$ and this is the only case where the assertion that $b_1 = 0$ cannot be made, rather b_1 is now arbitrary. The recurrence relation is

$$b_m = \frac{-b_{m-2}}{m(m-1)} \qquad (m \geqslant 2),$$

which leads to the linearly independent solutions

$$y_1 = b_0 x^{-1/2}\left(1 - \frac{x^2}{2!} + \frac{x^4}{4!} - \frac{x^6}{6!} + \dots\right) = b_0 x^{-1/2} \cos x$$

and

$$y_2 = b_1 x^{-1/2}\left(x - \frac{x^3}{3!} + \frac{x^5}{5!} - \frac{x^7}{7!} + \dots\right) = b_1 x^{-1/2} \sin x.$$

With $\varrho = \frac{1}{2}$ the usual condition $c_1 = 0$ holds and the only solution to emerge is

$$c_0 x^{1/2}\left(1 - \frac{x^2}{3!} + \frac{x^4}{5!} - \frac{x^6}{7!} + \ \cdots \right) = c_0 x^{-1/2} \sin x,$$

which is essentially the same as y_2. With $b_0 = b_1 = 1$ the standard forms of solution are obtained by including the factor $1/2^\nu \, \Gamma(\nu+1) = \sqrt{(2/\pi)}$; they are

$$J_{1/2}(x) = \sqrt{(2/\pi x)} \sin x \quad \text{and} \quad J_{-1/2}(x) = \sqrt{(2/\pi x)} \cos x.$$

These results illustrate a general theorem which asserts that if ν is half an odd integer then $J_\nu(x)$ may be expressed in a closed form which involves algebraic and circular functions of x.

The student is advised to look at some tables of Bessel functions to obtain an impression of how much information is available. When seen in its practical numerical forms the subject tends to lose much of the complexity which arises from a theoretical treatment, no matter how simple, and the student may come to regard work involving Bessel functions with but little more concern than is shown for more familiar functions.

Problems 5.10

1. Show that the series for $J_\nu(x)$, in (9.3), is convergent for all x.

2. Show that $$\int_0^\infty J_1(ax) \, dx = \frac{1}{a}.$$

3. Given that $$J_\nu' = J_{\nu-1} - \frac{\nu}{x} J_\nu \quad \text{show that}$$

$$2 \int_0^a \frac{\{J_1(x)\}^2}{x} \, dx = 1 - \{J_0(a)\}^2 - \{J_1(a)\}^2.$$

4. By means of the relation

$$\frac{2\nu}{x} J_\nu(x) = J_{\nu-1}(x) + J_{\nu+1}(x),$$

or otherwise, show that

$$J_{3/2}(x) + J_{-5/2}(x) = \sqrt{(2/\pi x^5)} \{4x \sin x + (3 - 2x^2) \cos x\}.$$

5. Show that $\dfrac{d}{dx} \{x^\nu J_\nu(x)\} = x^\nu J_{\nu-1}(x)$

and hence that

$$\int_0^a x^\nu J_{\nu-1}(x) \, dx = a^\nu J_\nu(a),$$

provided ν is real and positive.

CHAPTER 6

The Laplace Transformation

6.1. INTRODUCTION

A mathematical transformation is a special way of changing a function, the purpose being to produce a different function, the transform, which is more useful than the original for achieving certain results. The transformation due to Laplace is the best-known integral transform in applied mathematics and it is used mainly for the solution of differential equations.

An integral transform of a function $f(t)$ has the form

$$\int_a^b f(t)\, K(s,\, t)\, dt,$$

where K is a specified function of s and t, called the kernel of the transform. Between given values of the limits a and b, the integral is determined as a function of s.

It is probably true that upon first meeting the Laplace transform method for solving differential equations, most students find it less attractive than the conventional methods, such as the D operator approach, and may tend to conclude that it is inferior. This is an unfortunate attitude that should fade with experience, for the transform method is of special use when conventional methods are difficult to apply and it has valuable features in connection with many of the equations which arise in practice.

The systems to which the Laplace transform technique is

213

applied are usually described in terms of time as the independent variable. For this reason the functions considered in the theory and in examples are taken to be dependent upon t rather than on the more conventional x.

6.2. DEFINITION OF THE TRANSFORMATION

Let $f(t)$ be a function of t defined for all positive values of t. Multiply $f(t)$ by the kernel e^{-st}, where s is independent of t, and integrate the product with respect to t from zero to infinity. The result is a function of s, denoted by $\bar{f}(s)$ to indicate its relation to the original function $f(t)$. This operation is the *Laplace transformation* of $f(t)$ and the resulting transform $\bar{f}(s)$, or $\bar{f}$, is also denoted by $\mathcal{L}\{f(t)\}$ or simply by $\mathcal{L}(f)$.

$$\mathcal{L}\{f(t)\}, = \mathcal{L}(f) = \bar{f}(s) = \bar{f}, = \int_0^\infty e^{-st} f(t)\, dt. \qquad (2.1)$$

Conversely the original function $f(t)$ is called the *inverse transform* of $\bar{f}(s)$ and is denoted by $\mathcal{L}^{-1}\{\bar{f}(s)\}$.

$$f(t), = f = \mathcal{L}^{-1}(\bar{f}), = \mathcal{L}^{-1}\{\bar{f}(s)\}.$$

Note the slight variations in notation, any one of which is used without special comment. A function and its transform are said to make a transform pair; such a pair is unique.

The transform $\bar{f}(s)$ is a function of s, just as $f(t)$ is a function of t, but the range of values which s may take is liable to be restricted to ensure that the integral in (2.1) converges. The symbol p is often used instead of s.

The function $f(t)$ must satisfy certain conditions for its tiaplace transform to exist; these are not considered here for Lmay be assumed that they are satisfied by the functions to be sutdied and by most of those which arise in practice. A property of $\bar{f}(s)$ which should be apparent from the definition is that

$$\lim_{s \to \infty} \bar{f}(s) = 0,$$

a result needed sometimes in the solution of differential equations.

6.3. TRANSFORMS OF ELEMENTARY FUNCTIONS

The application of transform methods usually requires a knowledge of the transforms of elementary functions. Although these transforms and those of more complicated functions are readily available in standard tables, the technique of the method can be applied with ease only when several basic forms have been memorised. Some of the most important transforms are now derived from the definition (2.1); they are summarised in Table 2.

TABLE 2.   LAPLACE TRANSFORMS

$f(t)$	$\bar{f}(s)$	$f(t)$	$\bar{f}(s)$
1	$\dfrac{1}{s}$	$\cos at$	$\dfrac{s}{s^2+a^2}$
t	$\dfrac{1}{s^2}$	$\sin at$	$\dfrac{a}{s^2+a^2}$
$\dfrac{t^{n-1}}{(n-1)!}$, $n = 1, 2, \ldots$	$\dfrac{1}{s^n}$	$\cosh at$	$\dfrac{s}{s^2-a^2}$
e^{at}	$\dfrac{1}{s-a}$	$\sinh at$	$\dfrac{a}{s^2-a^2}$

THE TRANSFORM OF UNITY

$$\mathscr{L}(1) = \int_0^\infty e^{-st}\, dt = -\frac{1}{s} e^{-st} \Big|_0^\infty = \frac{1}{s}. \qquad (3.1)$$

The integral converges only for $s > 0$.

THE TRANSFORM OF t

$$\mathcal{L}(t) = \int_0^\infty te^{-st}\, dt = -\frac{t}{s} e^{-st} \Big|_0^\infty + \frac{1}{s} \int_0^\infty e^{-st}\, dt = \frac{1}{s^2} \qquad (s > 0).$$

(3.2)

THE TRANSFORM OF t^n (n a positive integer)

$$\mathcal{L}(t^n) = \int_0^\infty t^n e^{-st}\, dt = -\frac{t^n}{s} e^{-st} \Big|_0^\infty + \frac{n}{s} \int_0^\infty t^{n-1} e^{-st}\, dt,$$

$$= \frac{n}{s}\, \mathcal{L}(t^{n-1}) \quad \text{by (2.1).}$$

By induction

$$\mathcal{L}(t^{n-1}) = \frac{(n-1)}{s}\, \mathcal{L}(t^{n-2}), \quad \mathcal{L}(t^{n-2}) = \frac{(n-2)}{s}\, \mathcal{L}(t^{n-3})$$

and so on. Combine these results to obtain

$$\mathcal{L}(t^n) = \frac{n}{s}\, \frac{(n-1)}{s}\, \frac{(n-2)}{s} \cdots \frac{1}{s}\, \mathcal{L}(1) = \frac{n!}{s^{n+1}} \quad (s > 0). \quad (3.3)$$

THE TRANSFORM OF e^{at} (a constant)

$$\mathcal{L}(e^{at}) = \int_0^\infty e^{-(s-a)t}\, dt = \frac{-1}{(s-a)} e^{-(s-a)t} \Big|_0^\infty = \frac{1}{s-a}. \quad (3.4)$$

This integral converges only for $(s-a) > 0$.

The Laplace transformation has the important property of being a linear operation, that is the transform of a sum of functions is equal to the sum of their transforms. For example

$$\mathcal{L}\{af(t) + b\, g(t) + c\, h(t)\} = a\bar{f}(s) + b\, \bar{g}(s) + c\, \bar{h}(s),$$

where a, b and c are constants. This property is easily proved from the definition (2.1) and is now used to obtain the transforms of some trigonometric and hyperbolic functions.

THE TRANSFORM OF e^{iat}

By (3.4) $\quad \mathscr{L}(e^{iat}) = \dfrac{1}{s-ia} = \dfrac{s+ia}{s^2+a^2} = \dfrac{s}{s^2+a^2} + i\,\dfrac{a}{s^2+a^2}\,.$

The linearity property applied to the relation $e^{iat} = \cos at + i \sin at$ shows that

$$\mathscr{L}(e^{iat}) = \mathscr{L}(\cos at) + i\,\mathscr{L}(\sin at).$$

A comparison of real parts and of imaginary parts gives

$$\mathscr{L}(\cos at) = \frac{s}{s^2+a^2} \qquad (s > 0) \tag{3.5}$$

and

$$\mathscr{L}(\sin at) = \frac{a}{s^2+a^2} \qquad (s > 0). \tag{3.6}$$

THE TRANSFORM OF $\cosh at$

$$\mathscr{L}(\cosh at) = \frac{1}{2}\,\mathscr{L}(e^{at}+e^{-at}) = \frac{1}{2}\left(\frac{1}{s-a} + \frac{1}{s+a}\right)$$

$$= \frac{s}{s^2-a^2} \qquad (s > |a|). \tag{3.7}$$

Similarly

$$\mathscr{L}(\sinh at) = \frac{a}{s^2-a^2} \qquad (s > |a|). \tag{3.8}$$

In practice it is nearly always the transform $\bar{f}(s)$ that is known, and the main task in the application of transform methods is to derive the inverse $f(t)$, for this is usually the solution sought of a differential equation. The ability to do

5*

this is achieved through experience and by the confiden application of the theorems and techniques given in the res of this chapter. At this stage simple examples only can b given.

Example 3.1.

Obtain the inverse transform $f(t)$ when the transform $\bar{f}(s)$ is (a) $\dfrac{(a-b)}{(s-a)(s-b)}$, (b) $\dfrac{a^2}{s(s^2+a^2)}$, where a and b ar constants.

(a) The transform does not correspond with any of those i Table 2. If either factor in the denominator were absen then $\bar{f}(s)$ would have the form in (3.4). This sugges that $\bar{f}(s)$ should be expressed in partial fractions as

$$\bar{f}(s) = \frac{1}{(s-a)} - \frac{1}{(s-b)}, \quad \text{whence} \quad f(t) = e^{at} - e^{bt}.$$

(b) A similar argument applies, the known transforms beir those in (3.1) and (3.5). In partial fractions

$$\bar{f}(s) = \frac{1}{s} - \frac{s}{(s^2+a^2)}, \quad \text{whence} \quad f(t) = 1 - \cos at.$$

Problems 6.3

Derive the Laplace transforms of the following functions, where a, b are constants. When possible use the transforms already obtained.

1. $at^2 + bt + c$.   5. $\sin 2t \cos 2t$.   8. $\cosh t + \sinh t$.

2. $(t+a)^2$.   6. $2 \sin^2 t$.   9. te^t.

3. e^{at+b}.   7. $\cos(at+b)$.   10. $e^{at} \cos bt$.

4. $\cos 3t + \sin 3t$.

Derive the inverse transforms of the following functions of s.

11. $\dfrac{s+2}{s^2-4}$.   12. $\dfrac{2(s+1)}{s^3}$.   13. $\dfrac{5}{(s^2+4)(s^2+9)}$.

14. $\dfrac{5s}{(s^2-4)(s^2+1)}$.

16. $\dfrac{2s-5}{4s^2+25}$.

18. $\dfrac{s-4}{s^2+8}$.

15. $\dfrac{2s+1}{s(s+1)}$.

17. $\dfrac{9s+12}{9s^2-16}$.

19. By means of the gamma function, defined in section 5.8, show that $\mathcal{L}(\sqrt{t}) = \sqrt{\pi}/2s^{3/2}$.

20. Show that if $\bar{f}(s)$ is the transform of $f(t)$ then the transform of $f(at)$ is $\dfrac{1}{a}\,\bar{f}\!\left(\dfrac{s}{a}\right)$, where a is a constant.

6.4. THE FIRST SHIFT THEOREM

This important theorem shows how a function $f(t)$ is modified when each s in its transform is replaced by $s-a$, where a is a constant. The theorem states that

$$\text{if}\quad \bar{f}(s) = \mathcal{L}\{f(t)\}\quad \text{then}\quad \bar{f}(s-a) = \mathcal{L}\{e^{at}\,f(t)\},\quad (4.1)$$

showing that the change in the transform corresponds to multiplication of the original function by e^{at}. To prove this replace s by $s-a$ in definition (2.1) to obtain

$$\bar{f}(s-a) = \int_0^\infty e^{-(s-a)t}\,f(t)\,dt = \int_0^\infty e^{-st}\{e^{at}\,f(t)\}\,dt = \mathcal{L}\{e^{at}\,f(t)\}.$$

By means of the theorem the transforms in Table 2 may be extended and the inverses will involve functions of t which include an exponential factor; numbers **9** and **10** of Problems 6.3 are of this type. Since many of the functions which arise in applications have this form, an oscillatory function which decays exponentially for example, the application of (4.1) is often required.

Example 4.1.

Derive $f(t)$ from $\bar{f}(s) = \dfrac{s+3}{s^2+2s+5}$.

In a problem of this type, where the denominator, $h(s)$ say is a quadratic function of s with real, rational coefficients it is advisable first to test whether $h(s)$ can be factorised into real, rational factors. This is possible when the discriminant $b^2 - 4ac$ is a perfect square; in this example its value is -16 so factorisation is not advised. This is usually the case and the next step is to complete the square of $s^2 + 2s$ to obtain $s^2 + 2s + 5 \equiv (s+1)^2 + 4$. The transform is now

$$\bar{f}(s) = \frac{s+3}{(s+1)^2 + 4},$$

where $h(s)$ is of the form in (3.5) or (3.6) with s replaced by $s+1$. Finally, the conversion to $s+1$ is made in the numerator to obtain

$$\bar{f}(s) = \frac{(s+1)+2}{(s+1)^2 + 4} = \frac{s+1}{(s+1)^2 + 2^2} + \frac{2}{(s+1)^2 + 2^2}.$$

By means of the theorem and transforms (3.5,6) the inversion is made and the result is

$$f(t) = e^{-t} \cos 2t + e^{-t} \sin 2t = e^{-t} (\cos 2t + \sin 2t).$$

It is possible to make the inversion through the factorisation of $h(s)$, although this is not advised. With this method $\bar{f}(s)$ is expressed in partial fractions as

$$\bar{f}(s) = \frac{s+3}{(s+1+2i)(s+1-2i)} \equiv \frac{(1+i)}{2(s+1+2i)} + \frac{(1-i)}{2(s+1-2i)}.$$

By (3.4) the inverse is

$$f(t) = \tfrac{1}{2}(1+i)e^{-(1+2i)t} + \tfrac{1}{2}(1-i)e^{-(1-2i)t}$$
$$= \tfrac{1}{2}e^{-t}\{(e^{2it} + e^{-2it}) - i(e^{2it} - e^{-2it})\} = e^{-t}(\cos 2t + \sin 2t).$$

Clearly this approach is longer, more subject to error and less elegant than the method which uses the shift theorem.

Example 4.2.

Derive $f(t)$ from $\bar{f}(s) = \dfrac{5s+13}{s^2+5s+6}$.

Here $h(s) = s^2+5s+6 = (s+2)(s+3)$ is expressible in simple factors. In partial fractions $\bar{f}(s) = 3/(s+2)+2/(s+3)$, whence $f(t) = 3e^{-2t}+2e^{-3t}$. The inversion can also be made with the shift theorem, using the techniques of Example 4.1. In this way

$$\bar{f}(s) = \frac{5s+13}{\left(s+\frac{5}{2}\right)^2-\frac{1}{4}} = \frac{5\left(s+\frac{5}{2}\right)+\frac{1}{2}}{\left(s+\frac{5}{2}\right)^2-\frac{1}{4}} .$$

By the theorem and transforms (3.7,8)

$$f(t) = e^{-5t/2}(5\cosh t/2 + \sinh t/2),$$

which is a form equivalent to that derived by the first method.

Example 4.3.

Derive $f(t)$ from $\bar{f}(s) = \dfrac{1}{s^2+4s+4}$.

This is a critical case where $h(s)$ has repeated factors and $\bar{f}(s) = 1/(s+2)^2$. The theorem is applied to transform (3.2) and the inverse is $f(t) = te^{-2t}$.

These few examples show how the form of the inverse $f(t)$ depends on the form of a quadratic $h(s)$ in the transform $\bar{f}(s)$. There is no set way of inverting a given transform, but careful thought at the start usually ensures the choice of the best way. Note that the form of $f(t)$ depends on $h(s)$ in a way similar to that in which the solution of a reduced equation depends on the type of roots of the auxiliary equation; see section 4.2.

Problems 6.4

Obtain the inverse transforms of the following functions.

1. $\dfrac{6}{(s+1)^4}$.

2. $\dfrac{(n-1)!}{(s-a)^n}$.

3. $\dfrac{5(s-2)}{s^2-s-6}$.

4. $\dfrac{s-4}{s^2-8s+20}$.

5. $\dfrac{3s+5}{s^2-2s-3}$.

6. $\dfrac{2s+5}{s^2+6s+10}$.

7. $\dfrac{4(3s+2)}{4s^2+12s+9}$.

8. $\dfrac{s^2-1}{(s+2)^3}$.

9. $\dfrac{12s-21}{16s^2+24s+201}$.

10. $\dfrac{5s^2-20s+36}{(s-2)(s^2-4s+20)}$.

6.5. TRANSFORMS OF DERIVATIVES AND INTEGRALS

The application of transform methods to differential equations requires the transforms of the derivatives which occur in the equations. If $f(t)$ and its derivative $f'(t)$ satisfy certain conditions over the range $t \geqslant 0$ then the Laplace transform of $f'(t)$ is given by

$$\mathscr{L}\{f'(t)\} = s\mathscr{L}\{f(t)\} - f(0). \tag{5.1}$$

This result is derived from the definition (2.1) and by integration by parts:

$$\mathscr{L}\{f'(t)\} = \int_0^\infty e^{-st} f'(t)\, dt = e^{-st} f(t)\Big|_0^\infty + s\int_0^\infty e^{-st} f(t)\, dt$$
$$= -f(0) + s\mathscr{L}\{f(t)\}.$$

The conditions to be satisfied ensure that the term $e^{-st}f(t) \to 0$ as $t \to \infty$.

The transform of the second-order derivative is obtained by applying (5.1) to $f''(t)$:

$$\mathscr{L}\{f''(t)\} = s\mathscr{L}\{f'(t)\} - f'(0)$$
$$= s[s\mathscr{L}\{f(t)\} - f(0)] - f'(0), \quad \text{by (5.1)},$$

giving $\quad \mathscr{L}\{f''(t)\} = s^2\mathscr{L}\{f(t)\} - sf(0) - f'(0). \tag{5.2}$

The process may be applied to the third-order derivative and gives

$$\mathscr{L}\{f'''(t)\} = s^3\mathscr{L}\{f(t)\} - s^2 f(0) - s f'(0) - f''(0). \quad (5.3)$$

By induction the transform of the nth-order derivative is

$$\mathscr{L}\{f^{(n)}(t)\} = s^n\mathscr{L}\{f(t)\} - s^{n-1} f(0) - s^{n-2} f'(0) - \dots$$
$$\dots - s f^{(n-2)}(0) - f^{(n-1)}(0). \quad (5.4)$$

Although the above results are used mainly with differential equations they are also useful for deriving the transforms of certain functions.

Example 5.1.

Derive the transforms of $f(t) = $ (a) sin at, (b) t cos at by means of (5.2).

(a) $f(t) = $ sin at, $f(0) = 0$, $f'(t) = a$ cos at, $f'(0) = a$ and $f''(t) = -a^2$ sin $at = -a^2 f(t)$. So $\mathscr{L}(f'') = -a^2\mathscr{L}(f)$, $= s^2\mathscr{L}(f) - a$ by (5.2), and $\mathscr{L}(f) = \dfrac{a}{s^2 + a^2}$.

(b) $f(t) = t$ cos at, $f(0) = 0$, $f'(t) = $ cos $at - at$ sin at, $f'(0) = 1$ and $f''(t) = -2a$ sin $at - a^2 t$ cos $at = -2a$ sin $at - a^2 f(t)$. So $\mathscr{L}(f'') = -2a\mathscr{L}(\text{sin } at) - a^2\mathscr{L}(f)$, $= s^2\mathscr{L}(f) - 1$ by (5.2), and $\mathscr{L}(f) = \dfrac{s^2 - a^2}{(s^2 + a^2)^2}$ by the result of (a).

A differential equation which describes the current $i(t)$ in a circuit often contains a term for the charge $q(t)$ on a capacitor. Since $i = dq/dt$, q may be expressed as the integral of i with respect to t. Thus it is often necessary to transform the integral of a function as well as its derivatives. If $f(t)$ satisfies certain conditions then the transform of the integral of $f(t)$ from 0 to

t is given by

$$\mathcal{L}\left\{\int_0^t f(u)\,du\right\} = \frac{1}{s}\,\mathcal{L}\{f(t)\}. \tag{5.5}$$

As t is the upper limit of the integral it is usual to denote the variable in the integrand by another symbol, such as u.

Proof. Let $g(t) = \int_0^t f(u)\,du$ then $g'(t) = f(t)$ and

$$\mathcal{L}\{f(t)\} = \mathcal{L}\{g'(t)\},\ = s\mathcal{L}\{g(t)\} - g(0) \quad \text{by (5.1)}.$$

But as $g(0) = 0$

$$\mathcal{L}\{g(t)\} = \frac{1}{s}\,\mathcal{L}\{f(t)\}.$$

This result may be used to derive the inverses of certain functions and is useful when expressed in the form

$$\mathcal{L}^{-1}\left\{\frac{1}{s}\,\bar{f}(s)\right\} = \int_0^t f(u)\,du. \tag{5.6}$$

Example 5.2.

Derive the inverse transform of $1/s^2(s-a)$.
Consider $1/(s-a)$ to be $\bar{f}(s)$ in (5.6).
Now as $\mathcal{L}^{-1}\{1/(s-a)\} = e^{at}$ (5.6) gives

$$\mathcal{L}^{-1}\left\{\frac{1}{s}\,\bar{f}(s)\right\} = \int_0^t e^{au}\,du = \frac{1}{a}(e^{at}-1).$$

Now take $1/s(s-a)$ to be $\bar{f}(s)$ in (5.6) and repeat the process to obtain

$$\mathcal{L}^{-1}\{1/s^2(s-a)\} = \frac{1}{a}\int_0^t (e^{au}-1)\,du = \frac{1}{a^2}(e^{at}-at-1).$$

THE CONVOLUTION THEOREM

This theorem has applications in network theory when, for example, the transform of an output function is expressed as the product of the transform of an input function with that of a transfer or generalised impedance function. The theorem states that if $\mathscr{L}^{-1}\{\bar{f}(s)\} = f(t)$ and $\mathscr{L}^{-1}\{\bar{g}(s)\} = g(t)$ then the inverse transform of the product $\bar{f}\bar{g}$ is

$$\mathscr{L}^{-1}\{\bar{f}(s)\,\bar{g}(s)\} = \int_0^t f(u)\,g(t-u)\,du, \qquad (5.7)$$

denoted by f^*g. This relation between f and g is called the *convolution* of f and g; it obeys the commutative, associative and distributive laws of algebra.

Example 5.3.

Consider the transform in Example 5.2, taking $\bar{f} = 1/(s-a)$ and $\bar{g} = 1/s^2$, then by (5.7)

$$\mathscr{L}^{-1}\left\{\frac{1}{(s-a)} \cdot \frac{1}{s^2}\right\} = \int_0^t e^{au}(t-u)\,du =$$

$$\frac{t}{a}\,e^{au} - \frac{u}{a}\,e^{au} + \frac{1}{a^2}\,e^{au}\bigg|_0^t = (e^{at} - at - 1)/a^2,$$

as before.

6.6. APPLICATION TO ORDINARY DIFFERENTIAL EQUATIONS

The principles and methods given are now applied to the solution of ordinary linear differential equations with constant coefficients. For illustration take the general equation of second order

$$a\frac{d^2y}{dt^2}+b\frac{dy}{dt}+cy=f(t), \qquad (6.1)$$

where $y(t)$ is the unknown function to be found. In a specific equation $f(t)$ is a given function of t and, in this context, it is not used to denote an unknown function. The operation d/dt will be denoted by a prime rather than by a dot.

By the linearity property the transform of equation (6.1) is obtained by transforming each term:

$$a\{s^2\bar{y}(s)-sy(0)-y'(0)\}+b\{s\bar{y}(s)-y(0)\}+c\bar{y}(s)=\bar{f}(s). \qquad (6.2)$$

The transforms of the derivatives are derived from (5.1,2) and $\mathcal{L}\{y(t)\}$ is denoted by $\bar{y}(s)$. Equation (6.2) is called the *subsidiary equation* (S.E.) of (6.1) and is solved algebraically for $\bar{y}(s)$ giving

$$\bar{y}(s)=\frac{(as+b)y(0)+ay'(0)+\bar{f}(s)}{as^2+bs+c}. \qquad (6.3)$$

The solution is completed by taking the inverse transform of (6.3) to derive $y(t)$.

The initial values of y and y', when given, are usually inserted at the stage when the subsidiary equation is formed. The Laplace transform method is specially suited for the initial-value problems of applied mathematics, electric circuit and control theory in particular, in which differential equations are to be solved for times $t>0$ with given conditions at $t=0$.

Example 6.1.

Obtain the solution of $y''+5y'+6y=10\sin t$ which satisfies the initial conditions $y(0)=y'(0)=0$.

S.E. $$s^2\bar{y}+5s\bar{y}+6\bar{y}=\frac{10}{s^2+1},$$

whence $\quad \bar{y} = \dfrac{10}{(s^2+5s+6)(s^2+1)} = \dfrac{10}{(s+2)(s+3)(s^2+1)}$

$\qquad \equiv \dfrac{2}{s+2} - \dfrac{1}{s+3} + \dfrac{(-s+1)}{(s^2+1)} .$

By known transforms, $y(t) = 2e^{-2t} - e^{-3t} - \cos t + \sin t.$

Example 6.2.

A constant e.m.f. E is applied to a circuit of L, R, C components in series, the initial current and charge being zero. Obtain expressions for the current at time t.

The differential equation is $\quad L\dfrac{di}{dt} + Ri + \dfrac{q}{C} = E,$

where $\qquad\qquad\qquad i = \dfrac{dq}{dt} .$

S.E. $\qquad\qquad\qquad Ls\bar{i} + R\bar{i} + \dfrac{\bar{i}}{Cs} = \dfrac{E}{s}$

whence $\qquad\qquad \bar{i} = \dfrac{E}{L} \dfrac{1}{[s^2+(Rs/L)+(1/LC)]}$

$\qquad\qquad\qquad = \dfrac{E}{L} \dfrac{1}{(s+a)^2+b^2} ,$

where the symbols $a = R/2L$ and $b^2 = 1/LC - R^2/4L^2$ are introduced to simplify the algebra. By known transforms and theorem (4.1), three expressions for $i(t)$ arise which depend on the nature of b^2.

When $b^2 > 0$ $\qquad\qquad i = \dfrac{E}{bL} e^{-at} \sin bt.$

When $b^2 = 0$ $\qquad\qquad i = \dfrac{E}{L} te^{-at}.$

When $b^2 = -k^2 < 0$ $\quad i = \dfrac{E}{kL} e^{-at} \sinh kt.$

The transform method is applicable to simultaneous differential equations. The solution of a pair of such equations, with dependent variables $x(t)$ and $y(t)$, leads to a pair of subsidiary equations which involve $\bar{x}(s)$ and $\bar{y}(s)$. The elimination of either transform allows the inverse of the other to be obtained, whence the solution is completed by the method most suitable for the given problem.

Problems 6.6

Obtain particular solutions of the following differential equations by the Laplace transform method. Also derive some solutions by D operator methods and compare the two techniques.

1. $y' - y = -t,$ $y(0) = 1.$

2. $y'' - 4y = 2,$ $y(0) = 0,$ $y'(0) = 1.$

3. $y'' - 3y' + 2y = e^{3t},$ $y(0) = 0,$ $y'(0) = 1.$

4. $y'' + 2y' + y = e^{2t},$ $y(0) = y'(0) = 1.$

5. $y'' + 5y' + 6y = e^t,$ $y(0) = 0,$ $y'(0) = 1.$

6. $y'' + 2y' + 5y = 0,$ $y(0) = 2,$ $y'(0) = -4.$

7. $y'' + 3y' + 2y = \cos t,$ $y(0) = 0,$ $y'(0) = 1.$

8. $y'' - 2y' - 3y = \cos t - \sin t,$ $y(0) = 0,$ $y'(0) = 1.$

9. $x' - 7x + y = 0,$ $y' - 2x - 5y = 0;$ $x(0) = 0,$ $y(0) = -1.$

10. $x' - 2y' = t,$ $x'' + 2y = e^{-t};$ $x(0) = y(0) = 0,$ $x'(0) = 1.$

6.7. TECHNIQUES OF INVERSION

Most of the transforms which arise in practice as solutions to subsidiary equations are of the form $\bar{y}(s) = g(s)/h(s)$, where $g(s)$ and $h(s)$ are polynomials in s. If $f(t)$ in equation (6.1) is limited to the functions considered so far, then (6.3) shows that the degree of $g(s)$ is less than that of $h(s)$, a result that is true in general. Thus $\bar{y}(s)$ is normally in a form suitable for resolution into partial fractions and the student should be familiar with the rules which govern this process. In this section the process is developed in a slightly more advanced way which should remove some of the tedium associated with the conventional methods.

Express $\bar{y}(s)$ in the form

$$\bar{y}(s) = \frac{g(s)}{h(s)} = \frac{g(s)}{(s-a_1)(s-a_2)(s-a_3)\ldots(s-a_n)}, \quad (7.1)$$

where $g(s)$ and $h(s)$ have no common factor and $h(s)$ is reduced to its linear factors, some of which may be complex. The values $a_1, a_2, a_3, \ldots, a_n$ of s which make $h(s)$ zero are called the *poles* of $\bar{y}(s)$, a term from the language of complex-variable theory. When $h(s)$ has no repeated factors the poles are distinct and are said to be *simple* poles. When $h(s)$ contains repeated factors the related poles are called *multiple* poles. For example

$$\bar{y}(s) = \frac{s+2}{s(s-3)(s+1)^2} \quad \text{has simple poles at } s = 0,3 \text{ and a double pole at } s = -1.$$

$$\bar{y}(s) = \frac{2s-1}{s^2+4} = \frac{2s-1}{(s+2i)(s-2i)} \quad \text{has simple poles at } s = \pm 2i.$$

When $\bar{y}(s)$ has simple poles only the expansion of (7.1) in partial fractions is

$$\bar{y}(s) = \frac{A_1}{(s-a_1)} + \frac{A_2}{(s-a_2)} + \frac{A_3}{(s-a_3)} + \ldots + \frac{A_n}{(s-a_n)}, \quad (7.2)$$

where $A_1, A_2, A_3, \ldots, A_n$ are constants to be determined. To find A_1 multiply (7.2) by $(s-a_1)$:

$$(s-a_1)\,\bar{y}(s) = A_1 + A_2\frac{(s-a_1)}{(s-a_2)} + A_3\frac{(s-a_1)}{(s-a_3)}$$

$$+ \ldots + A_n\frac{(s-a_1)}{(s-a_n)}. \quad (7.3)$$

Taking the limit of each side of this equation as $s \to a_1$ leaves A_1 on the right and a finite limit on the left because $\bar{y}(s)$ contains the factor $(s-a_1)$ in $h(s)$ and this cancels before the limit is taken. The result of this process is summarised by

$$A_1 = \lim_{s \to a_1} (s-a_1)\bar{y}(s) \tag{7.4}$$

and in general by

$$A_r = \lim_{s \to a_r} (s-a_r)\bar{y}(s). \tag{7.5}$$

This way of resolving into partial fractions is called the "cover-up" method by some authors to indicate that the coefficient A_r associated with the fraction $1/(s-a_r)$ may be obtained by "covering-up" or neglecting $(s-a_r)$ in $\bar{y}(s)$ whilst a_r is substituted for s. By transform (3.4) each simple pole a_r will contribute a term $A_r e^{a_r t}$ to the inverse transform $y(t)$ of $\bar{y}(s)$. Thus the inverse of a function $\bar{y}(s)$ having n simple poles only is given by the inversion formula

$$y(t) = \sum_{r=1}^{n} \lim_{s \to a_r} (s-a_r)\bar{y}(s)e^{st}. \tag{7.6}$$

This is a much simplified version of the complex inversion formula, which provides a powerful direct method for inverting Laplace transforms by means of complex-variable theory.

Example 7.1.

Invert
$$\bar{y}(s) = \frac{s+6}{s(s-2)(s+3)}.$$

$$y(t) = \lim_{s \to 0} \frac{(s+6)e^{st}}{(s-2)(s+3)} + \lim_{s \to 2} \frac{(s+6)e^{st}}{s(s+3)} + \lim_{s \to -3} \frac{(s+6)e^{st}}{s(s-2)}$$

$$= -1 + \frac{4}{5} e^{2t} + \frac{1}{5} e^{-3t}.$$

In practice the result is derived immediately by inspection of $\bar{y}(s)$.

The above process may be extended to transforms which have multiple poles. Suppose the factor $(s-a_1)$ occurs twice in

$h(s)$, then in partial fractions

$$\bar{y}(s) = \frac{A_{11}}{(s-a_1)^2} + \frac{A_{12}}{(s-a_1)} + \frac{A_2}{(s-a_2)} + \ldots + \frac{A_n}{(s-a_n)}. \quad (7.7)$$

The constants A_r associated with the simple poles $a_2, \ldots, a_n$ are obtained from

$$A_r = \lim_{s \to a_n} (s-a_r)\bar{y}(s) \qquad (r = 2, \ldots, n) \qquad (7.8)$$

as in (7.5)

Multiply (7.7) by $(s-a_1)^2$ to obtain

$$(s-a_1)^2 \bar{y}(s) = A_{11} + A_{12}(s-a_1)$$
$$+ A_2 \frac{(s-a_1)^2}{(s-a_2)} + \ldots + A_n \frac{(s-a_1)^2}{(s-a_n)}, \quad (7.9)$$

whence A_{11} is given by

$$A_{11} = \lim_{s \to a_1} (s-a_1)^2 \bar{y}(s). \qquad (7.10)$$

Differentiate (7.9) with respect to s and then let $s \to a_1$ to obtain

$$A_{12} = \lim_{s \to a_1} \left\{ \frac{d}{ds} (s-a_1)^2 \bar{y}(s) \right\}. \qquad (7.11)$$

This result is easily extended to the case where $(s-a_1)$ occurs k times:

$$A_{1m} = \lim_{s \to a_1} \frac{1}{(m-1)!} \left\{ \frac{d^{m-1}}{ds^{m-1}} (s-a_1)^k \bar{y}(s) \right\} \qquad (m = 2, \ldots, k). \quad (7.12)$$

These methods for inversion and resolution into partial fractions should be used only when they are clearly superior to the conventional methods, which evaluate the constants by comparison of coefficients and by substitution of suitable values for s. In many problems a combination of both methods provides the easiest solution.

Example 7.2.

Obtain $\mathcal{L}^{-1}\{\bar{y}(s)\}$ when $\bar{y}(s)$ is

(a) $\dfrac{s-2}{(s-1)^2\,(s+2)\,(s-3)}$,   (b) $\dfrac{16(s+1)}{s(s-2)\,(s^2+4)}$.

(a)   Let $\bar{y}(s) \equiv \dfrac{A_{11}}{(s-1)^2} + \dfrac{A_{12}}{(s-1)} + \dfrac{A_2}{(s+2)} + \dfrac{A_3}{(s-3)}$

then by (7.5)

$$A_2 = \tfrac{4}{45}, \quad A_3 = \tfrac{1}{20} \quad \text{and by (7.10)} \; A_{11} = \tfrac{1}{6}.$$

By (7.11)

$$A_{12} = \lim_{s \to 1} \left\{ \frac{(s+2)(s-3)-(s-2)(2s-1)}{(s+2)^2\,(s-3)^2} \right\} = -\frac{5}{36}.$$

By transforms (3.2,4) and theorem (4.1)

$$y(t) = \frac{t}{6}\,e^t - \frac{5}{36}\,e^t + \frac{4}{45}\,e^{-2t} + \frac{e^{3t}}{20}.$$

(b) Here the factor (s^2+4) could be expressed as $(s+2i)(s-2i)$ and the inverse of $\bar{y}(s)$ would follow directly from (7.6). However, this method would involve complex algebra and this can be avoided in the following way. Use (7.5) on the simple poles 0, 2 to obtain

$$\bar{y}(s) \equiv -\frac{2}{s} + \frac{3}{s-2} + \frac{As+B}{s^2+4}.$$

Multiply this equation by $h(s)$:

$$16(s+1) \equiv -2(s-2)(s^2+4)+3s(s^2+4)+(As+B)\,s(s-2).$$

To find A equate the coefficients of s^3: $0 = -2+3+A$,
therefore $A = -1$.

To find B equate the coefficients of s: $16 = -8+12-2B$,
therefore $B = -6$.

By known transforms $y(t) = -2+3e^{2t}-\cos 2t-3\sin 2t$.

Problems 6.7

Obtain particular solutions of the following differential equations by the Laplace transform method.

1. $y'' - 3y' + 2y = e^{3t}$; $y(0) = y'(0) = 0$.

2. $y'' - 4y' + 3y = te^{-2t}$; $y(0) = y'(0) = 0$.

3. $y' + 2y = t^2$; $y(0) = 0$.

4. $y''' - 3y'' + 2y' = e^{3t}$; $y(0) = y'(0) = y''(0) = 0$.

5. $y'' - 5y' + 4y = e^{3t} \cosh 2t$; $y(0) = 0, \quad y'(0) = 1$.

6. Show that if $\mathcal{L}\{y(t)\} = \bar{y}(s) = g(s)/h(s)$ and $\bar{y}(s)$ has a simple pole at $s = a$, then the contribution of this pole to $y(t)$ is $[g(a)/h'(a)]e^{at}$. Use this result on $\bar{y}(s)$ in Example 7.2b by means of the four simple poles.

7. Solve the equations in Example 7.3 of Chapter 4 by transform methods, obtaining an expression for i in the form

$$i = \frac{2E}{5R} [5 - e^{-7Rt/12L} \{5 \cosh (5Rt/12L) + 4 \sinh (5Rt/12L)\}].$$

8. By means of the convolution theorem obtain $\mathcal{L}^{-1}\{1/(s^2 + 4)^2\}$ and hence solve the equation $y'' - 4y = 16t \cos 2t$; $y(0) = 1$, $y'(0) = 2$.

6.8. DIFFERENTIATION AND INTEGRATION OF TRANSFORMS

If $f(t)$ satisfies certain conditions then the derivative of its transform

$$\bar{f}(s) = \int_0^\infty e^{-st} f(t)\, dt = \mathcal{L}\{f(t)\}$$

may be obtained by differentiating under the integral sign with respect to s:

$$\bar{f}'(s) = -\int_0^\infty e^{-st}\{t f(t)\}\, dt, = -\mathcal{L}\{t f(t)\}. \tag{8.1}$$

Thus differentiation of a transform corresponds to multiplication of its inverse by $-t$. This result provides a way of extending known transforms.

16*

Example 8.1.

Obtain $\qquad \mathscr{L}^{-1}\left\{\dfrac{s^2}{(s^2+a^2)^2}\right\}.$

By (8.1) $\mathscr{L}(t\cos at) = -\dfrac{d}{ds}\left\{\dfrac{s}{s^2+a^2}\right\} = \dfrac{s^2-a^2}{(s^2+a^2)^2}$ and by (3.6)

$\dfrac{1}{a}\mathscr{L}(\sin at) = \dfrac{1}{s^2+a^2} = \dfrac{s^2+a^2}{(s^2+a^2)^2}$, whence by addition

$$\mathscr{L}^{-1}\left\{\dfrac{s^2}{(s^2+a^2)^2}\right\} = \dfrac{1}{2a}(\sin at + at\cos at).$$

Provided the limit of $f(t)/t$ exists as t approaches zero from the right then

$$\mathscr{L}\{f(t)/t\} = \int_s^\infty \bar{f}(s)\,ds. \qquad (8.2)$$

Thus integration of a transform from s to infinity corresponds to division of its inverse by t.

Example 8.2.

Given $\qquad \bar{f}(s) = \mathscr{L}(1-e^t) \quad$ obtain $\quad \mathscr{L}\left\{\dfrac{1-e^t}{t}\right\}.$

$\bar{f}(s) = \dfrac{1}{s} - \dfrac{1}{s-1}$; therefore $\quad \mathscr{L}\{f(t)/t\} = \int_s^\infty \left\{\dfrac{1}{s} - \dfrac{1}{s-1}\right\}ds$

$= \log\dfrac{s}{(s-1)}\Big|_s^\infty = \log\dfrac{(s-1)}{s}.$

Some ordinary differential equations with variable coefficients may be solved by Laplace transform methods, in particular those which have terms of the form $t^k y^{(n)}$, where k is a positive integer. The generalisation of (8.1) applied to such terms is

$$\mathscr{L}\{t^k f(t)\} = (-1)^k \frac{d^k}{ds^k}\{\bar{f}(s)\}. \tag{8.3}$$

The following two theorems may be needed:

The initial-value theorem: $\lim_{t \to 0} f(t) = \lim_{s \to \infty} s\bar{f}(s)$. $\qquad$ (8.4)

The final-value theorem: $\lim_{t \to \infty} f(t) = \lim_{s \to 0} s\bar{f}(s)$, $\qquad$ (8.5)

provided the poles of $s\bar{f}(s)$ are zero or have negative real parts.

Example 8.3.

Obtain the solution of Bessel's differential equation of zero order, $ty'' + y' + ty = 0$, which satisfies the condition $y(0) = 1$. Transform the equation

$$-\frac{d}{ds}\{s^2\bar{y} - s - y'(0)\} + (s\bar{y} - 1) - \frac{d\bar{y}}{ds} = 0;$$

$$-2s\bar{y} - s^2\frac{d\bar{y}}{ds} + 1 + s\bar{y} - 1 - \frac{d\bar{y}}{ds} = 0,$$

whence $\qquad\qquad (s^2 + 1)\dfrac{d\bar{y}}{ds} + s\bar{y} = 0.$

This is a first-order equation in $\bar{y}$ which, after a separation of the variables, is solved by quadratures to give $\bar{y} = C/\sqrt{(s^2+1)}$. To determine the constant C use theorem (8.4):

$$\lim_{t \to 0} y(t), = 1, = \lim_{s \to \infty} \frac{Cs}{\sqrt{(s^2+1)}} = \lim_{s \to \infty} \frac{C}{\sqrt{(1+1/s^2)}} = C,$$

whence $\qquad\qquad \bar{y} = (s^2+1)^{-1/2}.$

This does not correspond with a known transform but it may be expanded by the binomial theorem as an infinite series of terms in powers of s which upon inversion gives an infinite series in powers of t.

$$\bar{y} = \frac{1}{s} - \frac{1}{2s^3} + \frac{3}{2^2 2! \, s^5} - \frac{15}{2^3 3! \, s^7} + \ldots$$

$$y(t) = 1 - \frac{t^2}{2^2} + \frac{t^4}{2^2 4^2} - \frac{t^6}{2^2 4^2 6^2} + \ldots,$$

which is the Bessel function of the first kind of order zero, denoted by $J_0(t)$ and defined in section 5.10.

Thus $\qquad \mathscr{L}^{-1}\{(s^2+1)^{-1/2}\} = J_0(t);$

also $\qquad \mathscr{L}^{-1}\{(s^2+a^2)^{-1/2}\} = J_0(at).$

Problems 6.8

1. Solve $ty'' + (1+2t)y' + 2y = 0; \quad y(0) = 1.$

2. Show that the solution of the equation

$$ty'' + (1-t)y' + (n-1)y = 0; \quad y(0) = 1,$$

has the transform $(s-1)^{n-1}/s^n$ and derive the solution for $n = 3$.

3. Solve $ty'' + y' + 9ty = 0; \quad y(0) = 2.$

4. Solve $ty'' + (1+4t)y' + (2+5t)y = 0; \quad y(0) = 3.$

5. The solution of $ty'' + 2(t-1)y' + 2(t-1)y = 4e^{-t}\cos t$ is required which satisfies $y(0) = y(\pi) = 0$. Show that the transform $\bar{y}(s)$ of the solution $y(t)$ satisfies the equation

$$(s^2+2s+2)\frac{d\bar{y}}{ds} + 4(s+1)\bar{y} = \frac{-4(s+1)}{(s^2+2s+2)}$$

and thence obtain $y(t)$.

6. Obtain the solution of the equation

$$ty'' + (t-2)y' - y = 1 - e^{-t}$$

which satisfies the conditions $y(0) = 0$ and $y(\infty) = -1$.

6.9. THE UNIT STEP FUNCTION

The unit step function is denoted by $H(t-a)$, after Heaviside, or by $u_a(t)$ and is defined by

$$H(t-a) = u_a(t) = \begin{cases} 0 & \text{when} \quad t < a \\ 1 & \text{when} \quad t > a \end{cases} \qquad (a \geqslant 0). \quad (9.1)$$

The function, which is discontinuous at $t = a$, is shown in Fig. 6.1.

When $a = 0$, $\quad H(t) = u_0(t) = \begin{cases} 0 & \text{when} \quad t < 0 \\ 1 & \text{when} \quad t > 0 \end{cases}$, $\quad$ (9.2)

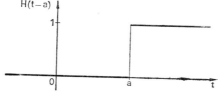

FIG. 6.1. Unit step function $H(t-a)$

and the function takes the simple form shown in Fig. 6.2. The Laplace transform of the unit step function is

$$\mathscr{L}\{H(t-a)\} = \int_0^\infty e^{-st}H(t-a)\, dt = \int_0^a e^{-st}0\, dt + \int_a^\infty e^{-st}1\, dt$$

$$= -\frac{1}{s}e^{-st}\Big|_a^\infty = \frac{e^{-as}}{s} \qquad (s > 0). \qquad (9.3)$$

FIG. 6.2. Unit step function $H(t)$.

When $a = 0$ this result gives $\mathscr{L}\{H(t)\} = 1/s$, in accordance with transform (3.1). Many functions which occur in practice, such as pulses and certain periodic functions, may be represented as a combination of step functions.

Example 9.1.

The rectangular pulse of amplitude A shown in Fig. 6.3 is defined by a pair of step functions as

$f(t) = A\{H(t-a) - H(t-b)\}$ and its transform is

$\bar{f}(s) = \dfrac{A}{s}(e^{-as} - e^{-bs})$.

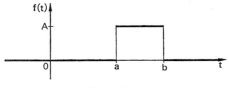

FIG. 6.3.

Example 9.2.

The pulse shown in Fig. 6.4 is defined by

$f(t) = t\{H(t) - H(t-a)\}$ and its transform is

$$\bar{f}(s) = \frac{1}{s^2} - \int_a^\infty te^{-st}\,dt = \frac{1}{s^2} + \frac{1}{s^2}\left[ste^{-st} + e^{-st}\right]_a^\infty = \frac{1}{s^2} - \frac{e^{-as}}{s^2}(1+as)$$

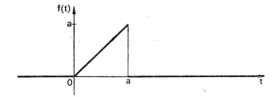

FIG. 6.4.

Example 9.3.

The square wave of period $2a$ and amplitude A shown in Fig. 6.5 is defined by an infinite series of step functions as $f(t) = A\{H(t) - 2H(t-a) + 2H(t-2a) - 2H(t-3a) + \ \ldots\}$

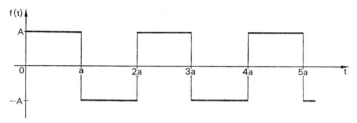

Fig. 6.5.

whence $\bar{f}(s) = \dfrac{A}{s}\{1 - 2e^{-as} + 2e^{-2as} - 2e^{-3as} + \ \ldots\}$ which may

be expressed as $\bar{f}(s) = \dfrac{A}{s}\{1 - 2e^{-as}(1 - e^{-as} + e^{-2as} - \ \ldots)\}.$

The terms between the inner brackets constitute a geometric series with common ratio $-e^{-as}$ and the sum to infinity of this series is $1/(1 + e^{-as})$ $(s > 0)$, so the transform becomes

$$\bar{f}(s) = \frac{A}{s}\left\{1 - \frac{2e^{-as}}{1 + e^{-as}}\right\} = \frac{A}{s}\left\{\frac{1 - e^{-as}}{1 + e^{-as}}\right\} = \frac{A}{s}\tanh\frac{as}{2}.$$

Problems 6.9

1–5. Obtain the Laplace transforms of the functions defined graphically in Figs. 6.6–6.10 below.

Obtain the inverse transforms of the following functions.

6. $e^{-\pi s}/s.$

7. $2(e^{-2s} - 3e^{-3s})/s.$

8. $(1 + e^{-s} + e^{-2s} + e^{-3s} + \ \ldots)/s.$

9. $1/s(1 - e^{-s}).$

10. $e^{-as}(1 + as)/s^2.$ Use result (5.6).

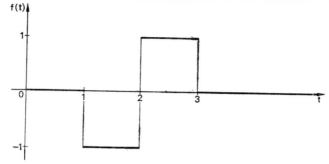

FIG. 6.6.

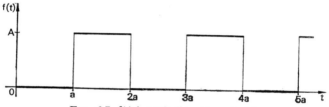

FIG. 6.7. $f(t)$ is periodic with period $2a$

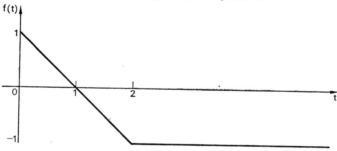

FIG. 6.8.

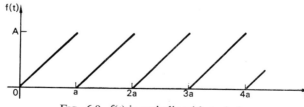

FIG. 6.9. $f(t)$ is periodic with period a

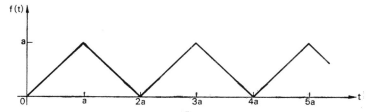

Fig. 6.10. $f(t)$ is periodic with period $2a$

6.10. THE SECOND SHIFT THEOREM

The first shift theorem shows how $f(t)$ is modified when each s in its transform is replaced by $s-a$. The second shift theorem shows how $f(s)$ is modified when each t in its inverse is replaced by $t-a$. In order to make this statement correspond with the symbolic form of the theorem, $f(t)$ is put in the clearly equivalent form $f(t) H(t)$, $t > 0$. The theorem states that

$$\text{if} \quad f(t), \quad \equiv f(t) H(t), \quad = \mathcal{L}^{-1}\{\bar{f}(s)\}$$
$$\text{then} \quad f(t-a) H(t-a) = \mathcal{L}^{-1}\{e^{-as}\bar{f}(s)\}. \tag{10.1}$$

To derive this let $\bar{f}(s) = \mathcal{L}\{f(u)\} = \int_0^\infty e^{-su} f(u)\, du,$

then $\qquad e^{-as}\, \bar{f}(s) = \int_0^\infty e^{-s(u+a)} f(u)\, du.$

Substitute t for $u+a$ to obtain

$$e^{-as}\, \bar{f}(s) = \int_a^\infty e^{-st} f(t-a)\, dt.$$

In order to convert the lower limit from a to 0, replace $f(t-a)$ by the function which is zero in the range $0 \le t < a$ and which equals $f(t-a)$ when $t > a$, namely $f(t-a) H(t-a)$, then

$$e^{-as}\, \bar{f}(s) = \int_0^\infty e^{-st} f(t-a) H(t-a)\, dt = \mathcal{L}\{f(t-a) H(t-a)\}$$

as required. This theorem is important for the analysis of systems which involve delayed time functions.

Example 10.1.

The transform of a unit rectangular pulse of duration b seconds which starts at $t = 0$ is $\mathcal{L}\{H(t) - H(t - b)\} = \dfrac{1}{s}(1 - e^{-bs})$. The transform of the same pulse delayed to start at $t = a$ is, by the theorem, $\mathcal{L}\{H(t - a) - H[t - (a + b)]\} = \dfrac{e^{-as}}{s}(1 - e^{-bs})$.

Example 10.2.

Obtain $\mathcal{L}^{-1}(e^{-4s}/s^2)$.

Since $\mathcal{L}^{-1}(1/s^2) = t$ the theorem gives the required inverse as $(t - 4)H(t - 4)$.

Example 10.3.

A constant e.m.f. E is applied during the interval between $t = a$ and $t = b$ to a circuit of R, C components in series, the capacitor being initially uncharged. Obtain an expression for the current at time t.

The differential equation is

$$Ri + \frac{q}{C} = E\{H(t - a) - H(t - b)\}; \qquad i = \frac{dq}{dt}.$$

The subsidiary equation is

$$R\bar{i} + \frac{\bar{i}}{Cs} = \frac{E}{s}(e^{-as} - e^{-bs}),$$

$$\text{whence} \quad \bar{i} = \frac{E}{R}\frac{(e^{-as} - e^{-bs})}{(s + 1/RC)}$$

and, by (10.1),

$$i(t) = \frac{E}{R} \{e^{-(t-a)/RC} H(t-a) - e^{-(t-b)/RC} H(t-b)\}.$$

Analysis of this expression shows that:

when $\quad 0 \leqslant t < a \quad\quad i = 0,$

when $\quad a < t < b \quad\quad i = \dfrac{E}{R} e^{-(t-a)/RC}$

and when $\quad t > b \quad\quad i = \dfrac{E}{R} \{e^{-(t-a)/RC} - e^{-(t-b)/RC}\},$

as shown in Fig. 6.11.

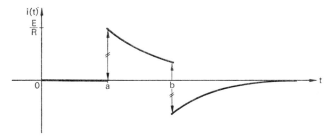

FIG. 6.11.

Example 10.4.

Given that $y = y' = 0$ when $t = 0$, solve the equation $y'' + y = f(t)$, where $f(t) = n+1$ in the interval $n\pi < t < (n+1)\pi$ and $n = 0, 1, 2, \ldots$. Sketch the solution.

In terms of a series of step functions

$$f(t) = 1 + H(t-\pi) + H(t-2\pi) + H(t-3\pi) + \ldots$$

and its transform is

$$\bar{f}(s) = \frac{1}{s}(1 + e^{-\pi s} + e^{-2\pi s} + e^{-3\pi s} + \ldots).$$

The subsidiary equation is

$$s^2\bar{y} + \bar{y} = \bar{f}(s), \quad \text{whence} \quad \bar{y} = \frac{\bar{f}(s)}{(s^2+1)}$$

or $\quad \bar{y} = \left\{\dfrac{1}{s} - \dfrac{s}{s^2+1}\right\}(1 + e^{-\pi s} + e^{-2\pi s} + e^{-3\pi s} + \ldots).$

Since $\quad \mathcal{L}^{-1}\left\{\dfrac{1}{s} - \dfrac{s}{s^2+1}\right\} = 1 - \cos t, \quad$ by (10.1)

$$y = (1 - \cos t) + \{1 - \cos(t-\pi)\} H(t-\pi)$$
$$+ \{1 - \cos(t-2\pi)\} H(t-2\pi) + \ldots. \tag{i}$$

R.H.S. of

$$\begin{aligned}
\text{(i)} &= 1 - \cos t && \text{when} \quad 0 < t < \pi, \\
&= (1 - \cos t) + (1 + \cos t) = 2 && \text{when} \quad \pi < t < 2\pi, \\
&= 2 + (1 - \cos t) = 3 - \cos t && \text{when} \quad 2\pi < t < 3\pi, \\
&= 3 - \cos t + (1 + \cos t) = 4 && \text{when} \quad 3\pi < t < 4\pi
\end{aligned}$$

and, in general,

$$y = (n+1) - \tfrac{1}{2}(\cos n\pi + 1)\cos t \quad \text{when} \quad n\pi < t < (n+1)\pi,$$

as indicated in Fig. 6.12.

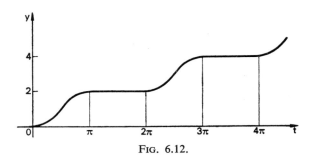

Fig. 6.12.

Problems 6.10

Obtain the inverse transforms of the following functions.

1. $\dfrac{2e^{-4s}}{(s-1)^3}.$
2. $\dfrac{3e^{-2\pi s/3}}{(s^2+9)}.$
3. $\dfrac{(s+3)e^{-\pi s}}{(s^2+4s+5)}.$

4. $\dfrac{(s+3)e^{-2s}}{(s^2+6s+5)}$.　　**5.**(a) $\dfrac{2+se^{-2\pi s}}{(s^2+4)}$ $(t > 2\pi)$,

(b) $\dfrac{1}{s(s^2+1)(1-e^{-\pi s})}$ $(\pi < t < 2\pi)$.

6. Obtain and sketch the solution of $y''+\pi^2 y = f(t)$, where

$$f(t) = \begin{Bmatrix} \pi^2, & t < 1 \\ 0, & t > 1 \end{Bmatrix} \quad \text{and} \quad y(0) = 1,\ y'(0) = 0.$$

7. Solve the equation $y''+4y = f(t)$, where $f(t) = \begin{Bmatrix} 4, & t < \pi \\ 0, & t > \pi \end{Bmatrix}$ and $y(0) = 0,\ y'(0) = 2$.

8. An e.m.f. $E(t) = \begin{Bmatrix} t, & 0 \leqslant t \leqslant 1 \\ 1, & t > 1 \end{Bmatrix}$ is applied to a series $L,\ R,\ C$ circuit for which the charge and current are zero at $t = 0$. Taking a critical case, given by $L = C = 1$ and $R = 2$, obtain an expression for the current $i(t)$.

9. Obtain the solution of $y''+y = f(t)$, where $f(t) = \begin{Bmatrix} t, & 0 \leqslant t \leqslant \pi \\ \pi, & t > \pi \end{Bmatrix}$, which satisfies $y(0) = y'(0) = 0$. Sketch the solution for values of $t \geqslant \pi$.

10. By means of the Convolution theorem show that

(a) $\mathscr{L}^{-1}\{1/(s^2+1)^2\} = (\sin t - t \cos t)/2$,

(b) $\mathscr{L}^{-1}\{s/(s^2+1)^2\} = (t \sin t)/2$.

Solve the equation $y''+y = f(t)$, where $f(t) = \begin{Bmatrix} \sin t, & 0 \leqslant t \leqslant \pi/2 \\ 1, & t > \pi/2 \end{Bmatrix}$ and $y(0) = y'(0) = 0$.
Show that $y = (\pi+4)/4$ when $t = (2n+1)\pi$, $n = 0, 1, 2, \ldots$.

11. Obtain the solution of $y'' - 4y' + 20y = 100f(t)$, where $f(t) = \begin{Bmatrix} t, & 0 \leqslant t \leqslant a \\ a, & t \geqslant a \end{Bmatrix}$, which satisfies the conditions $y = y' = 0$ when $t = 0$. Show that if $a = \pi$ and $t > \pi$ then

$$y = 5\pi + e^{2t}(\cos 4t + \tfrac{3}{4} \sin 4t)(e^{-2\pi} - 1).$$

12. Consider the third-order equation $y''' - 3y'' + 2y' = f(t)$, where $f(t) = n$ in the interval $n\pi < t < (n+1)\pi$, $n = 0, 1, 2, \ldots$ and $y(0) = y'(0) = y''(0) = 0$. Obtain the solution $y(t)$ for the three intervals $0 < t < \pi$, $\pi < t < 2\pi$, $2\pi < t < 3\pi$ and describe its behaviour near to $t = \pi$ as t increases through this value.

6.11. PERIODIC FUNCTIONS

The Laplace transform of a periodic function $f(t)$, period T, defined for positive values of t is

$$\mathscr{L}\{f(t)\} = \frac{1}{(1-e^{-sT})} \int_0^T e^{-st} f(t)\, dt. \qquad (11.1)$$

To derive this expression divide the range of integration into intervals of length equal to the period and express the transform of $f(t)$ in the form

$$\mathscr{L}\{f(t)\} = \int_0^\infty e^{-st} f(t)\, dt = \int_0^T e^{-st} f(t)\, dt$$
$$+ \int_T^{2T} e^{-st} f(t)\, dt + \int_{2T}^{3T} e^{-st} f(t)dt + \ldots .$$

In order to convert all the limits on the right to 0 and T substitute $t = u+T$ in the second integral, $t = u+2T$ in the third integral and so on to obtain

$$\mathscr{L}\{f(t)\} = \int_0^T e^{-st} f(t)\, dt + e^{-sT} \int_0^T e^{-su} f(u+T)\, du$$
$$+ e^{-2sT} \int_0^T e^{-su} f(u+2T)du + \ldots .$$

Since f is periodic $f(u+T) = f(u+2T) = \ldots = f(u)$, so all integrands have the same form. As a definite integral is a function of its limits, and not of the variable, all the integrals are equal and

$$\mathscr{L}\{f(t)\} = (1+e^{-sT}+e^{-2sT}+ \ldots) \int_0^T e^{-st} f(t)\, dt.$$

The sum of the infinite geometric series being $1/(1-e^{-sT})$, $(s > 0)$, the required result follows.

Example 11.1.

By means of (11.1) obtain the transform of the square wave considered in Example 9.3.

The range of integration is from 0 to $T = 2a$ and $f(t) = \pm A$ for the first and second halves of the range respectively. By (11.1)

$$\bar{f}(s) = \frac{A}{(1-e^{-2as})}\left\{\int_0^a e^{-st}\,dt - \int_a^{2a} e^{-st}\,dt\right\}$$

$$= \frac{A}{s(1-e^{-2as})}(e^{-2as}-2e^{-as}+1)$$

$$= \frac{A(1-e^{-as})^2}{s(1-e^{-2as})} = \frac{A(1-e^{-as})}{s(1+e^{-as})} = \frac{A}{s}\tanh\frac{as}{2}.$$

Example 11.2.

Obtain the transform of the following function, period $2\pi/a$, which represents the output of a half-wave rectifier.

$$f(t) = \begin{cases} \sin at & \text{when} \quad 0 \leqslant t \leqslant \pi/a, \\ 0 & \text{when} \quad \pi/a \leqslant t \leqslant 2\pi/a. \end{cases}$$

By (11.1)

$$\bar{f}(s) = \frac{1}{(1-e^{-2\pi s/a})}\int_0^{\pi/a} e^{-st}\sin at\,dt = \frac{a}{(1-e^{-\pi s/a})(s^2+a^2)}.$$

Note that there is no contribution to $\bar{f}(s)$ when $f(t) = 0$.

Problems 6.11

1. Use result (11.1) to derive the transforms of the periodic functions defined in numbers **2. 4** and **5** of Problems 6.9.

2. By means of result (11.1) obtain the transforms of $\cos at$ and $\sin at$.

HGM: – METCS 17

3. Obtain the transform of the function $f(t) = A \,|\sin at\,|$, which represents the output of a full-wave rectifier.

4. Obtain the transform of the function $f(t) = e^t$, $0 < t < 2\pi$, extended periodically with period 2π.

5. Solve the equation $y'' + y = f(t)$, where $f(t)$ is the saw-tooth wave defined by $f(t) = At/a$, $0 < t < a$, extended periodically with period a; $y(0) = y'(0) = 0$. If $a = \pi$ show that, for $0 < t < n\pi$,

$$y = \frac{A}{\pi}\left[t - \sin t - \pi\left\{ (n-1) + \frac{1}{2}(1 + \cos n\pi)\cos t \right\} \right],$$

where n is a positive integer.

6. Obtain an expression for the solution in the range $0 < t < n\pi/a$, $n = 1, 2, 3, \ldots$, of the equation $y'' + a^2 y = f(t)$, where $f(t) = A \,|\sin at\,|$ as in problem **3** and $y = y' = 0$ at $t = 0$.

6.12. THE UNIT IMPULSE FUNCTION

Properties of the impulse function can be shown by considering limiting forms of certain other functions. Take, as one example, the function φ defined as

$$\varphi(t) = \begin{Bmatrix} 1/\varepsilon & 0 \leqslant t \leqslant \varepsilon \\ 0 & t < 0, t > \varepsilon \end{Bmatrix} \quad (\varepsilon > 0),$$

and illustrated in Fig. 6.13. As $\varepsilon \to 0$ the dimensions of the rectangular shaded region change in such a way that its area is always unity. This property may be expressed by the integral

$$\int_0^\infty \varphi(t)\, dt = 1.$$

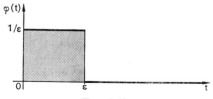

FIG. 6.13.

The unit impulse function $\delta(t)$ is defined to possess the properties of $\varphi(t)$ as $\varepsilon \to 0$, which implies that $\delta(t) = 0$ when $t < 0$ and $t > 0$ except for a vanishingly small interval to the right of $t = 0$. It follows that

$$\int_0^\infty \delta(t)\, dt = 1.$$

The impulse function has the following properties, which should seem reasonable from the definition. If $f(t)$ is a continuous function then

$$\int_0^\infty \delta(t)\, f(t)\, dt = f(0) \tag{12.1}$$

and

$$\int_0^\infty \delta(t-a)\, f(t)\, dt = f(a), \tag{12.2}$$

where the notation $\delta(t-a)$ means that the impulse occurs at $t = a$. By (12.1) the Laplace transform of $\delta(t)$ is

$$\mathscr{L}\{\delta(t)\} = \int_0^\infty e^{-st}\, \delta(t)\, dt = 1.$$

Similarly, by (12.2),

$$\mathscr{L}\{\delta(t-a)\} = \int_0^\infty e^{-st}\, \delta(t-a)\, dt = e^{-as}.$$

Example 12.1.

An impulsive e.m.f. E is applied at $t = 0$ to an inductance L and a capacitor C in series, the initial charge and current being zero. Obtain an expression for the charge at time t. The differential equation for the charge q is

$$L\frac{d^2q}{dt^2} + \frac{q}{C} = E\,\delta(t); \qquad i = \frac{dq}{dt}.$$

The S.E. is $Ls^2\bar{q} + \bar{q}/C = E$, whence $\bar{q} = (E/L)\,1/[(s^2 + 1/LC)]$ and

$$q = E\sqrt{\frac{C}{L}}\sin\left(\frac{t}{\sqrt{LC}}\right).$$

This result is consistent with the given condition $q(0) = 0$. However, the expression for q shows that

$$\frac{dq}{dt} = i = \frac{E}{L}\cos\left(\frac{t}{\sqrt{LC}}\right)$$

so $i(0) = E/L$, which is inconsistent with the given condition $i(0) = 0$. The impulse function is an ideal hardly realised in practice and such a result could be interpreted as a change in the value of i from zero to E/L in an extremely short time, instantaneously in effect. This simple explanation of the apparent inconsistency may be justified by a more detailed analysis using the theory of impulsive motion.

Careful consideration should be given to the solution of problems which involve impulse functions for, although the Laplace transform procedure gives valid results, certain relations, such as (5.5), may cease to be strictly true and may need to be modified.

Problems 6.12

1. Show by simple considerations that $\int_0^t \delta(t-a)\,dt = H(t-a)$.

2. An impulsive e.m.f. E is applied at $t = 0$ to a circuit of L, R, C components in series. Given that $q(0) = i(0) = 0$, and the notation $a = R/2L$, $b^2 = (1/LC - a^2) > 0$, show that the current at time t is given by

$$i = \frac{E}{bL}\,e^{-at}(b\cos bt - a\sin bt).$$

CHAPTER 7

Line and Multiple Integrals

7.1. INTRODUCTION

In this chapter various ways are considered in which a function of several variables may be integrated. The values of a function of position in three-dimensional space can be summed over a set of points along a space curve or over a set on a surface or in a volume. When suitable limits of these sums are defined there arise integrals which are called line, surface and volume integrals respectively.

The theory is explained in terms of space coordinates because the processes of integration are most easily pictured in this way, but the principles involved are applicable to problems which may not be expressed in terms of space variables. Many of the integrals may be conceived as representing areas or volumes or masses, for example, and such associations are often helpful, although the attempt to connect all integrals with familiar ideas should not be carried too far. Some integrals cannot be treated easily in this way and the student's ability to evaluate them should not be frustrated by the search for a connection which may not be apparent or even exist.

In order to preserve some uniformity of notation the letters f, F and $\mathcal{G}$ are used to denote functions of one, two and three variables respectively.

7.2. DEFINITION OF A LINE INTEGRAL

Consider the familiar integral $\int f(x)\,dx$, where f is a continuous function of x. The contribution to the integral from the small interval between $x = x_0$ and $x = x_0 + \delta x$ is nearly equal to $f(x_0)\,\delta x$. Now consider the integral $\int F(x, y)\,dx$, where F is a continuous function of x and y. The contribution to this integral from the same interval is approximately $F(x_0, y)\,\delta x$, but this depends on y as well as on the increment δx. Thus a definite integral of the form $\int_a^b F(x, y)\,dx$ is defined only when corresponding values of x and y are given over the range between the limits. In practice the relation given between the two variables is usually functional, $y = f(x)$, and this defines a curve called the path of integration; the related integral is called a *line* or *curvilinear integral* and is denoted by $\int_C F(x, y)\,dx$. Similar remarks apply to the line integral $\int_C F(x, y)\,dy$ for which $x = g(y)$ is a function of y and the limits are values of y. In general $\int_C F\{x, f(x)\}\,dx \neq \int_C F\{g(y), y\}\,dy$ for the same path C.

Example 2.1.

Evaluate (a) $\int_C (x+y)\,dx$, (b) $\int_C (x+y)\,dy$, where C is the arc of the parabola $y = x^2$ between $(0, 0)$ and $(2, 4)$.

(a) $\displaystyle \int_C (x+y)\,dx = \int_0^2 (x+x^2)\,dx = \frac{x^2}{2} + \frac{x^3}{3}\bigg|_0^2 = 14/3.$

(b) $\displaystyle \int_C (x+y)\,dy = \int_0^4 (\sqrt{y}+y)\,dy = \frac{2}{3}\,y^{3/2} + \frac{y^2}{2}\bigg|_0^4 = 40/3.$

Or, expressing y in terms of x,

$$\int_{C}^{\cdot} (x+y)\,dy = \int_{0}^{2} (x+x^2)\,2x\,dx = \frac{2}{3}\,x^3 + \frac{x^4}{2}\bigg|_{0}^{2} = 40/3.$$

A line integral may be evaluated in either of the two directions along a curve; for example the arc AB of the curve in Fig. 7.1 may be described from A to B or from B to A and the two integrals are related by

$$\int_{AB} F(x,\,y)\,dx = -\int_{BA} F(x,\,y)\,dx,$$

for

$$\int_{AB} F\,dx = \int_{a}^{b} F\,dx = -\int_{b}^{a} F\,dx = -\int_{BA} F\,dx.$$

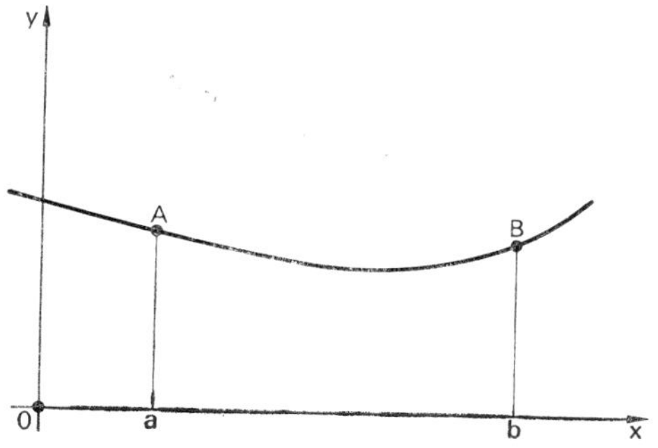

Fig. 7.1.

Should the curve be cut by an ordinate in more than one point, the line integral is formed from the sum of integrals for each part of the curve. Thus in Fig. 7.2

$$\int_{AB} F(x,\,y)\,dx = \int_{AK} F\{x,\,f_1(x)\}\,dx + \int_{KB} F\{x,\,f_2(x)\}\,dx$$

$$= \int_{a}^{k} F\{x,\,f_1(x)\}\,dx + \int_{k}^{b} F\{x,\,f_2(x)\}\,dx,$$

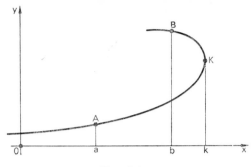

FIG. 7.2.

where $y = f_1(x)$ and $y = f_2(x)$ are the equations of the arcs AK, KB respectively.

When a curve is the boundary of an area in the plane the directions along the curve are described as being in the positive or negative sense. The positive sense is defined to be such that a person who follows the curve in this direction has the enclosed area to his left. The notation $\oint_C$ is used when the path of integration is a closed circuit.

Example 2.2.

Show that the area enclosed by a curve is given by $-\oint_C y\,dx$, where C is traversed in the positive sense.

In Fig. 7.3

$$\text{area}\quad S = \int_a^b f_2(x)\,dx - \int_a^b f_1(x)\,dx = \int_a^b \{f_2(x) - f_1(x)\}\,dx.$$

The line integral

$$\oint_C y(x)\,dx = \int_a^b f_1(x)\,dx + \int_b^a f_2(x)\,dx = \int_a^b \{f_1(x) - f_2(x)\}\,dx = -S.$$

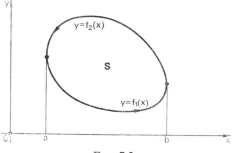

Fig. 7.3.

A curvilinear integral often occurs in the form $\int_C F(x, y)\, ds$, where s is the path length of the curve C measured from an initial point. Such an integral could arise, for example, when F defines a field of force, whose magnitude and direction depend on position. Figure 7.4 represents a curve C drawn in a field of electric force $E(x, y)$, the direction of which is confined to the x–y plane. At any point N on the curve E is resolved into a component $E_t(x, y)$, tangential to the curve, and a normal component not shown; E is also resolved into compo-

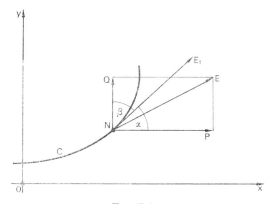

Fig. 7.4.

nents $P(x, y)$ and $Q(x, y)$ parallel to the axes. The work done by E on a charge which moves along C is proportional to $\int_C E_t\, ds = \int_C (P \cos \alpha + Q \cos \beta)\, ds = \int_C (P\, dx + Q\, dy)$, a form taken by many line integrals.

The analysis given above is restricted to functions of two variables and the paths of integration are confined to a plane only for ease of explanation. Most of the ideas can be extended to three dimensions, for which line integrals have general forms such as

$$\int_C \mathcal{G}(x, y, z)\, ds \quad \text{and} \quad \int_C (P\, dx + Q\, dy + R\, dz).$$

7.3. EVALUATION OF LINE INTEGRALS

A line integral of the form $\int_C F(x, y)\, ds$ may be evaluated with respect to x by the substitution $ds = \sqrt{\{1 + (dy/dx)^2\}}\, dx$, since the differentials are related by $(ds)^2 = (dx)^2 + (dy)^2$. If the evaluation is with respect to y then the substitution $ds = \sqrt{\{1 + (dx/dy)^2\}}\, dy$ is made.

Example 3.1.

Evaluate $\int_C \dfrac{y}{x}\, ds$ along $y = x^2$ between $x = 0$ and $x = \frac{1}{2}$.

$$\int_C = \int_0^{1/2} x \sqrt{(1 + 4x^2)}\, dx = \tfrac{1}{12}(1 + 4x^2)^{3/2} \Big|_0^{1/2} = \big(2\sqrt{2} - 1\big)/12.$$

When the curve C is defined in terms of a parameter t by the equations $x = x(t)$, $y = y(t)$, the integral may be evaluated with respect to t by the substitution $ds = \sqrt{\{(dx/dt)^2 + (dy/dt)^2\}}\, dt$. Thus the equations $x = t$, $y = t^2$ applied to the integral of Example 3.1 lead to expressions in terms of t similar to those used in the solution. Many of the line integrals which occur in applications are of the form $\int_C \{P(x, y)\,dx + Q(x, y)\,dy\}$ and the evaluation of this is usually straightforward.

Example 3.2.

Evaluate $\int_C \{x^2y\,dx + (x + 2y^2)\,dy\}$ along $y = x^2$ from the origin to $(1, 1)$.

In terms of x

$$\int_C = \int_0^1 \{x^4 + (x + 2x^4)2x\}\,dx = \int_0^1 (4x^5 + x^4 + 2x^2)\,dx$$

$$= \frac{2}{3}\,x^6 + \frac{x^5}{5} + \frac{2}{3}\,x^3 \Big|_0^1 = \frac{23}{15}.$$

If $F(x, y) \equiv 1$ or $\mathcal{G}(x, y, z) \equiv 1$ then the line integrals $\int_C F\,ds$ and $\int_C \mathcal{G}\,ds$ simply measure the arc length of C between the limits.

Problems 7.3

1. Evaluate the integral in Example 3.2, between the same points, along (a) $y = x$, (b) $x = y^2$.

2. Evaluate $\oint_C \{(x - y)^2\,dx + 2xy\,dy\}$, where C is the boundary of the square $y = 0$, $x = 1$, $y = 1$, $x = 0$ described in the positive sense.

3. Evaluate $\oint_C (2xe^y\,dx + x\,dy)$, where C is the boundary of the region formed by $y = 0$, $x = 2$ and $y = x^2$ described in the negative sense.

4. Evaluate $\oint_C \{(3x - 2y + 1)\,dx + (5x - y - 4)\,dy\}$, C being the triangle with vertices at the origin, $(2, 0)$, $(2, 2)$ described in the positive sense.

5. Evaluate $\int_C y\,ds$, C being the arc of $y = 2\sqrt{x}$ between $x = 0$ and $x = 3$.

6. Evaluate $\int_C \{(x+y)\,dx+(x-y)\,dy\}$, where C is the arc of $x = 2t^2+t-1$, $y = t^2+1$ between $(0, 2)$ and $(2, 2)$.

7. Evaluate $\int_C \{x^2z\,dx+(x-y)\,dy-xyz\,dz\}$, where C is the arc of $y = x^2$, in the plane $z = 1$, between $(0, 0, 1)$ and $(2, 4, 1)$.

8. Evaluate $\int_C \{3x^2 \sin y\,dx+xy^2 \cos z\,dy+4yz^2dz\}$, where C is the line $x = y = z$ between the origin and (π, π, π).

9. Evaluate $\int_C (x \sin \pi y^2\,dx-4xy^2\,dy)$ first with respect to x and then with respect to y. C is the arc of $y^2 = 4x$ between $(4, -4)$ and $(4, 4)$.

10. Evaluate $\oint_C (y\,dx-x\,dy)$, where C is the circle of radius 2 centred at $(2, 0)$ and described in the positive sense. (Use plane polar coordinates; $x = r \cos \theta$, $y = r \sin \theta$.)

11. A unit charge moves along the parabola $y = x^2-3x+2$ from $(0, 2)$ to $(3, 2)$ in a field of electric force whose component along the curve is $E_t = 3(x^2-y-2)-x$. Calculate the work done on the charge.

12. Evaluate

$$\oint_C \left\{ \frac{-y\,dx}{x^2+y^2} + \frac{x\,dy}{x^2+y^2} \right\},$$

where C is (a) the square formed by $x = \pm 1$, $y = \pm 1$, (b) the square formed by $x = 1$, $x = 2$, $y = 1$, $y = 2$.

7.4. DEPENDENCE ON THE PATH OF INTEGRATION

Figure 7.5 shows two points A, B in the x–y plane joined by two different curves, ALB and AMB. The value of a line integral depends, in general, on the curve which connects the initial and terminal points. Thus, in general,

$$\int_{ALB} (P\,dx+Q\,dy) \neq \int_{AMB} (P\,dx+Q\,dy).$$

An exception to this rule occurs when $P\,dx+Q\,dy$ is an exact, or a perfect, differential $d\varphi$, say, where φ is a *one-valued* function of x and y. In this case the line integral

$$\int_{AB} (P\,dx+Q\,dy), = \int_{AB} d\varphi = \varphi_B-\varphi_A, \qquad (4.1)$$

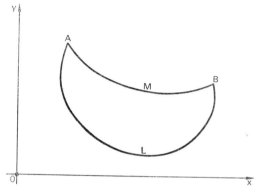

FIG. 7.5.

depends only on the values of φ at the two points A, B and is therefore independent of a connecting curve.

The necessary and sufficient condition for $P\,dx + Q\,dy$ to be an exact differential is that

$$\partial P/\partial y = \partial Q/\partial x. \qquad (4.2)$$

Proof of necessity.

If $\qquad P\,dx + Q\,dy = d\varphi(x, y), = \dfrac{\partial \varphi}{\partial x}\,dx + \dfrac{\partial \varphi}{\partial y}\,dy, \qquad$ then

$P = \partial \varphi / \partial x$ and $Q = \partial \varphi / \partial y$. The elimination of φ between these two relations gives $\partial P/\partial y = \partial Q/\partial x$ since

$$\partial^2 \varphi / \partial y\,\partial x = \partial^2 \varphi / \partial x\,\partial y.$$

The line integral of an exact differential of a one-valued function round a closed curve is zero.

Proof. Consider the circuit $ALBMA$ in Fig. 7.5.

$$\oint_C d\varphi = \int_{ALB} d\varphi + \int_{BMA} d\varphi = \int_{ALB} d\varphi - \int_{AMB} d\varphi = 0, \quad \text{by (4.1).}$$

These results may be extended to line integrals of the form $\int_{AB} (P\,dx + Q\,dy + R\,dz)$, where A and B are points in space.

The conditions for $P\,dx + Q\,dy + R\,dz$ to be an exact differential, $d\varphi(x, y, z)$, are that

$$\partial P/\partial y = \partial Q/\partial x, \quad \partial P/\partial z = \partial R/\partial x \quad \text{and} \quad \partial Q/\partial z = \partial R/\partial y.$$
$$(4.3)$$

Example 4.1.

Evaluate $\int_C \{2y(x-y)\,dx + x(x-4y)\,dy\}$ between the origin and $(3, 1)$.

Here $\partial P/\partial y = \partial Q/\partial x = 2x - 4y$, which confirms the implication that the integral is independent of the path C. As $P = \partial\varphi/\partial x$ and $Q = \partial\varphi/\partial y$, two expressions for φ may be derived by taking the partial integral of P and of Q with respect to x and y respectively.

$\varphi = \int 2y(x-y)\,dx = x^2y - 2xy^2 + f(y)$, since y is held constant.

$\varphi = \int x(x-4y)\,dy = x^2y - 2xy^2 + g(x)$, since x is held constant.

Comparison of these two expressions for φ shows that $f(y) = g(x)$, which is true only if $f = g = $ a constant.

Thus

$$\int_C (P\,dx + Q\,dy) = \int_C d\varphi = \varphi \Big|_{(0,\,0)}^{(3,\,1)} = x^2y - 2xy^2 \Big|_{(0,\,0)}^{(3,\,1)} = 3.$$

The integral could be evaluated in the usual way by taking any path between the points given, say from the origin to $(3, 0)$ and then to $(3, 1)$.

Example 4.2.

Evaluate $\int_C \{(3x^2 - 2y)dx + 2(y - x)dy\}$ between $(2, 1)$ and $(3, 2)$.

$\partial P/\partial y = \partial Q/\partial x = -2$. Proceeding as above,

$\varphi = \int (3x^2 - 2y)\,dx = x^3 - 2xy + f(y)$ and

$\varphi = \int 2(y-x)\,dy = y^2 - 2xy + g(x)$. These expressions for φ are identical if $f(y) = y^2$ and $g(x) = x^3$, each plus an arbitrary constant of no interest. Thus $\int_C d\varphi = x^3 - 2xy + y^2 \Big|_{(2,\,1)}^{(3,\,2)} = 14$.

Problems 7.4

1. Show that the line integral $\int_C \{(y^2+4)\,dx + 2(xy+1)\,dy\}$ is independent of the integration path joining any two points in the x–y plane. By choosing a convenient path evaluate the integral between the points $(1, 2)$ and $(3, 4)$. Verify the result by another method.

2. Determine the value of k for which the line integral

$$\int_C \left\{ \frac{(1-ky^2)\,dx}{(1+xy)^2} + \frac{(1-kx^2)\,dy}{(1+xy)^2} \right\}$$

is independent of the curve C connecting two points in the x–y plane. Using this value of k evaluate the integral along any convenient path which connects the origin to $(1, 2)$.

3. Evaluate $\int_C xe^{xy}\{y^3(2+xy)\,dx + xy^2(3+xy)\,dy\}$, where C connects the origin to $(1, 1)$ and is defined in the following ways:

    (a) The x axis from the origin to $(1, 0)$ and then the straight line from $(1, 0)$ to $(1, 1)$.

    (b) $y = x$.

    (c) $y = x^3$.

Also make the evaluation without reference to a specific path.

4. Evaluate $\oint_C [\{2z(x+y)+1\}\,dx + 2\{z(x+y)+1\}\,dy + \{(x+y)^2+3\}\,dz]$,

where C is the plane triangle with vertices at the origin, $(2, 2, 2)$ and $(0, 0, 2)$, described in the positive sense as viewed from the point $(1, 0, 1)$. Verify the result by another method.

5. Apply condition (4.2) to the integral in problem **12** of Problems 7.3 and explain the inconsistency between the results of (a) and (b). (A change to polar coordinates may assist the analysis.)

7.5. DEFINITION OF A DOUBLE INTEGRAL

In Fig. 7.6 the meaning of a line integral which has the form $\int_C F(x, y) \, ds$ is shown graphically. Such an integral may be defined as the limit of the sum $\sum_{s_A}^{s_B} F(x, y) \, \delta s$, as $\delta s \to 0$, taken along the curve AB in the x–y plane. The values of the function $F(x, y)$ at points (x, y) are measured parallel to the z axis and $z = F(x, y)$ defines a surface which is curved in general. The curve AB is shown projected parallel to the z axis thus generating a surface, which meets the surface $z = F(x, y)$ in the space curve GH. The area of $ABHG$ is equal approximately to a sum of elemental areas $z \, \delta s$, one of which is shown, and the limiting value of this sum as $\delta s \to 0$ is the line integral $\int_{AB} F(x, y) \, ds$.

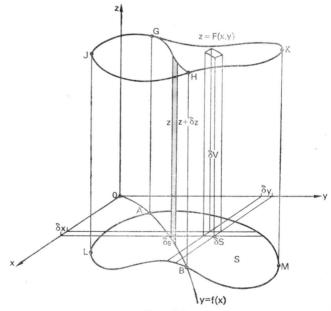

Fig. 7.6.

Example 5.1.

By means of a line integral evaluate the area of the plane surface bounded by the z axis, the line $y = x$ and the projection of this line on the plane $6x + 3y + 2z = 6$.

The required surface is shown shaded in Fig. 7.7. From the given equation $z = F(x, y) = 3(2-2x-y)/2$, $= 3(2-3x)/2$ as $y = x$ along the path of integration; also $ds = \sqrt{2}\,dx$.

$$\int_{OB} F(x, y)\,ds = \frac{3}{\sqrt{2}} \int_{0}^{2/3} (2-3x)\,dx$$

$$= \frac{3}{\sqrt{2}} \left(2x - \frac{3}{2} x^2\right) \Big|_{0}^{2/3} = \sqrt{2},$$

which is verified easily from the dimensions of $\triangle OBG$.

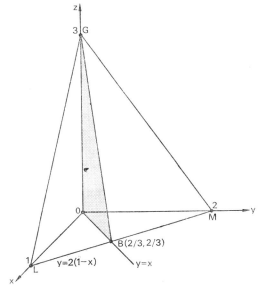

Fig. 7.7.

The above analysis shows how a line integral may be considered as a limiting process which measures an area. A similar limiting process may be used to evaluate a volume. Figure 7.6 shows in the x–y plane a curve $ALBMA$ which encloses an area S. The cylinder generated by lines parallel to the z axis which pass through the curve meets the surface $z = F(x, y)$ in the space curve $GJHKG$. The volume V to be evaluated is enclosed by the plane $z = 0$, the surface $z = F(x, y)$ and the cylinder on the boundary of S. Consider V to be divided into many elemental volumes by sets of planes parallel to the x–z and y–z planes. One such element δV is shown as a solid figure which is almost equivalent to a cuboid with base area $\delta S = \delta x \delta y$ and of height z, where z is evaluated at some point within δS. Thus V is given approximately by $\sum z \delta S = \sum F(x, y) \, \delta x \, \delta y$, where the summation extends over S. The limit of this sum as $\delta S \rightarrow 0$ is called the *double integral* of $F(x, y)$ over the *field of integration* S and is denoted by

$$\iint\limits_{S} F(x, y) \, dS \quad \text{or by} \quad \iint\limits_{S} F(x, y) \, dx \, dy.$$

If $F(x, y) \equiv 1$ then the double integral $\iint\limits_{S} dx \, dy$ is simply the volume of the cylinder with unit height which stands on S and therefore it measures the area of S. If $F(x, y)$ is negative then the double integral is negative because the surface $z = F(x, y)$ lies below the x–y plane. When the purpose of a double integral is to measure *actual* volumes, those which lie below the coordinate plane on which S is defined are assigned positive values.

7.6. EVALUATION OF DOUBLE INTEGRALS

Figure 7.8 shows a field of integration S bounded by the curve $ACBDA$ and contained between the lines $x = a$, $x = b$, $y = c$, $y = d$ which are tangents to the curve at A, B, C, D. Arc ACB is defined by $y = f_1(x)$ and arc ADB by $y = f_2(x)$.

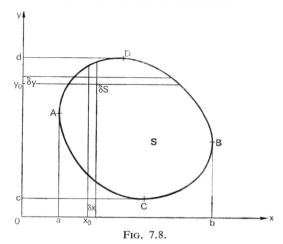

FIG. 7.8.

The volume of the region formed between S, the planes $x = x_0$, $x = x_0 + \delta x$ and the surface $z = F(x, y)$ is measured approximately by

$$\delta x \sum_{y=f_1(x_0)}^{y=f_2(x_0)} F(x_0, y) \, \delta y$$

where the sum is formed from increments δy in y along the strip of width δx, x being treated as a constant x_0. The limit of this sum as $\delta y \to 0$ is

$$\int_{f_1(x_0)}^{f_2(x_0)} F(x_0, y) \, dy$$

which is a function of x_0, say $h(x_0)$. The required volume V is given approximately by the sum of elemental volumes standing on all strips of width δx in S, that is by

$$\sum_{x=a}^{x=b} h(x) \, \delta x,$$

where x varies between the boundary limits of S parallel to Ox.

The limit of this sum as $\delta x \to 0$ is

$$\int_a^b h(x) \, dx.$$

In this way V may be evaluated by two integrations, the first with respect to y, for which the limits are generally functions of x, and the second with respect to x between constant limits. This process for evaluating a double integral may be expressed as

$$\iint_S F(x, y) \, dx \, dy = \int_a^b \left\{ \int_{f_1(x)}^{f_2(x)} F(x, y) \, dy \right\} dx$$

$$\text{or} \quad \int_a^b dx \int_{f_1(x)}^{f_2(x)} F(x, y) \, dy, \qquad (6.1)$$

which is called a repeated integral. Being functions of x, the limits for the inner integral imply that x is to be treated as a constant during its evaluation with respect to y.

Example 6.1.

By means of a double integral evaluate the volume of the tetrahedron $OLMG$, shown in Fig. 7.7, formed between the coordinate planes and the plane $6x + 3y + 2z = 6$.

In the notation of the theory the volume is given by

$$V = \iint_S z \, dx \, dy, \quad \text{where} \quad z = 3(2 - 2x - y)/2$$

$$\text{and} \quad S \quad \text{is} \quad \Delta OLM;$$

$_1(x) = 0, f_2(x) = 2(1 - x), a = 0$ and $b = 1$. Upon substitution

$$V = \int_0^1 dx \int_0^{2(1-x)} \frac{3}{2}(2 - 2x - y) \, dy = \frac{3}{2} \int_0^1 \left[2y - 2xy - \frac{y^2}{2} \right]_0^{2(1-x)} dx$$

$$= 3 \int_0^1 (1 - x)^2 \, dx = -(1 - x)^3 \Big|_0^1 = 1,$$

which is verified easily from the dimensions of the tetrahedron.

Example 6.2.

The axis of a circular cylinder, radius *a*, is the *z* axis. Determine the value of *a* for two units of volume to be enclosed in the positive octant by the plane $z = 0$, the cylinder and the curved surface $z = xy$.

The field of integration is bounded by the lines $x = 0$, $y = 0$ and the circle $x^2 + y^2 = a^2$ in the positive quadrant of the plane $z = 0$. To cover the field *y* needs to vary from its general value on the boundary $y = 0$, that is 0, along a strip where *x* is constant to its general value on the circle, that is $\sqrt{(a^2 - x^2)}$; *x* then varies between the constant limits 0 and *a*. In this way the volume is given by

$$V = \iint_S z \, dx \, dy = \int_0^a \int_0^{\sqrt{(a^2 - x^2)}} xy \, dx \, dy = \int_0^a x \left[\frac{y^2}{2} \right]_0^{\sqrt{(a^2 - x^2)}} dx$$

$$= \frac{1}{2} \int_0^a x(a^2 - x^2) \, dx = -\frac{1}{8} (a^2 - x^2)^2 \Big|_0^a = \frac{a^4}{8}.$$

For *V* to be two units $a = 2$.

Problems 7.6

Evaluate the double integrals in **1–5** and sketch their fields of integration.

1. $\int_0^1 \int_{-x}^{2x} (3 + x^2 + y^2) \, dx \, dy.$ **2.** $\int_0^3 \int_0^{\sqrt{(9 - x^2)}} x \, dx \, dy.$

3. $\int_0^2 dx \int_0^1 xy \, dy.$ **4.** $\int_0^1 \int_x^{\sqrt{x}} (x^2 - y^2) \, dx \, dy.$

5. $\int_0^2 \int_x^{6-x} (x + 2y) \, dx \, dy.$

6. Evaluate $\iint\limits_{S} (x-y)\, dx\, dy$, where S is the region enclosed by arcs of the parabolas $y = x^2$ and $y^2 = x$. Explain how the value of the integral may be obtained by inspection only.

7. Evaluate $\iint\limits_{S} xy\, dx\, dy$, where S is the region bounded by the semi-circle $x^2+y^2 = a^2$, $y \geqslant 0$, and the x axis. Explain how the value of this integral may be obtained by inspection only.

8. Evaluate $\iint\limits_{S} (x^2+y^2)\, dx\, dy$, where S is the square region with vertices at $(3, 0)$, $(0, 3)$, $(-3, 0)$ and $(0, -3)$. Explain the validity of considering the value of the integral over S to be four times the value of the integral over that part of S in the positive quadrant.

9. Evaluate $\iint\limits_{S} (4x-3y)\, dx\, dy$, where S is the region bounded by the parabola $y = 4-x^2$ and the x axis. Explain the invalidity of considering the value of the integral over S to be twice the value of the integral over that half of S in the positive quadrant.

10. Evaluate $\iint\limits_{S} [x/(x+y)]\, dx\, dy$, where S is the triangular region with vertices at $(a, 0)$, $(a, 2a)$ and $(0, a)$. Determine the area of S for which the value of the integral is $\log 27$.

11. Evaluate the volume of the region in the positive octant bounded by the coordinate planes, the cylinder $x^2+y^2 = 16$ and the plane $z = x+y$.

12. Evaluate the volume of the region enclosed by the parabolic cylinder $4y = x^2$ and the planes $z = y-1$, $z = 0$.

13. Evaluate the volume of the region in the positive octant bounded by the coordinate planes, the parabolic cylinder $z = 1-x^2$ and the plane $x+y = 1$.

14. Evaluate the volume of the region common to the cylinders $x^2+y^2 = a^2$ and $x^2+z^2 = a^2$.

15. Evaluate the volume of the region enclosed by the cylinder $x^2+y^2 = 4x$, the plane $z = 0$ and the surface $z = xy$.

16. Evaluate the volume of the region in the positive octant enclosed by the coordinate planes, the parabolic cylinder $z = x^2$ and the elliptical cylinder $4x^2+y^2 = 1$.

17. Evaluate $\iint\limits_{S} xy^2\, dx\, dy$, where S is the smaller segment of the circle $x^2+y^2 = 4$ formed by $x+y = 2$.

18. Evaluate the volume of the region in the positive octant formed between the paraboloid $y^2 + z^2 = 4x$, the parabolic cylinder $y^2 = x$ and the planes $z = 0$, $x = 3$.

19. Evaluate the volume of the region enclosed by the paraboloid $1 - z = x^2 + (y/2)^2$ and the x–y plane.

20. Evaluate, correct to one decimal place, the volume of the region in the positive octant enclosed by the plane $z = 0$, the surfaces formed by the projections of the curves $y = e^x$ and $y = e^{-x}$ normal to this plane, and by the surfaces $z = xy$, $x = 2$.

7.7. REVERSAL OF INTEGRATION ORDER

A double integral may be evaluated by performing the two integrations in the reverse order to that considered. Let arc CAD in Fig. 7.8 be defined by $x = g_1(y)$ and arc CBD by $x = g_2(y)$. The volume V may be constructed by first treating y as a constant, y_0, and by summing terms like $F(x, y_0)\delta x$ along a strip of width δy between limits $g_1(y_0)$ and $g_2(y_0)$. The elemental volumes standing on strips of width δy are then summed with respect to y between constant limits c and d. The limiting forms of these sums as $\delta x \to 0$ and as $\delta y \to 0$ may again be expressed as a repeated integral. Thus a double integral may be evaluated in the equivalent form to (6.1):

$$\iint_S F(x, y)\, dx\, dy = \int_c^d dy \int_{g_1(y)}^{g_2(y)} F(x, y)\, dx,$$

where y is treated as a constant during the evaluation of the inner integral.

Example 7.1.

Evaluate the volume of tetrahedron $OLMG$ in Fig. 7.7 by reversing the previous order of integration.

In this case $g_1(y) = 0$, $g_2(y) = 1-y/2$, $c = 0$ and $d = 2$.

$$V = \int_0^2 dy \int_0^{1-y/2} \tfrac{3}{2}(2-2x-y)\,dx = \tfrac{3}{2} \int_0^2 \left[2x-x^2-yx\right]_0^{1-y/2} dy$$

$$= \tfrac{3}{2} \int_0^2 (1-y/2)^2\,dy = \left[-(1-y/2)^3\right]_0^2 = 1, \quad \text{as before.}$$

A reversal of integration order may require the field of integration to be divided in a suitable way.

Example 7.2.

Reverse the integration order of the repeated integral
$\int_0^1 dx \int_{3x}^{4-x^2} F(x, y)\,dy$.

In this type of problem it is important to have a clear idea of the integration field, so a sketch is often advisable. Figure 7.9 shows how S is defined by the limits of the integrals. The inner integral implies that, whilst x is held constant, y varies from a value on the straight line $y = 3x$ to a value on the parabola $y = 4-x^2$. This information alone defines the region enclosed by *ABCDA*. The outer integral shows that x may vary only from 0 to 1 and this restricts S to the shaded part of the region. When the integration order is reversed the inner integral is evaluated with respect to x whilst y is held constant. At any point on *OB* $x = 0$, the lower limit, but on *OAB* x is defined by $y/3$ on *OA* and by $+\sqrt{(4-y)}$ on *AB*. Thus S is divided by a line through A parallel to *Ox* and the given repeated integral is equivalent to

$$\int_0^3 dy \int_0^{y/3} F(x, y)\,dx + \int_3^4 dy \int_0^{+\sqrt{(4-y)}} F(x, y)\,dx.$$

The order in which differentials, such as dx and dy, appear in $dx\,dy$ within a double integral should not be significant, for usually the limits indicate the variable with respect to which the first integration is to be performed. Some texts state a convention whereby the order in which the differentials are put implies

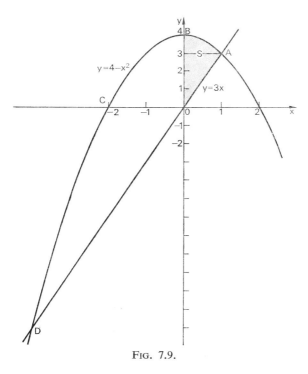

Fig. 7.9.

the order of integration. This is not necessary and here dx usually precedes dy for alphabetic reasons only. However, when both pairs of limits are constants an indication is needed of the variables to which they apply. Here this is given by associating each differential with a separate integral sign.

Problems 7.7

1. Evaluate the integrals in **1–5** of Problems 7.6 with the order of integration reversed.

2. Evaluate $\displaystyle\int_0^3 \int_{\sqrt{(y/3)}}^1 e^{x^2}\, dx\, dy$.   **3.** Evaluate $\displaystyle\int_0^1 \int_x^{2-x} \frac{x}{y}\, dx\, dy$.

4. Sketch the integration field of $\displaystyle\int_0^1 \int_{x+1}^{2+\sqrt{(1-x)}} (1/y)\, dx\, dy$ and evaluate the integral by changing the order of integration.

5. By means of a reversal in the integration order show that

$$\int_0^\infty \int_{\sqrt{x}}^{3\sqrt{x}} \frac{y}{\sqrt{(1+y^4)^5}}\, dx\, dy = 2^2/3^3.$$

6. Express the sum $\displaystyle\int_0^1 \int_0^{\sqrt[3]{y}} F(x, y)\, dx\, dy + \int_1^2 \int_0^{2-y} F(x, y)\, dx\, dy$ as one double integral and evaluate it for $F = 3x^2 + y$.

7.8. TRIPLE INTEGRALS

The process of repeated integration for the evaluation of double integrals is easily extended to triple integrals, for which the function to be integrated is of the form $\mathcal{G}(x, y, z)$ and the region of integration V is a closed volume in space. A triple integral is expresed as $\iiint_V \mathcal{G}(x, y, z)dV$ and may be evaluated as the repeated integral

$$\int_a^b dx \int_{f_1(x)}^{f_2(x)} dy \int_{F_1(x, y)}^{F_2(x, y)} \mathcal{G}(x, y, z)\, dz$$

or in terms of five other arrangements of integrals depending upon the order of integration. The limits of the innermost integral represent surfaces in space and it is evaluated by keeping x and y constant; the result is a function of x and y which leads to a double integral of the form (6.1).

The double integral (6.1) may be expressed as the triple integral

$$\iiint_V dx\,dy\,dz = \int_a^b dx \int_{f_1(x)}^{f_2(x)} dy \int_0^{F(x,y)} dz,$$

for the innermost integral is simply $F(x, y)$ and this becomes the integrand of the second integral to give the usual form of double integral. Thus in this special case where $\mathscr{G}(x, y, z) \equiv 1$ the triple integral evaluates the volume of the region of integration.

Example 8.1.

Evaluate the triple integral of the function $\mathscr{G}(x, y, z) = (1+x+y+z)^{-3}$ throughout the region bounded by the coordinate planes and the plane $x+y+z = 2$.

$$\iiint_V \mathscr{G}(x, y, z)\,dx\,dy\,dz = \int_0^2 dx \int_0^{2-x} dy \int_0^{2-x-y} (1+x+y+z)^{-3}\,dz$$

$$= \int_0^2 dx \int_0^{2-x} \left[-\tfrac{1}{2}(1+x+y+z)^{-2} \right]_0^{2-x-y} dy$$

$$= -\tfrac{1}{2} \int_0^2 dx \int_0^{2-x} \left\{ \tfrac{1}{9} - (1+x+y)^{-2} \right\} dy$$

$$= -\tfrac{1}{2} \int_0^2 \left[y/9 + (1+x+y)^{-1} \right]_0^{2-x} dx$$

$$= -\tfrac{1}{2} \int_0^2 \left\{ (2-x)/9 + \tfrac{1}{3} - (1+x)^{-1} \right\} dx$$

$$= -\tfrac{1}{2} \left[5x/9 - x^2/18 - \log(1+x) \right]_0^2$$

$$= (9 \log 3 - 8)/18, \quad \doteqdot \tfrac{1}{10}.$$

This result could be interpreted as the mass of the region V were it filled by a substance whose density varies from point to point as $\mathscr{J}(x, y, z)$.

Problems 7.8

1. Evaluate $\int\limits_0^1 dx \int\limits_2^3 dy \int\limits_{-1}^1 (x+y+z)\, dz$.

2. Evaluate $\int\limits_{-1}^1 dz \int\limits_0^z dy \int\limits_{y-z}^{y+z} (x+y+z)\, dx$.

3. Evaluate the volume of the region enclosed by the planes $x = 0$, $y = 0$, $z = 1$ and $z = x+y$.

4. Evaluate the mass of a cube which has edge of length a and density equal to the square of the distance from a vertex.

5. Evaluate the volume of the region enclosed in the positive octant by the planes $x+y = 2$, $z = y$ and the surface $z = ye^x$.

6. Evaluate the mass of the region, with density equal to $|xyz|$, enclosed by the paraboloid $9-z = x^2+y^2$ and the plane $z = 0$.

7. Evaluate the mass of the region, with density $|x|$, common to the paraboloids $4-z = x^2+y^2$ and $z = x^2+y^2$.

8. Evaluate $\int\limits_0^4 dw \int\limits_0^{1/w} dx \int\limits_1^{1/wx} dy \int\limits_0^{wx+\log y} \dfrac{w^3 xe^z\, dz}{y}$.

7.9. CHANGE OF VARIABLES IN MULTIPLE INTEGRALS

The variables in terms of which a multiple integral is expressed are often not the most suitable for the evaluation of the repeated integrals and it may be necessary to change to other variables, in particular to those which describe polar coordinate systems.

Consider the double integral $\iint\limits_S F(x, y)\, dS$, where S is a region of the x–y plane, and introduce new variables u, v related to x, y by the functions $x = x(u, v)$ and $y = y(u, v)$. Provided certain conditions are satisfied, these relations yield the inverse functions $u = u(x, y)$ and $v = v(x, y)$; also on area S in the x-y plane is transformable into a corresponding unique area $\bar{S}$ in the u-v plane, and conversely. The transformation of the integrand $F(x, y)$ to a function of u, v is simply $F\{x(u, v), y(u, v)\}$, denoted here by $\bar{F}(u, v)$. The main task is to express dS in terms of u and v.

Figure 7.10 shows a region S of the x-y plane and the corresponding region $\bar{S}$ of the u-v plane. $\bar{S}$ is divided into elemental

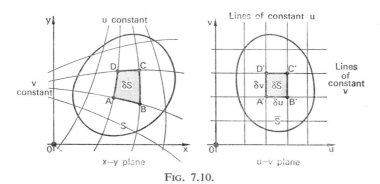

FIG. 7.10.

areas $\delta\bar{S} = \delta u \, \delta v$ by sets of lines on which u is constant and by sets on which v is constant. Through the functional relations between the four variables, the two sets of lines in the u-v plane are transformed into two sets of curves, possibly straight lines, in the x-y plane and $\delta\bar{S}$ is transformed into an elemental area δS. In general the shape of δS is not rectangular and may not be considered as $\delta x \, \delta y$, but this is immaterial because the double integral over S does not depend on the shape of the element δS. The increments δx, δy in x, y due to increments δu, δv in u, v are given approximately by

$$\delta x \doteqdot \frac{\partial x}{\partial u} \, \delta u + \frac{\partial x}{\partial v} \, \delta v \quad \text{and} \quad \delta y \doteqdot \frac{\partial y}{\partial u} \, \delta u + \frac{\partial y}{\partial v} \, \delta v.$$

Let point $A'(u, v)$ be transformed into point $A(x, y)$; then it follows that the points $B'(u+\delta u, v)$, $C'(u+\delta u, v+\delta v)$ and $D'(u, v+\delta v)$ are transformed into points whose positions are given approximately by

$$B\left(x+\frac{\partial x}{\partial u}\delta u, y+\frac{\partial y}{\partial u}\delta u\right),$$

$$C\left(x+\frac{\partial x}{\partial u}\delta u+\frac{\partial x}{\partial v}\delta v, y+\frac{\partial y}{\partial u}\delta u+\frac{\partial y}{\partial v}\delta v\right) \quad \text{and}$$

$$D\left(x+\frac{\partial x}{\partial v}\delta v, y+\frac{\partial y}{\partial v}\delta v\right).$$

As δu and δv both tend to zero, the shape of $ABCD$ approximates more closely to that of a parallelogram and its area is nearly twice that of triangle ABD. Thus the area of δS is approximately equal to the numerical value of

$$\begin{vmatrix} 1 & 1 & 1 \\ x & x+\dfrac{\partial x}{\partial u}\delta u & x+\dfrac{\partial x}{\partial v}\delta v \\ y & y+\dfrac{\partial y}{\partial u}\delta u & y+\dfrac{\partial y}{\partial v}\delta v \end{vmatrix} = \begin{vmatrix} 1 & 0 & 0 \\ x & \dfrac{\partial x}{\partial u}\delta u & \dfrac{\partial x}{\partial v}\delta v \\ y & \dfrac{\partial y}{\partial u}\delta u & \dfrac{\partial y}{\partial v}\delta v \end{vmatrix}$$

$$= \begin{vmatrix} \dfrac{\partial x}{\partial u} & \dfrac{\partial x}{\partial v} \\ \dfrac{\partial y}{\partial u} & \dfrac{\partial y}{\partial v} \end{vmatrix} \delta u\, \delta v.$$

A determinant of this form is denoted by $\partial(x, y)/\partial(u, v)$ or by $J(u, v)$ and is called the Jacobian of (x, y) with respect to (u, v). In the limit, as $\delta u, \delta v \to 0$, $\delta S \to dS = |J|\, du\, dv$ and the transformation of the double integral is

$$\iint_S F(x, y)\, dS = \iint_{\bar{S}} \bar{F}(u, v)\, |J|\, du\, dv. \tag{9.1}$$

By similar reasoning the triple integral $\iiint\limits_{V} \mathcal{G}(x, y, z)\, dV$ evaluated through the volume V in (x, y, z) space may be transformed through the relations $x = x(u, v, w)$, $y = y(u, v, w)$, $z = z(u, v, w)$ and the transformation is given by

$$\iiint\limits_{V} \mathcal{G}(x, y, z)\, dV = \iiint\limits_{\bar{V}} \bar{\mathcal{G}}(u, v, w)\, |J|\, du\, dv\, dw,$$

where $J = \partial(x, y, z)/\partial(u, v, w)$ is the Jacobian of (x, y, z) with respect to (u, v, w).

Jacobians have the important property that if the two variables in each partial derivative are interchanged, so that $\partial x/\partial u$ becomes $\partial u/\partial x$ and so on, then the value of the Jacobian is inverted. Sometimes the value of a Jacobian $J(u, v)$, say, is more easily obtained by inverting the value of

$$J(x, y) = \partial(u, v)/\partial(x, y).$$

Example 9.1.

Sketch the integration field S of the double integral

$$\int\limits_{0}^{1} dx \int\limits_{0}^{1-x} (x+y)\, e^{y/(x+y)}\, dy$$

and obtain its value by means of the transformation $x+y = u$, $y = uv$.

The region S of the x–y plane is shown in Fig. 7.11 together with the region $\bar{S}$ of the u–v plane into which S is transformed. The definition of $\bar{S}$ is derived by the following reasoning.

The transformation is $x+y = u$ (i), $y = uv$ (ii). From (i) and (ii) $x = u(1-v)$ (iii), which implies that when $x = 0$, $u = 0$ or $v = 1$. In other words the line $x = 0$ in the x–y plane is equivalent to the lines $u = 0$, $v = 1$ in the u–v plane. Similarly, (ii) implies that $y = 0$ is equivalent to the lines $u = 0$,

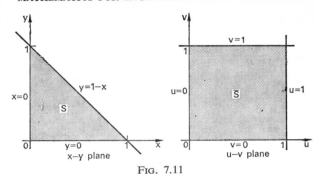

FIG. 7.11

$v = 0$. From (ii) and (iii) the line $y = 1-x$ is equivalent to $u = 1$.

The Jacobian $J(u,v) = \begin{vmatrix} (1-v) & -u \\ v & u \end{vmatrix} = u$. As a point of interest, since $u = x+y$ and $v = y/(x+y)$, $J(x,y) = \begin{vmatrix} 1 & 1 \\ -y/(x+y)^2 & x/(x+y)^2 \end{vmatrix} = 1/(x+y) = 1/u = 1/J(u,v)$ and, in this example, the evaluation of $J(u,v)$ via the inverse of $J(x,y)$ is not easier than the direct method. Conditions for the validity of a transformation are that $J(u,v) \neq 0$ and $J(x,y) \neq 0$ at all points *within* S; also $x = x(u,v)$ and $y = y(u,v)$ must be single valued. Clearly these conditions are satisfied in this example. By (9.1) the integral over $\bar{S}$ is

$$\int_0^1 \int_0^1 u^2 e^v \, du \, dv = \left[\frac{u^3}{3}\right]_0^1 \left[e^v\right]_0^1 = (e-1)/3.$$

Problems 7.9

Evaluate the following multiple integrals, over or through the regions described, by means of the transformations given. In each case sketch the region of integration S in terms of the original variables and the region $\bar{S}$ in terms of the new variables. Verify, as in Example 9.1, that the condition on the Jacobians is satisfied and show that a selected point within S is mapped onto the interior of $\bar{S}$.

1. $\iint\limits_{S} (x^2+y^2) \, dx \, dy.$ S is the square with vertices at the origin, $(1, -1)$, $(2, 0)$ and $(1, 1)$. $u = x+y$, $v = x-y$.

2. $\iint\limits_{S} e^{y/(x+y)} \, dx \, dy.$ S is the triangle with vertices at the origin, $(1, 0)$ and $(0, 1)$. $u = x+y$, $y = uv$.

3. $\iint\limits_{S} e^{(y-x)/(y+x)} \, dx \, dy.$ S is the triangle with vertices at the origin, $(1, 0)$ and $(0, 1)$. $u = x+y$, $v = y-x$.

4. $\iint\limits_{S} \dfrac{(x+y)}{x^2} \, e^{x+y} \, dx \, dy.$ S is the triangle with vertices at the origin, $(2, 0)$ and $(1, 1)$. $u = x+y$, $v = y/x$.

5. $\iint\limits_{S} ye^{-(x+y)} \sin\left\{ \dfrac{\pi y^2}{(x+y)^2} \right\} dx \, dy.$ S is the infinite positive quadrant. $u = x+y$, $v = y$.

6. $\iint\limits_{S} e^{-(x+y)^2} \sinh\left(\dfrac{x}{x+y} \right) dx \, dy.$ S is the infinite positive quadrant. $u = x$, $v = x+y$.

Give the result correct to three decimal places.

7. $\iint\limits_{S} \dfrac{x^2 y}{(x+y)^3} \, dx \, dy.$ S is the triangle with vertices at the origin, $(2, 0)$ and $(0, 2)$. $x = uv$, $y = u(1-v)$.

8. $\iint\limits_{S} \dfrac{1}{\sqrt{\{1+(y-x)^2\}}} \, dx \, dy.$ S is the square with vertices at the origin, $(1, 0)$, $(1, 1)$ and $(0, 1)$. $u = y+x$, $v = y-x$.

9. $\iiint\limits_{S} (x+y+z)^3 xyz \, dx \, dy \, dz.$ S is the tetrahedron formed by the coordinate planes and $x+y+z = 1$.

$$u = x+y+z, \quad uv = y+z, \quad uvw = z.$$

10. $\iiint\limits_{S} e^{(x+y+z)^3} \, dx \, dy \, dz.$ S and the transformation as in problem 9.

7.10. POLAR COORDINATES

The position of a point in the x–y plane is defined by plane polar coordinates (r, θ) through the relations $x = r \cos \theta$, $y = r \sin \theta$ and these give the inverse functions $r = +\sqrt{x^2+y^2}$, $\theta = \tan^{-1} y/x$. Figure 7.12 implies that lines in the $r-\theta$

plane, not shown, on which r is constant are transformed into circles in the x–y plane and that lines on which θ is constant are transformed into rays which meet at the origin. Clearly an element of area δS formed by increments δr, $\delta \theta$ in r, θ is

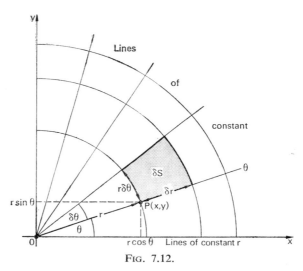

FIG. 7.12.

approximately equal to $r \, \delta r \, \delta \theta$ and, in the limit as δr, $\delta \theta \to 0$, $\delta S \to dS = r \, dr \, d\theta$. This result is implied by the Jacobian

$$J(r, \theta) = \begin{vmatrix} \cos \theta & -r \sin \theta \\ \sin \theta & r \cos \theta \end{vmatrix} = r.$$

Thus
$$\iint\limits_{S} F(x, y) \, dS = \iint\limits_{\bar{S}} \bar{F}(r, \theta) r \, dr \, d\theta. \tag{10.1}$$

Cylindrical polar coordinates (r, θ, z) define the position of a point in space by the plane polar coordinates (r, θ), which give the projection of the point on the x–y plane, and by the coordinate z which measures its distance from the x–y plane. Figure 7.13 shows that an element of volume δV

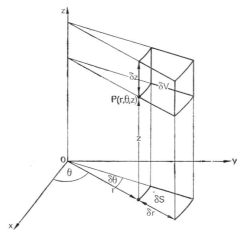

Fɪɢ. 7.13.

formed by increments δr, $\delta\theta$ and δz is given approximately by $\delta V = \delta S\, \delta z \doteq r\, \delta r\, \delta\theta\, \delta z$, whence $dV = r\, dr\, d\theta\, dz$ and

$$\iiint\limits_{V} \mathcal{G}(x, y, z)\, dV = \iiint\limits_{\bar{V}} \bar{\mathcal{G}}(r, \theta, z) r\, dr\, d\theta\, dz. \quad (10.2)$$

Figure 7.14 shows how spherical polar coordinates (r, θ, φ) define the position of a point P in space. The relations between the variables are

$$x = r \sin \theta \cos \varphi, \quad y = r \sin \theta \sin \varphi \quad \text{and} \quad z = r \cos \theta.$$

Note that r is the distance from the origin to P and that θ is measured from the z axis; some conventions retain θ in the x–y plane and measure φ from the z axis. The element of volume δV is approximately equal to $r^2 \sin \theta\, \delta r\, \delta\theta\, \delta\varphi$, whence $dV = r^2 \sin \theta\, dr\, d\theta\, d\varphi$ and

$$\iiint\limits_{V} \mathcal{G}(x, y, z)\, dV = \iiint\limits_{\bar{V}} \bar{\mathcal{G}}(r, \theta, \varphi) r^2 \sin \theta\, dr\, d\theta\, d\varphi. \quad (10.3)$$

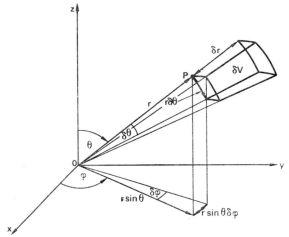

FIG. 7.14.

Example 10.1.

Evaluate the volume common to the cylinder $x^2 + y^2 = b^2$ and the sphere $x^2 + y^2 + z^2 = a^2$, $a > b$.

The field of integration S is enclosed by the circle $x^2 + y^2 = b^2$ in the x–y plane. Due to the relative symmetry of the sphere and cylinder the double integral may be taken over the positive quadrant only; the result multiplied by 8 gives the required volume V.

By (10.1), with $F(x, y) = z$,

$$V = 2 \iint_S z \, dx \, dy = 2 \iint_S \sqrt{(a^2 - x^2 - y^2)} \, dx \, dy$$

$$= 8 \int_0^{\pi/2} \int_0^b \sqrt{(a^2 - r^2)} r \, dr \, d\theta = 8 \left[-\frac{1}{3} (a^2 - r^2)^{3/2} \right]_0^b \left[\theta \right]_0^{\pi/2}$$

$$= \frac{4\pi}{3} \{a^3 - (a^2 - b^2)^{3/2}\}.$$

If $a = b$ then $V = \frac{4}{3}\pi a^3$ is the volume of the sphere.

Example 10.2.

Evaluate the volume of a sphere, radius a, by means of spherical polar coordinates.

Let the sphere be centred at the origin; then by (10.3), with $\mathcal{J}(x, y, z) = 1$,

$$V = \iiint_V dV = \int_0^a r^2\, dr \int_0^\pi \sin \theta\, d\theta \int_0^{2\pi} d\varphi$$

$$= \left[\frac{r^3}{3} \right]_0^a \left[-\cos \theta \right]_0^\pi \left[\varphi \right]_0^{2\pi} = \frac{4\pi a^3}{3}.$$

Problems 7.10

1. Express $\int_{-2}^{2} \int_0^{\sqrt{4-x^2}} (3x - 2y)\, dx\, dy$ in plane polar coordinates and obtain its value. Also evaluate the double integral without changing the variables.

2. Evaluate $\int_0^3 \int_0^{\sqrt{(9-x^2)}} \frac{x}{\sqrt{(x^2+y^2)}}\, dx\, dy$.

3. Evaluate the mass of a circular cylinder, radius a and height $3a$, which has a density equal to the distance from its axis.

4. Evaluate the volume of the region contained by the planes $z = x$, $z = 2x$ and the cylinder $x^2 + y^2 = 2x$.

5. Evaluate the triple integral of the function $1/(x^2 + y^2 + z^2)$ through the region enclosed by the sphere $x^2 + y^2 + z^2 = a^2$.

6. Evaluate the volume of the region common to the sphere $x^2 + y^2 + z^2 = 2$ and the cone $x^2 + y^2 = z^2$.

7. Evaluate the double integral of the function $\log(x^2 + y^2)$ over the field bounded by $x = 0$, $y = x$ and $x^2 + y^2 = 4$ in the positive quadrant.

8. Evaluate the mass of a hemisphere which has unit radius and a density equal to four times the perpendicular distance from the plane face.

9. Evaluate the double integral of the function xy/re^r, where $r^2 = x^2+y^2$, over that part of the circular field $x^2+y^2 = 1$ in the positive quadrant.

10. Evaluate the volume common to the sphere $x^2+y^2+z^2 = 9$ and the circular cylinder $x^2+y^2 = 3y$.

11. Evaluate $\iint\limits_{S} (x^2/\sqrt{x^2+y^2})\, dx\, dy$, where S is the triangular field with vertices at the origin, $(3, 0)$ and $(3, 3)$.

12. Evaluate the triple integral of the function $\sqrt{(x^2+y^2)^{-3}}$ throughout the region enclosed by the plane $z = 0$, the cone $2(x^2+y^2) = z^2$ and the spheres of radii 2 and 4 centred at the origin.

13. Express $\int\limits_{0}^{a} \int\limits_{y}^{a} \{x^2/(x^2+y^2)\}\, dx\, dy$ in plane polar coordinates and obtain its value. Also consider the double integral in its original form and obtain its value by changing the order of integration.

14. The region between the loop of the curve $r = 4\cos 2\theta$, $-\pi/4 \leqslant \theta \leqslant \pi/4$, and the circle $r = 8\cos\theta$ has unit area density. Evaluate its moment of inertia about the axis through the origin normal to the plane of the region.

15. Verify that the Jacobian $\partial(x, y, z)/\partial(r, \theta, \varphi)$ of the spherical polar transformation is $r^2 \sin\theta$.

16. A right circular cone of unit height has semi-vertical angle $60°$. Evaluate the mass of the cone if the density at any point is six times the distance of the point from the vertex.

17. Evaluate the triple integral of the function $\sqrt{(x^2+y^2+z^2)^{-3}}$ through the region between concentric spheres of radii a and $2a$ centred at the origin.

18. Evaluate the volume of the region enclosed between the paraboloid

$$z = x^2+y^2 \quad \text{and the cone} \quad z^2 = x^2+y^2.$$

19. The lemniscate $r^2 = \cos 2\theta$ is projected normally to the x–y plane to form a pair of cylinders. Evaluate the volume of the region common to the cylinders and the sphere of unit radius centred at the origin.

20. Evaluate the volume of the region enclosed by the cylinder $x^2+y^2 = 16$ and the planes $z = 6-x$, $z = 2x-10$.

7.11. SURFACE INTEGRALS

In Fig. 7.15 S is part of a surface $z = F(x, y)$, curved in general, shown projected parallel to $0z$ on to the region R of the x–y plane. A point P on S is projected on to the point $N(x, y)$ in R,

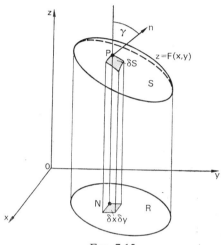

FIG. 7.15.

where an element of area $\delta x \delta y$ is formed from increments δx, δy in x, y. The projection of $\delta x \, \delta y$ on S is the element δS; its projection on the tangent plane to S at P is denoted by $\delta S'$, but this is not shown. Let $\cos \alpha$, $\cos \beta$, $\cos \gamma$ be the direction cosines, defined in section 3 of Chapter 8, of the normal n to S at P, then the area of $\delta S'$ is $\delta x \delta y / |\cos \gamma|$. As δx and δy tend to zero, the ratio $\delta S / \delta S'$ tends to unity so $\delta S'$ becomes $dS' = dS = dxdy / |\cos \gamma|$. The direction cosines ($\cos \alpha$, $\cos \beta$, $\cos \gamma$) of a normal to a surface $z = F(x, y)$ are proportional to $(p, q, -1)$ respectively, where $p = \partial z / \partial x$ and $q = \partial z / \partial y$; see Example 5.3 of Chapter 8. Since $\cos^2 \alpha + \cos^2 \beta + \cos^2 \gamma = 1$ it follows that $1 = k^2(p^2 + q^2 + 1)$, k being the constant of proportionality, and $\cos \gamma = -k = -1/\sqrt{p^2+q^2+1}$ whence $dS = \sqrt{p^2+q^2+1}\ dx\ dy$.

The integral of a function $\mathcal{J}(x, y, z)$ taken over a portion S of one side of a surface $z = F(x, y)$ is called a *surface integral* and is denoted by

$$\iint\limits_{S} \mathcal{J}(x, y, z)\ dS.$$

Such an integral may be evaluated as a repeated integral of the form

$$\iint\limits_{R} \mathcal{J}\{x, y, F(x, y)\}\ \frac{dx\ dy}{|\cos \gamma|}$$

$$= \iint\limits_{R} G(x, y)\ \sqrt{(p^2+q^2+1)}\ dx\ dy, \qquad (11.1)$$

where G is the function to which $\mathcal{J}$ is reduced by substituting $F(x, y)$ for z. The integral could also be evaluated by taking R as the projection of S on the plane $y = 0$ or on $x = 0$. If $\mathcal{J} \equiv 1$ then the integral measures the area of S.

This definition of a surface integral does not depend on a particular side of S because the numerical value only of the direction cosine of the normal is taken. This independence on the side of S is consistent with some applications of surface integrals which may, for example, be concerned with surface areas or with the mass of a lamina whose density varies from point to point as $\mathcal{J}(x, y, z)$.

Example 11.1.

Evaluate the surface area A of a sphere, radius a.

Let the sphere be centred at the origin; then a possible field of integration R is enclosed by the circle $x^2+y^2 = a^2$ on the plane $z = 0$. By (11.1), with $\mathcal{J} = 1$, the surface area is

$$A = 2 \iint_R \sqrt{(p^2 + q^2 + 1)} \, dx \, dy.$$

Since $x^2 + y^2 + z^2 = a^2$, $2zp = -2x$ and $2zq = -2y$ whence $p^2 + q^2 + 1 = x^2/z^2 + y^2/z^2 + 1 = a^2/z^2$.
This gives

$$A = 2a \iint_R \frac{dx \, dy}{z} = 2a \iint_R \frac{dx \, dy}{\sqrt{(a^2 - x^2 - y^2)}}$$

$$= 8a \int_0^{\pi/2} \int_0^a \frac{r \, dr \, d\theta}{\sqrt{(a^2 - r^2)}} = 8a \left[\theta\right]_0^{\pi/2} \left[-\sqrt{(a^2 - r^2)}\right]_0^a = 4\pi a^2.$$

A surface integral may occur in a form such as
$\iint_S \mathcal{J}(x, y, z) \, dx \, dy$.
This is defined to be $\iint_R G(x, y) \, dx \, dy$ on that side of S for which $\cos \gamma$ is positive and to be $-\iint_R G(x, y) \, dx \, dy$ on the side for which $\cos \gamma$ is negative. S may have to be divided into parts on which $\cos \gamma$ has different signs. The normal to a specified side of a surface is drawn from that side immediately into the surrounding space.

Example 11.2.

Evaluate $\iint_S xyz \, dy \, dz$ where S is the surface of the sphere $x^2 + y^2 + z^2 = a^2$ in the positive octant, the required side of S facing the origin.

All over S the direction cosine $\cos \alpha$ of the normal to the required side is negative. The field of integration is in the positive quadrant of the plane $x = 0$, being bounded by $y = 0$,

$z = 0$ and $y^2 + z^2 = a^2$, so the surface integral is

$$\iint_S xyz \, dy \, dz = -\int_0^a z \, dz \int_0^{\sqrt{(a^2-z^2)}} y \sqrt{(a^2 - y^2 - z^2)} \, dy$$

$$= \frac{1}{3} \int_0^a z \left[(a^2 - y^2 - z^2)^{3/2} \right]_0^{\sqrt{(a^2-z^2)}} dz$$

$$= -\frac{1}{3} \int_0^a z(a^2 - z^2)^{3/2} \, dz = \frac{1}{15} \left[(a^2 - z^2)^{5/2} \right]_0^a$$

$$= -\frac{a^5}{15}.$$

In **7** and **8** of Problems 7.11 below, take the side of S which faces away from the origin.

Problems 7.11

1. Evaluate the surface area of a sphere, radius a, not by the method of Example 11.1 but more directly by consideration of an elemental surface area in spherical polar coordinates; see Fig. 7.14.

2. Evaluate $\iint_S \{z/(x^2 - y^2)\} \, dS$, where S is that portion of the surface of a sphere, centred at the origin with radius a, defined by $\pi/3 \leqslant \theta \leqslant \pi/2$, $-\pi/6 \leqslant \varphi \leqslant \pi/6$.

3. Evaluate $\iint_S (xy + yz + xz) \, dS$, where S is the entire surface of the tetrahedron defined in Example 6.1.

4. Evaluate $\iint_S [z/\sqrt{(4x^2 + 4y^2 + 1)^3}] \, dS$, where S is the surface of the paraboloid $4 - z = x^2 + y^2$, $0 \leqslant z \leqslant 4$.

5. Evaluate $\iint_S x(xy - 2yz + 3z) \, dS$, where S is the entire surface of the region, in the positive octant, enclosed by the coordinate planes, the plane $z = b$ and the cylinder $x^2 + y^2 = a^2$.

6. Evaluate $\iint_S x(y \sin z - z \sin y)\, dS$, where S is the entire surface of the cube defined by the coordinate planes and the planes $x = \pi$, $y = \pi$, $z = \pi$.

7. Evaluate $\iint_S \mathcal{I}\, dx\, dz$ and $\iint_S \mathcal{I}\, dx\, dy$, where $\mathcal{I} = xy^2z$ and S is the surface in the positive octant of the paraboloid $9 - z = x^2 + 4y^2$.

8. Evaluate $\iint_S (x + 2y + 3z)\, dx\, dz$, where S is the curved surface of the hemisphere $x^2 + y^2 + z^2 = a^2$, $0 \leqslant y \leqslant a$. Explain why the term $(x + 3z)$ in the integrand contributes nothing to the result.

9. Evaluate $\iint_S \{(x + y + z)\, dy\, dz + 2xy\, dx\, dz - (3x + z)\, dx\, dy\}$, where S is the entire outer surface of the tetrahedron formed by the coordinate planes and the plane $z = 2(1 - x - y)$.

10. Evaluate $\iint_S \{xe^x(y - z)\, dy\, dz + (x^2 + y^2 + z^2)\, dx\, dz - 3xy \sinh z\, dx\, dy\}$ where S is the entire outer surface of the cube defined by the coordinate planes and the planes $x = 1$, $y = 1$, $z = 1$.

7.12. INTEGRAL THEOREMS

The four types of integral defined in this chapter are related to one another by the three principal theorems indicated in the following scheme.

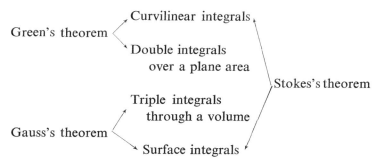

The theorems are stated here without proofs, though these are straightforward and are given in many texts. The functions P, Q, R and their first derivatives, which appear in the integrals, are continuous over the relevant regions and on their boundaries.

GREEN'S THEOREM. *If $P(x, y)$, $Q(x, y)$ are functions defined within and on the boundary C of a closed plane region S then*

$$\iint_S \left(\frac{\partial Q}{\partial x} - \frac{\partial P}{\partial y} \right) dx\,dy = \oint_C (P\,dx + Q\,dy), \qquad (12.1)$$

where the line integral is evaluated in the positive sense.

The theorem applies to cases where C is cut by lines parallel to the axes in more than two points and to the case where closed regions which are not part of S lie inside C.

The theorem shows that the line integral round every closed path in the plane is zero if and only if $\partial P/\partial y = \partial Q/\partial x$. This is the condition for $P\,dx + Q\,dy$ to be an exact differential and is in accordance with the results of section 4. As a special case let $P = -y$ and $Q = x$, then the theorem gives $2\iint_S dx\,dy = \oint_C (-y\,dx + x\,dy)$, showing that the area within C is $\frac{1}{2} \oint_C (x\,dy - y\,dx)$.

Example 12.1.

Evaluate $\oint_C (y^2\,dx - x\,dy)$ by means of Green's theorem, where C is the boundary of the square defined by $x = 0$, $x = 2$, $y = 0$ and $y = 2$.

The equivalent double integral is

$$\int_0^2 \int_0^2 (-1 - 2y)\,dx\,dy = -\int_0^2 dx \int_0^2 (2y + 1)\,dy$$

$$= -\left[x \right]_0^2 \left[y^2 + y \right]_0^2 = -12.$$

The student should verify that direct evaluation of the line integral gives the same result.

GAUSS'S THEOREM. *If P, Q, R are each functions of (x, y, z) defined inside and on the surface S which encloses a region V then*

$$\iiint_V \left(\frac{\partial P}{\partial x} + \frac{\partial Q}{\partial y} + \frac{\partial R}{\partial z} \right) dx\, dy\, dz$$

$$= \iint_S (P\, dy\, dz + Q\, dz\, dx + R\, dx\, dy), \tag{12.2}$$

where the surface integral is evaluated on the outside of S.

Example 12.2.

By means of Gauss's theorem evaluate

$$\iint_S (2xy^2\, dy\, dz + y^3\, dz\, dx - y^2z\, dx\, dy),$$

where S is the closed surface formed from the circular cylinder $x^2 + y^2 = a^2$ and the discs which the cylinder intercepts on the planes $z = 0$, $z = a$.

The equivalent triple integral is

$$I = \iiint_V (2y^2 + 3y^2 - y^2)\, dx\, dy\, dz = 4 \iiint_V y^2\, dx\, dy\, dz.$$

In cylindrical polars

$$I = 4 \int_0^a r^3\, dr \int_0^{2\pi} \sin^2 \theta\, d\theta \int_0^a dz$$

$$= \left[r^4 \right]_0^a \tfrac{1}{2} \left[\theta - \tfrac{1}{2} \sin 2\theta \right]_0^{2\pi} \left[z \right]_0^a = \pi a^5.$$

STOKES'S THEOREM. *If P, Q, R are each functions of (x, y, z) defined on an open surface S and on its bounding rim C then*

$$\iint_S \left\{ \left(\frac{\partial R}{\partial y} - \frac{\partial Q}{\partial z} \right) dy\, dz + \left(\frac{\partial P}{\partial z} - \frac{\partial R}{\partial x} \right) dz\, dx \right.$$
$$\left. + \left(\frac{\partial Q}{\partial x} - \frac{\partial P}{\partial y} \right) dx\, dy \right\} = \oint_C (P\, dx + Q\, dy + R\, dz), \quad (12.3)$$

where the sense of description of C is so related to the required side of S that a person traversing C and keeping on the correct side of S has that side to his left.

Green's theorem is the two-dimensional form of Stokes's theorem.

Example 12.3.

By means of Stokes's theorem evaluate

$$\oint_C (y\, dx + z\, dy + x\, dz),$$

where C is the boundary of the triangle with vertices at $(1, 0, 0)$, $(0, 2, 0)$, $(0, 0, 3)$ described clockwise when viewed from the origin, as shown in Fig. 7.16. The equivalent surface

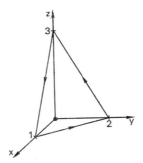

FIG. 7.16.

integral is $I = -\iint\limits_{S} (dy\,dz + dz\,dx + dx\,dy)$. This is the sum of three double integrals each being a surface integral of the type described at the end of section 11, though simplified by having $\mathcal{G} \equiv 1$. Each integral is merely the area of the projection of S on to the relevant coordinate plane.

Thus $\qquad I = -\left(\dfrac{2\times3}{2} + \dfrac{3\times1}{2} + \dfrac{1\times2}{2}\right) = -\dfrac{11}{2}$,

which may be verified by evaluation of the line integral.

The theorems of Gauss and Stokes are very concise when expressed in their vector forms, given in section 8 of the following chapter.

Problems 7.12

1. By means of Green's theorem evaluate the curvilinear integrals in **2, 3, 4, 10** and **12**(b) of Problems 7.3.

2. By means of Gauss's theorem evaluate the surface integrals in **9** and **10** of Problems 7.11.

3. By means of Stokes's theorem evaluate the curvilinear integral in **4** of Problems 7.4.

4. Evaluate $\oint\limits_{C} \{xyz\,dx + (x+y+z)\,dy + (2x - 3y^2 + 4z^3)\,dz\}$, where C lies on the plane $x = 4$ and is the intersection in the positive octant of that plane with the planes $y = 0$, $z = 0$ and the paraboloid $x = y^2 + z^2$, described in the positive sense when viewed from the origin. Also evaluate the curvilinear integral by an equivalent surface integral according to Stokes's theorem. Take the open surface of which C is the boundary to be (a) that part of the paraboloid defined by $0 \leqslant x \leqslant 4$, in the positive octant, and sections of the planes $y = 0$, $z = 0$ and (b) the region enclosed by C on the plane $x = 4$.

5. Evaluate $\oint\limits_{C} \{xe^y z^2(2\,dx + x\,dy) - (x^2 + y)\,dz\}$, where C is the intersection of the ellipsoid $x^2 + y^2/4 + z^2/9 = 1$ and the plane $y = 0$, described in the positive sense when viewed from a point on the positive y axis. Verify the value of the line integral by Stokes's theorem, taking the open surface related to C to be that half of the ellipsoid defined by $y \geqslant 0$. Awkward integrals which may arise should not be evaluated by formal integration but by consideration of the integrands in relation to the fields of integration and their symmetry. Also evaluate the line integral by Green's theorem applied to the region enclosed by C on the plane $y = 0$.

CHAPTER 8

Vector Analysis

8.1. INTRODUCTION

The analysis and solution of many problems which arise in physics and engineering, as well as in more abstract fields, are achieved more easily by using the concept of a vector, that is a quantity which has a direction associated with it. The construction of a mathematical system which permits both magnitudes and directions to be considered at the same time may seem strange and somewhat unreal to the student at first but, after some practice and careful thought, he should come to realise that vector analysis is a natural means of expression for many problems because its conciseness allows important relationships between various quantities to be kept distinct. In this way the mathematical expressions themselves almost form a picture of the physical ideas they describe and represent.

To attempt a reasonable review of vector analysis in one chapter is a formidable task. The three sections which follow this introduction give the basic ideas needed for an algebra of vectors and an indication through examples and problems of several applications. Probably the greatest power of vector analysis lies in its application to the theory of fields, electromagnetic, gravitational and hydromechanical for example. For this type of analysis the methods of the differential and integral calculus are applied to the basic vector concepts and a vector is considered more as a continuous function of position in space rather than as a discrete quantity. The rest of the chapter deals with this aspect, the true vector analysis, but the treatment is concerned with a proper understanding of fundamental ideas rather than with applications.

8.2. DEFINITIONS

Quantities that possess magnitude only are completely defined by a real number and the unit to which the number refers. Such quantities are called *scalars* and familiar examples are volume, mass, temperature, charge and electric potential, as well as numbers themselves which are not necessarily related to physical concepts. Scalars are combined in various ways by the familiar operations of elementary algebra.

Many quantities require for their definition a real number and a direction. Provided they satisfy certain conditions such quantities are called *vectors* and examples are velocity, acceleration, electric and magnetic intensity. Vectors are combined according to the laws of vector algebra, which take account of the directional property in ways which are consistent with how directed quantities combine in practice. Although vectors are said to be "added" and "multiplied" the processes are not simply those of scalar algebra, from which the terminology and much of the notation comes, and care should be taken not to identify familiar ideas with the new ones of vector algebra.

A vector may be represented by or considered as a directed line segment AB, shown in Fig. 8.1. The vector which the segment represents may be denoted by $\overrightarrow{AB}$ or by a single letter in boldface type, such as $\mathbf{v}$; in manuscript a vector quantity is indicated by underlining a letter, as in $\underline{v}$. The end points A and B of the segment shown are called the initial and terminal points respectively. The magnitude of $\mathbf{v}$, always positive, is indicated by the length of the segment and may be denoted in various ways, such as by $\left|\overrightarrow{AB}\right|$, $|\mathbf{v}|$ or v. The direction of $\mathbf{v}$ is defined in relation to a frame of reference, which is taken here to be a rectangular Cartesian system for which the x, y and z axes form a right-handed set. An essential feature of a vector is its independence of any coordinate system.

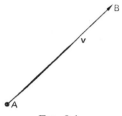

FIG. 8.1.

A vector which has unit magnitude is called a *unit vector*. The letters **i, j, k** are reserved for unit vectors which are parallel to and in the positive sense of the x, y, z axes respectively. A unit vector is often denoted by a circumflex, as in $\hat{\mathbf{v}}$. The *null* or *zero vector* is denoted by **0** and is defined to have zero magnitude and no specific direction.

Vectors **a** and **b** are said to be *equal* if their magnitudes are equal and if they have the same direction. Equality of vectors is expressed by **a** = **b** and inequality by **a** ≠ **b**. A vector which has the same magnitude as but opposite direction to **a** is denoted by −**a**. For most vectors the definition of equality means that equal vectors need not have the same initial point and, further, that a vector may be translated, that is displaced without rotation, and yet still may be considered as the same vector. Vectors which can be treated in this way are called *free vectors*. Some vectors are required to act along a given line and are called *line vectors*; equality of such vectors includes this extra condition. Vectors which have fixed initial points are called *position vectors* and, by means of their terminal points, they define the position of a point relative to the fixed point. When the location of the initial point of a position vector is not specified it may be assumed to be at the origin. The nature of a vector should be apparent from the context in which it appears.

The product of a vector **v** and a scalar c is a vector c**v** with magnitude $|c|\,v$ and with direction either the same as or opposite to that of **v** according as c is positive or negative; if $c = 0$ then c**v** is the null vector. Thus any vector **v** may be expressed as the product of its magnitude v and a unit vector $\hat{\mathbf{v}}$ in the same direction: $\mathbf{v} = v\hat{\mathbf{v}}$. The multiplication of vectors by scalars obeys the rules of scalar algebra.

A plane area of S sq. units may be represented by a *vector area* **S** of magnitude S and direction $\hat{\mathbf{n}}$ normal to the area. The sign of a vector area depends on the sense in which the boundary of the area is described when viewed from an external point such as P, shown in Fig. 8.2. If the sense of descrip-

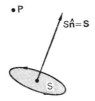

Fig. 8.2.

tion is anticlockwise then the positive direction of the vector area is that along which a right-handed screw would be translated by a rotation in the same sense.

Figure 8.3 shows a point $P(x, y, z)$ in three-dimensional space defined by the position vector $\overrightarrow{OP}$ relative to the origin O of the reference frame. The position vector $\overrightarrow{OP}$ is denoted more often by the radius vector **r**, which is reserved here entirely for this purpose. The coordinates (x, y, z) of P are called the (scalar) *components* of its position vector **r** and these are the orthogonal projections, with regard to sign, of OP on to the coordinate axes. The corresponding *vector components* of **r** are x**i**, y**j** and z**k**.

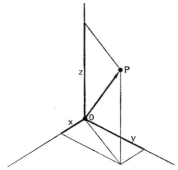

FIG. 8.3.

More generally Fig. 8.4 shows a vector $\mathbf{a} = \overrightarrow{AB}$ with end points A, B defined by (x_1, y_1, z_1), (x_2, y_2, z_2) respectively. The components of $\mathbf{a}$ with respect to the coordinate system are the scalars

$$a_x = x_2 - x_1, \, a_y = y_2 - y_1, \, a_z = z_2 - z_1. \tag{2.1}$$

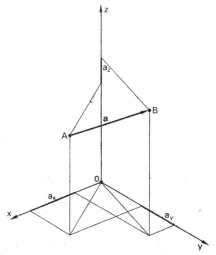

FIG. 8.4.

The corresponding vector components of $\mathbf{a}$ are $a_x\mathbf{i}$, $a_y\mathbf{j}$ and $a_z\mathbf{k}$. By definition the magnitude of $\mathbf{a}$ is the length AB which, by (2.1) and the theorem of Pythagoras, is given by

$$|\mathbf{a}| = \sqrt{a_x^2 + a_y^2 + a_z^2}. \tag{2.2}$$

The components of $\mathbf{a}$ are independent of the initial point A, for if $\mathbf{a}$ is translated then corresponding coordinates of A and B are changed by the same amount. In particular if $\mathbf{a}$ is translated so that A moves to the origin O then $a_x = x$, $a_y = y$, $a_z = z$ and $|\mathbf{a}| = \sqrt{x^2 + y^2 + z^2}$, where (x, y, z) are the coordinates of B. Two vectors $\mathbf{a}$ and $\mathbf{b}$ are equal if and only if their corresponding components are equal. Thus the equation $\mathbf{a} = \mathbf{b}$ is equivalent to the three conditions $a_x = b_x$, $a_y = b_y$, $a_z = b_z$.

8.3. ADDITION OF VECTORS

The *sum* or resultant of two vectors $\mathbf{a}$ and $\mathbf{b}$, denoted by $\mathbf{a}+\mathbf{b}$, is the vector formed between the initial point of $\mathbf{a}$ and the terminal point of $\mathbf{b}$ when $\mathbf{b}$ has been translated so that its initial point is at the terminal point of $\mathbf{a}$, as shown in Fig. 8.5. Clearly the same result is obtained by placing the initial point of $\mathbf{a}$ on the terminal point of $\mathbf{b}$; each way of forming the sum is illustrated by the parallelogram.

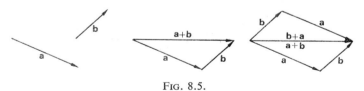

FIG. 8.5.

Thus vector addition is defined in accordance with the familiar parallelogram law of forces and is consistent with the way in which displacements are compounded. Some directed quantities do not conform with this law and therefore are not definable as vectors. It follows from the definition that vector addition is commutative and associative, that is

$$\mathbf{a}+\mathbf{b} = \mathbf{b}+\mathbf{a} \tag{3.1}$$

and $$(\mathbf{a}+\mathbf{b})+\mathbf{c} = \mathbf{a}+(\mathbf{b}+\mathbf{c}). \tag{3.2}$$

The difference of two vectors **a** and **b**, denoted by $\mathbf{a}-\mathbf{b}$, is defined as the sum $\mathbf{a}+(-\mathbf{b})$ and is illustrated in Fig. 8.6.

By the definitions of vector equality and addition a vector **a** is given in terms of its vector components by

$$\mathbf{a} = a_x\mathbf{i}+a_y\mathbf{j}+a_z\mathbf{k}, \tag{3.3}$$

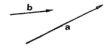

FIG. 8.6.

as shown in Fig. 8.7. The radius vector **r** is given by

$$\mathbf{r} = x\mathbf{i}+y\mathbf{j}+z\mathbf{k}. \tag{3.4}$$

The sum of two vectors is derived by adding their respective vector components.

$$\mathbf{a}+\mathbf{b} = (a_x+b_x)\mathbf{i}+(a_y+b_y)\mathbf{j}+(a_z+b_z)\mathbf{k}. \tag{3.5}$$

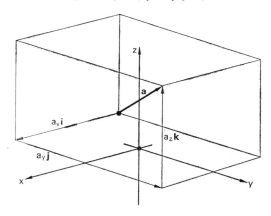

FIG. 8.7.

Let the direction of a vector **a** make angles α, β, γ with the positive directions of the x, y, z axes respectively. The cosines of these angles are called the direction cosines of **a** and are given by

$$\cos \alpha = a_x/a, \quad \cos \beta = a_y/a, \quad \cos \gamma = a_z/a; \quad (3.6)$$

by (2.2) the sum of their squares is unity:

$$\cos^2 \alpha + \cos^2 \beta + \cos^2 \gamma = 1. \quad (3.7)$$

If **a** is a unit vector then $a = 1$ and the direction cosines are equal to the components of $\hat{\mathbf{a}}$.

Let a vector **a** make an angle θ with a vector **b**. the quantity $a_b = a \cos \theta$ is called the component of **a** in the direction of **b** and is the orthogonal projection of **a** on the line along **b**, signed according to the relative senses of **a** and **b**. The component a_b is now expressed in terms of the components of **a**.

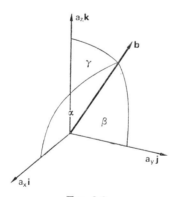

FIG. 8.8.

Figure 8.8 shows the vector components of **a**, parallel to the coordinate axes, and vector **b** with the angles α, β, γ which its direction makes with the axes. Clearly the components of $a_x\mathbf{i}$, $a_y\mathbf{j}$ and $a_z\mathbf{k}$ in the direction of **b** are $a_x \cos \alpha$, $a_y \cos \beta$ and

$a_z \cos \gamma$ respectively, so the total component a_b of **a** is the sum of the partial components:

$$a_b = a_x \cos \alpha + a_y \cos \beta + a_z \cos \gamma. \qquad (3.8)$$

Figure 8.9 shows an unlimited straight line L which is parallel to a vector **b**. The position vector of a fixed point A on the line is **a** whilst **r** is the position vector of any point P on the line.

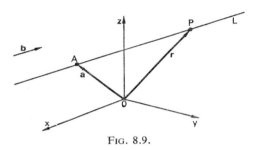

FIG. 8.9.

The vector $\overrightarrow{AP}$ can be represented by $t\mathbf{b}$, where t is a scalar of unlimited range. The vector equation of the line is

$$\mathbf{r} = \mathbf{a} + t\mathbf{b} \qquad (3.9)$$

for, as t varies, the locus of P is the line through A parallel to **b**

Problems 8.3

1. Evaluate the magnitude of the vector whose scalar components are (a) (3, 0, 4), (b) (3, −2, 6).

2. $ABCD$ is a parallelogram and the position vectors of A, B, C are **a, b, c**. Obtain the position vector of D in terms of the vectors given.

3. Show by means of a sketch that $(\mathbf{a}+\mathbf{b})+\mathbf{c} = \mathbf{a}+(\mathbf{b}+\mathbf{c})$.

4. Given $\mathbf{a} = \mathbf{i}-2\mathbf{j}+3\mathbf{k}$, $\mathbf{b} = 4\mathbf{i}+\mathbf{j}-3\mathbf{k}$ and $\mathbf{c} = 2\mathbf{i}-\mathbf{j}-\mathbf{k}$, evaluate the magnitudes of (a) $\mathbf{a}+\mathbf{b}+\mathbf{c}$, (b) $2\mathbf{a}-\mathbf{b}+3\mathbf{c}$, (c) $3\mathbf{a}+2\mathbf{b}-\mathbf{c}$.

5. Obtain a unit vector which is parallel to and in the same sense as the sum of $\mathbf{a} = 4\mathbf{i}+3\mathbf{j}+3\mathbf{k}$ and $\mathbf{b} = -\mathbf{i}+3\mathbf{j}-5\mathbf{k}$.

6. Show that the directed line segments $\mathbf{a} = 3\mathbf{i} - 4\mathbf{j} + \mathbf{k}$, $\mathbf{b} = 2\mathbf{i} + 3\mathbf{j} - 5\mathbf{k}$ and $\mathbf{c} = \mathbf{i} - 7\mathbf{j} + 6\mathbf{k}$ can form the sides of a triangle and evaluate the lengths of its medians. Show that the vectors which represent the medians can form a triangle.

7. Obtain in the form $v\hat{\mathbf{v}}$ the vector $\mathbf{v}$ which has its initial point at $(1, 2, 3)$ and its terminal point at $(-3, -2, -1)$.

8. Obtain the direction cosines of the vector $\overrightarrow{PQ}$ which joins $P(2, 1, 4)$ to $Q(5, -3, 2)$.

9. Evaluate the orthogonal projection of $\mathbf{a} = \mathbf{i} + 2\mathbf{j} - 4\mathbf{k}$ on $\mathbf{b} = 6\mathbf{i} - 2\mathbf{j} - 3\mathbf{k}$ and find the angle between $\mathbf{a}$ and $\mathbf{b}$.

10. Obtain the vector equation to the straight line which passes through the point $P(1, 1, 1)$ and is (a) parallel to the z axis, (b) parallel to the position vector of P, (c) parallel to the difference between $5\mathbf{i} + \mathbf{j} - 3\mathbf{k}$ and $2\mathbf{i} + 3\mathbf{j} - \mathbf{k}$.

8.4. PRODUCTS OF VECTORS

Vectors are combined in the product sense in two distinct ways which are consistent with how certain combinations of directed quantities occur in practice.

SCALAR PRODUCT

The scalar or dot product of two vectors $\mathbf{a}$ and $\mathbf{b}$, written as $\mathbf{a} \cdot \mathbf{b}$, is a *scalar* quantity defined as the product of the magnitudes of the vectors and the cosine of the angle θ $(\leqslant \pi)$ between their directions. Since $\cos \theta = \cos (-\theta)$, the sense in which θ is measured is immaterial and, as $ab = ba$, the product $\mathbf{a} \cdot \mathbf{b}$ is commutative.

$$\mathbf{a} \cdot \mathbf{b} = ab \cos \theta = \mathbf{b} \cdot \mathbf{a}. \qquad (4.1)$$

Figure 8.10 shows how $\mathbf{a} \cdot \mathbf{b}$ may be considered as the product of a with the component of b in the direction of $\mathbf{a}$, or as the product of b with the component of a in the direction of $\mathbf{b}$. The two vectors are shown with their initial points coincident but, by the definition of vector equality, they need not be coincident. If $\mathbf{a}$ and $\mathbf{b}$ are parallel and in the same sense then $\mathbf{a} \cdot \mathbf{b} = ab$, but if in the opposite sense then $\mathbf{a} \cdot \mathbf{b} = -ab$. The scalar pro-

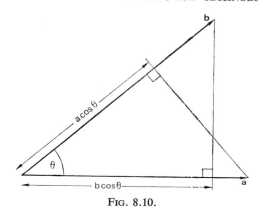

FIG. 8.10.

duct of a vector **a** with itself is called the square of **a**, or its self-product, and is written as $\mathbf{a}^2$.

Thus $$\mathbf{a} \cdot \mathbf{a} = \mathbf{a}^2 = a^2, \tag{4.2}$$

whence it follows that the square of a unit vector is unity and, in particular, that

$$\mathbf{i} \cdot \mathbf{i} = \mathbf{j} \cdot \mathbf{j} = \mathbf{k} \cdot \mathbf{k} = 1. \tag{4.3}$$

The definition of the scalar product implies that two non-zero vectors are perpendicular if and only if their scalar product is zero. Since the unit vectors **i**, **j**, **k** are perpendicular,

$$\mathbf{i} \cdot \mathbf{j} = \mathbf{j} \cdot \mathbf{k} = \mathbf{k} \cdot \mathbf{i} = 0. \tag{4.4}$$

The distributive law of multiplication applies to scalar products, thus

$$\mathbf{a} \cdot (\mathbf{b} + \mathbf{c}) = \mathbf{a} \cdot \mathbf{b} + \mathbf{a} \cdot \mathbf{c}. \tag{4.5}$$

By results (4.3, 4, 5) $\mathbf{a} \cdot \mathbf{b}$ in terms of the components of **a** and **b** is

$$\mathbf{a} \cdot \mathbf{b} = a_x b_x + a_y b_y + a_z b_z. \tag{4.6}$$

From this and (4.1) $\cos \theta$ is given by the useful form

$$\cos \theta = (a_x b_x + a_y b_y + a_z b_z)/ab. \tag{4.7}$$

If **b** is a unit vector then (4.6) is equivalent to (3.8).

Example 4.1.

Find the projection of $\mathbf{a} = \mathbf{i} + 2\mathbf{j} + \mathbf{k}$ on $\mathbf{b} = 4\mathbf{i} - 4\mathbf{j} + 7\mathbf{k}$. Figure 8.10 shows that the projection is $a \cos \theta$, which is given by $\mathbf{a} \cdot \hat{\mathbf{b}} = (\mathbf{i} + 2\mathbf{j} + \mathbf{k}) \cdot (4\mathbf{i} - 4\mathbf{j} + 7\mathbf{k})/9 = \frac{1}{3}$.

Example 4.2.

Find the work done by a force $\mathbf{F} = 3\mathbf{i} + 2\mathbf{j} - \mathbf{k}$ whose point of application moves along the vector $\mathbf{d} = 2\mathbf{i} - \mathbf{j} - 2\mathbf{k}$.

Work = (magnitude of force in direction of motion) (distance moved)

$$= (F \cos \theta)(d) = \mathbf{F} \cdot \mathbf{d} = (3\mathbf{i} + 2\mathbf{j} - \mathbf{k}) \cdot (2\mathbf{i} - \mathbf{j} - 2\mathbf{k}) = 6.$$

Example 4.3.

Derive the vector equation of the plane which is perpendicular to $\mathbf{a}$ and which contains the terminal point of $\mathbf{b}$, as shown in Fig. 8.11.

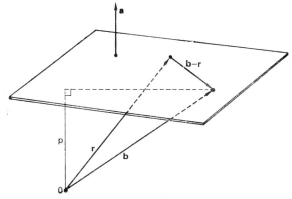

FIG. 8.11.

Let $\mathbf{r}$ be the position vector of any point (x, y, z) in the plane then, since $(\mathbf{b} - \mathbf{r})$ is perpendicular to $\mathbf{a}$, the required equation is

$$(\mathbf{b} - \mathbf{r}) \cdot \mathbf{a} = 0.$$

The perpendicular distance p from the origin to the plane is the projection of $\mathbf{b}$ on the direction of $\mathbf{a}$, that is $\mathbf{b} \cdot \hat{\mathbf{a}}$.

Problems 8.4a

1. Evaluate the projection of the vector $3\mathbf{i} + 4\mathbf{j} - \mathbf{k}$ on the vector with initial point $(1, 2, 3)$ and terminal point $(-2, 1, -4)$.

2. Show, with the aid of a sketch, that $\mathbf{a} \cdot (\mathbf{b} + \mathbf{c}) = \mathbf{a} \cdot \mathbf{b} + \mathbf{a} \cdot \mathbf{c}$.

3. Evaluate the angle between $\mathbf{a} = 3\mathbf{i} - 4\mathbf{j} + \mathbf{k}$ and $\mathbf{b} = 7\mathbf{i} + 3\mathbf{j} - \mathbf{k}$.

4. Obtain a unit vector which is parallel to the x–z plane and is perpendicular to the vector $\overrightarrow{AB}$, where A is $(2, 1, 2)$ and B is $(-3, 4, -1)$.

5. Obtain a unit vector which makes an angle of $60°$ with $(0, 0, 1)$ and is perpendicular to $(1, 0, 0)$, these two points being the terminal points of position vectors referred to the origin.

6. The position vectors of the vertices of a triangle are $\mathbf{0}, \mathbf{a}, \mathbf{b}$. Show that the area S of the triangle is given by $4S^2 = a^2 b^2 - (\mathbf{a} \cdot \mathbf{b})^2$.

7. By means of the scalar product prove, for a plane triangle, the cosine rule in the form $c^2 = a^2 + b^2 - 2ab \cos C$.

8. Evaluate c for the vectors $c\mathbf{i} + 2\mathbf{j} - 3\mathbf{k}$ and $2\mathbf{i} - \mathbf{j} + 2\mathbf{k}$ to be (a) perpendicular, (b) inclined at $60°$. Give the value of c for (b) to the nearest integer.

9. Calculate the work done by the force $\mathbf{F} = 3\mathbf{i} + 7\mathbf{j} - \mathbf{k}$ when its point of application moves along the straight line from $(2, 3, 1)$ to $(6, 8, -1)$.

10. Obtain in terms of x, y and z the equation to the plane which contains the point $(2, 1, 4)$ and is perpendicular to the vector $\mathbf{i} + 2\mathbf{j} + 3\mathbf{k}$. Evaluate the perpendicular distance from the origin to the plane.

VECTOR PRODUCT

The vector or cross product of two vectors $\mathbf{a}$ and $\mathbf{b}$, written as $\mathbf{a} \times \mathbf{b}$, is a vector quantity whose magnitude is defined as the product of the magnitudes of the vectors and the sine of the angle θ ($\leqslant \pi$) between their directions. The direction of $\mathbf{a} \times \mathbf{b}$ is defined to be parallel to a unit vector $\mathbf{n}$ which is perpendicular to the plane of $\mathbf{a}$ and $\mathbf{b}$ in such a sense that $\mathbf{a}$, $\mathbf{b}$, $\hat{\mathbf{n}}$, in this order, form a right-handed set. This simply means that a rotation through θ from $\mathbf{a}$ to $\mathbf{b}$ would drive a right-handed screw in the direction of $\hat{\mathbf{n}}$, as shown in Fig. 8.12. The angle θ is measured

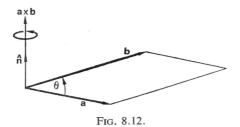

FIG. 8.12.

in the positive sense from the direction of the first vector in the product to that of the second. By the above definition the direction of $\mathbf{b} \times \mathbf{a}$ is such that $\mathbf{b}$, $\mathbf{a}$, $\hat{\mathbf{n}}$ form a right-handed set and this direction is opposite to that of $\mathbf{a} \times \mathbf{b}$, showing that the vector product is not commutative.

$$\mathbf{a} \times \mathbf{b} = ab \sin \theta \hat{\mathbf{n}} = -\mathbf{b} \times \mathbf{a}. \qquad (4.8)$$

Figure 8.12 shows that $|\mathbf{a} \times \mathbf{b}| = ab \sin \theta$ is the area of a parallelogram with $\mathbf{a}$, $\mathbf{b}$ as adjacent sides. The vector product of a vector $\mathbf{a}$ with itself or with $-\mathbf{a}$ is zero; in particular

$$\mathbf{i} \times \mathbf{i} = \mathbf{j} \times \mathbf{j} = \mathbf{k} \times \mathbf{k} = \mathbf{0}. \qquad (4.9)$$

The definition of the vector product implies that two non-zero vectors are parallel, though not necessarily in the same sense, if and only if their vector product is zero. If $\mathbf{a}$ and $\mathbf{b}$ are perpendicular then $|\mathbf{a} \times \mathbf{b}| = ab$, so, for the unit vectors $\mathbf{i}$, $\mathbf{j}$ and $\mathbf{k}$,

$$\mathbf{i} \times \mathbf{j} = \mathbf{k}, \quad \mathbf{j} \times \mathbf{k} = \mathbf{i}, \quad \mathbf{k} \times \mathbf{i} = \mathbf{j}. \qquad (4.10)$$

The distributive law of multiplication applies to vector products, thus

$$\mathbf{a} \times (\mathbf{b} + \mathbf{c}) = \mathbf{a} \times \mathbf{b} + \mathbf{a} \times \mathbf{c}. \tag{4.11}$$

By results (4.9, 10, 11) $\mathbf{a} \times \mathbf{b}$ in terms of the components of $\mathbf{a}$ and $\mathbf{b}$ is

$$\mathbf{a} \times \mathbf{b} = (a_y b_z - a_z b_y)\mathbf{i} + (a_z b_x - a_x b_z)\mathbf{j} + (a_x b_y - a_y b_x)\mathbf{k}, \tag{4.12}$$

which is much easier to use in the determinant form

$$\mathbf{a} \times \mathbf{b} = \begin{vmatrix} \mathbf{i} & \mathbf{j} & \mathbf{k} \\ a_x & a_y & a_z \\ b_x & b_y & b_z \end{vmatrix}. \tag{4.13}$$

Now $|\mathbf{a} \times \mathbf{b}|^2 = \mathbf{a}^2 \mathbf{b}^2 \sin^2 \theta = \mathbf{a}^2 \mathbf{b}^2 (1 - \cos^2 \theta) = \mathbf{a}^2 \mathbf{b}^2 - (\mathbf{a} \cdot \mathbf{b})^2$, so that

$$|\mathbf{a} \times \mathbf{b}| = \sqrt{\mathbf{a}^2 \mathbf{b}^2 - (\mathbf{a} \cdot \mathbf{b})^2} \tag{4.14}$$

which is an easy form to use when the magnitude only of a vector product is required.

Example 4.4.

Evaluate the area of the triangle with vertices at

$$A(1, 2, 3), \quad B(-2, 1, -4), \quad C(-3, -2, -2).$$
$$\overrightarrow{AB} = \overrightarrow{OB} - \overrightarrow{OA} = -3\mathbf{i} - \mathbf{j} - 7\mathbf{k}, \quad = \mathbf{a}.$$
$$\overrightarrow{AC} = \overrightarrow{OC} - \overrightarrow{OA} = -4\mathbf{i} - 4\mathbf{j} - 5\mathbf{k}, \quad = \mathbf{b}.$$

The area of $\varDelta ABC$ is

$$\tfrac{1}{2}|\mathbf{a} \times \mathbf{b}| = \tfrac{1}{2}\sqrt{(59)(57) - (51)^2} = \sqrt{(762)}/2, \quad \text{by (4.14)}.$$

Example 4.5.

Obtain an expression for the moment m of a force $\mathbf{F}$ about a point A, shown in Fig. 8.13, where B is the point of application and L the line of action.

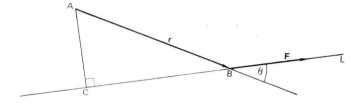

Fig. 8.13.

Let $\mathbf{r}$ be the position vector of B relative to A and let θ be the angle between $\mathbf{r}$ and $\mathbf{F}$. The moment is defined by

$$m = F.AC = F.AB \sin \theta = Fr \sin \theta = |\mathbf{r} \times \mathbf{F}|.$$

The vector $\mathbf{m} = \mathbf{r} \times \mathbf{F}$ is called the vector moment of $\mathbf{F}$ about A and its direction is parallel to the axis about which the turning effect occurs.

Problems 8.4b

1. Given $\mathbf{a} = 2\mathbf{i} - 3\mathbf{j} + \mathbf{k}$ and $\mathbf{b} = \mathbf{i} + 2\mathbf{j} - 2\mathbf{k}$ obtain $\mathbf{a} \times \mathbf{b}$ and $\mathbf{b} \times \mathbf{a}$. From either result obtain $(\mathbf{a} - \mathbf{b}) \times (\mathbf{a} + \mathbf{b})$.

2. Evaluate the area of the triangle which has vertices at the origin, $(1, 2, 3)$ and $(-3, -2, -1)$.

3. Obtain a unit vector which is perpendicular to the plane containing $\mathbf{r}_1 = 3\mathbf{i} + 6\mathbf{j} - 2\mathbf{k}$ and $\mathbf{r}_2 = \mathbf{i} - 3\mathbf{j} - 4\mathbf{k}$.

4. Use the scalar and vector products of $\mathbf{i} \cos \theta + \mathbf{j} \sin \theta$ with $\mathbf{i} \cos \varphi + \mathbf{j} \sin \varphi$ to show that $\cos (\theta - \varphi) = \cos \theta \cos \varphi + \sin \theta \sin \varphi$ and $\sin (\theta - \varphi) = \sin \theta \cos \varphi - \cos \theta \sin \varphi$.

5. Obtain a unit vector which is normal to the plane containing the points $(1, 3, 5)$, $(-2, 0, 6)$ and $(2, 2, 0)$.

6. The terminal points of the position vectors $\mathbf{a}, \mathbf{b}, \mathbf{c}$ define the vertices of a triangle. Show that the area of the triangle is $\frac{1}{2} |\mathbf{a} \times \mathbf{b} + \mathbf{b} \times \mathbf{c} + \mathbf{c} \times \mathbf{a}|$ and deduce a condition for three points to be collinear.

7. Let each face of a tetrahedron be associated with a vector whose magnitude is the area of the face and whose direction is parallel to an outward drawn normal to that face. Show that the sum of the four vectors is the null vector.

8. Given the vectors $\mathbf{a} = \mathbf{i} + 2\mathbf{j} + 3\mathbf{k}$, $\mathbf{b} = -2\mathbf{i} + \mathbf{j} + 4\mathbf{k}$ and $\mathbf{c} = 3\mathbf{i} - \mathbf{j} + \mathbf{k}$ obtain (a) $\mathbf{a} \times (\mathbf{b} \times \mathbf{c})$, (b) $(\mathbf{a} \times \mathbf{b}) \times \mathbf{c}$, (c) $(\mathbf{c} \times \mathbf{a}) \cdot (\mathbf{b} \times \mathbf{a})$, (d) $\{\mathbf{b} \times (\mathbf{b} \times \mathbf{c})\} \cdot \mathbf{a}$.

SCALAR TRIPLE PRODUCT

Products of vectors that have three or more factors occur in applications, the most important being the scalar triple product

$$
\mathbf{a} \cdot (\mathbf{b} \times \mathbf{c}) = (a_x \mathbf{i} + a_y \mathbf{j} + a_z \mathbf{k}) \cdot \begin{vmatrix} \mathbf{i} & \mathbf{j} & \mathbf{k} \\ b_x & b_y & b_z \\ c_x & c_y & c_z \end{vmatrix},
$$
$$
= \begin{vmatrix} a_x & a_y & a_z \\ b_x & b_y & b_z \\ c_x & c_y & c_z \end{vmatrix} \tag{4.15}
$$

by the form of (4.6). As the interchange of two pairs of rows in a determinant does not affect its value, the following relations hold.

$$
\mathbf{a} \cdot (\mathbf{b} \times \mathbf{c}) = \mathbf{b} \cdot (\mathbf{c} \times \mathbf{a}) = \mathbf{c} \cdot (\mathbf{a} \times \mathbf{b})
$$

and, as a scalar product is commutative,

$$
(\mathbf{b} \times \mathbf{c}) \cdot \mathbf{a} = (\mathbf{c} \times \mathbf{a}) \cdot \mathbf{b} = (\mathbf{a} \times \mathbf{b}) \cdot \mathbf{c}.
$$

Thus a scalar triple product has six equivalent forms, in which cyclic order of the factors is preserved, and these show that the dot and cross may be interchanged. For this reason the product $\mathbf{a} \cdot (\mathbf{b} \times \mathbf{c})$ is denoted also by $[\mathbf{a} \ \mathbf{b} \ \mathbf{c}]$.

Figure 8.14 shows how $\mathbf{a} \cdot (\mathbf{b} \times \mathbf{c})$ measures the volume of a parallelepiped which has $\mathbf{a}, \mathbf{b}, \mathbf{c}$ for its edges. The volume is the product of a base area and the perpendicular distance h to the surface above, that is

$$
|\mathbf{b} \times \mathbf{c}| \, h = |\mathbf{b} \times \mathbf{c}| \, \mathbf{a} \cdot \frac{(\mathbf{b} \times \mathbf{c})}{|\mathbf{b} \times \mathbf{c}|} = \mathbf{a} \cdot (\mathbf{b} \times \mathbf{c}).
$$

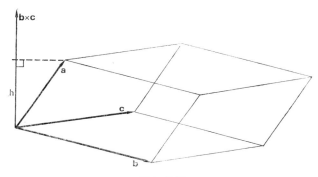

FIG. 8.14.

The volume is zero if **a**, **b**, **c** lie in the same plane, so the condition for three non-zero vectors to be coplanar is that any one of their scalar triple products should vanish.

VECTOR TRIPLE PRODUCT

This has the form $\mathbf{a} \times (\mathbf{b} \times \mathbf{c})$ and, since $(\mathbf{b} \times \mathbf{c})$ is normal to the plane containing **b** and **c**, its direction is in the plane of **b** and **c**. A useful expansion is

$$\mathbf{a} \times (\mathbf{b} \times \mathbf{c}) = (\mathbf{a} \cdot \mathbf{c})\mathbf{b} - (\mathbf{a} \cdot \mathbf{b})\mathbf{c}, \qquad (4.16)$$

but note that

$$(\mathbf{a} \times \mathbf{b}) \times \mathbf{c} = (\mathbf{a} \cdot \mathbf{c})\mathbf{b} - (\mathbf{b} \cdot \mathbf{c})\mathbf{a}.$$

Problems 8.4c

1. Determine the value of a for the three vectors with components $(3, 4, -5)$, $(1, -2, 3)$ and $(3a, 2, -1)$ to be coplanar.

2. By forming a suitable scalar triple product solve for x the equation $\mathbf{a}x + \mathbf{b}y + \mathbf{c}z = \mathbf{d}$, where the four vectors are constant and **a**, **b**, **c** are not coplanar.

3. Obtain the vector **v** which satisfies the conditions

$$\mathbf{a} \times \mathbf{v} + \mathbf{b} = \mathbf{0}, \quad \mathbf{c} \cdot \mathbf{v} = s,$$

where s is a scalar and $\mathbf{a} \cdot \mathbf{c} \neq 0$.

4. Given that $a \times b = c \times b$, $a \cdot d = 0$ and $b \cdot d \neq 0$, obtain an expression for a in terms of b, c and d.

5. Manipulate the equation $v \times a + sv = b$ so that v becomes the subject, the vectors a, b and the scalar s being non-zero. (*Hint*: Form separately the scalar product and the vector product of a with the equation and combine these with the equation itself.)

8.5. DIFFERENTIATION OF VECTORS

When a vector v is a function of a scalar variable t the rate of change of v with respect to t is called the derivative, denoted by dv/dt, and is defined by the limit

$$\frac{dv}{dt} = \lim_{\delta t \to 0} \frac{v(t + \delta t) - v(t)}{\delta t}, \tag{5.1}$$

which is a vector. Derivatives of higher order are defined in a similar way.

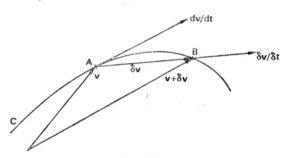

Fig. 8.15.

The meaning of the derivative is illustrated in Fig. 8.15 which shows $v(t)$ and $v(t + \delta t)$, $= v + \delta v$, related by the law of vector addition. As t changes, the terminal point of v describes a curve C. The average rate of change of v with t between any two points A, B on C is $\delta v/\delta t$ and its direction is along $\overline{AB}$. Definition (5.1) shows that the direction of the derivative at A is parallel to the tangent to C at A.

In general the components of $\mathbf{v}(t)$ are all functions of t, so the first-order derivative is given by

$$\frac{d\mathbf{v}}{dt} = \mathbf{i}\frac{dv_x}{dt} + \mathbf{j}\frac{dv_y}{dt} + \mathbf{k}\frac{dv_z}{dt}$$

and those of higher order by expressions of similar form. If $\mathbf{v}$ is a function of two or more scalar independent variables, $\mathbf{v}(x, y, z)$, say, then the partial derivatives of $\mathbf{v}$ are defined in the usual way. The total differential of $\mathbf{v}$ is

$$d\mathbf{v} = \frac{\partial \mathbf{v}}{\partial x}\,dx + \frac{\partial \mathbf{v}}{\partial y}\,dy + \frac{\partial \mathbf{v}}{\partial z}\,dz.$$

In particular the total differential of the position vector $\mathbf{r} = x\mathbf{i} + y\mathbf{j} + z\mathbf{k}$ is

$$d\mathbf{r} = \mathbf{i}\,dx + \mathbf{j}\,dy + \mathbf{k}\,dz.$$

Example 5.1.

If $\mathbf{r} = t^2\mathbf{i} + \sin t\mathbf{j} + t\mathbf{k}$ defines the position of a point in space at time t then its velocity is $d\mathbf{r}/dt = 2t\mathbf{i} + \cos t\mathbf{j} + \mathbf{k}$ and its acceleration is $d^2\mathbf{r}/dt^2 = 2\mathbf{i} - \sin t\mathbf{j}$.

DERIVATIVES OF SUMS AND PRODUCTS

In the expansions given below the vectors $\mathbf{a}$, $\mathbf{b}$ and $\mathbf{c}$ are differentiable functions of a scalar t and φ is a differentiable scalar function of t. The order of the factors in some of the products is significant.

$$\frac{d}{dt}(\mathbf{a} + \mathbf{b}) = \frac{d\mathbf{a}}{dt} + \frac{d\mathbf{b}}{dt}.$$

$$\frac{d}{dt}(\varphi\mathbf{a}) = \varphi\frac{d\mathbf{a}}{dt} + \mathbf{a}\frac{d\varphi}{dt}.$$

$$\frac{d}{dt}(\mathbf{a} \cdot \mathbf{b}) = \mathbf{a} \cdot \frac{d\mathbf{b}}{dt} + \mathbf{b} \cdot \frac{d\mathbf{a}}{dt}.$$

$$\frac{d}{dt}(\mathbf{a} \times \mathbf{b}) = \mathbf{a} \times \frac{d\mathbf{b}}{dt} + \frac{d\mathbf{a}}{dt} \times \mathbf{b}.$$

$$\frac{d}{dt}(\mathbf{a} \cdot \mathbf{b} \times \mathbf{c}) = \mathbf{a} \cdot \mathbf{b} \times \frac{d\mathbf{c}}{dt} + \mathbf{a} \cdot \frac{d\mathbf{b}}{dt} \times \mathbf{c} + \frac{d\mathbf{a}}{dt} \cdot \mathbf{b} \times \mathbf{c}.$$

$$\frac{d}{dt}\{\mathbf{a} \times (\mathbf{b} \times \mathbf{c})\} = \mathbf{a} \times \left(\mathbf{b} \times \frac{d\mathbf{c}}{dt}\right) + \mathbf{a} \times \left(\frac{d\mathbf{b}}{dt} \times \mathbf{c}\right) + \frac{d\mathbf{a}}{dt} \times (\mathbf{b} \times \mathbf{c}).$$

THE UNIT TANGENT VECTOR

In Fig. 8.16 points P, Q are shown, with position vectors $\mathbf{r}$, $\mathbf{r} + \delta\mathbf{r}$, on a curve C. The length of arc from a fixed point on C to P is s and arc PQ is of length δs. As $\delta s \to 0$ the ratio

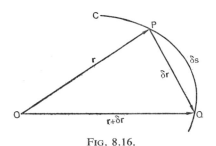

FIG. 8.16.

(chord PQ)/(arc PQ) $= \delta r/\delta s$ approaches unity and the direction of $\delta \mathbf{r}/\delta s$ tends to that of the tangent at P. Thus the vector $\dfrac{d\mathbf{r}}{ds} = \lim\limits_{\delta s \to 0} \dfrac{\delta \mathbf{r}}{\delta s}$ is called a unit tangent vector and is denoted here by $\mathbf{T}$. In components

$$\mathbf{T} = \frac{d\mathbf{r}}{ds} = \mathbf{i}\frac{dx}{ds} + \mathbf{j}\frac{dy}{ds} + \mathbf{k}\frac{dz}{ds}$$

and, as this is a unit vector, its direction cosines are dx/ds, dy/ds, dz/ds. If $\mathbf{r}$ is expressed in terms of a parameter t then a unit tangent vector is given also by $(d\mathbf{r}/dt)/(|d\mathbf{r}/dt|)$.

Example 5.2.

Obtain a unit tangent vector to the parabola $y = x^2$ at $(2, 4)$. The position vector of any point on the parabola is given in terms of a parameter t by $\mathbf{r} = t\mathbf{i} + t^2\mathbf{j}$; at $(2, 4)$ $t = 2$. $d\mathbf{r}/dt = \mathbf{i} + 2t\mathbf{j}$, $= \mathbf{i} + 4\mathbf{j}$ at $(2, 4)$, and $\mathbf{T} = (\mathbf{i} + 4\mathbf{j})/\sqrt{17}$. A slightly different approach is to consider $\mathbf{T} = d\mathbf{r}/ds$ as $(d\mathbf{r}/dx)(dx/ds)$; then $\mathbf{r} = x\mathbf{i} + x^2\mathbf{j}$, $d\mathbf{r}/dx = \mathbf{i} + 2x\mathbf{j}$ and $ds/dx = \sqrt{1 + (dy/dx)^2} = \sqrt{(1 + 4x^2)}$ give the above value for $\mathbf{T}$ at $(2, 4)$.

Example 5.3

A surface is defined by $z = F(x, y)$; $\partial z/\partial x$ is denoted by p and $\partial z/\partial y$ by q. Show that the vector $p\mathbf{i} + q\mathbf{j} - \mathbf{k}$ is perpendicular to the tangent plane to the surface at any point.

Now
$$dz = \frac{\partial z}{\partial x}\,dx + \frac{\partial z}{\partial y}\,dy,$$

whence $p\,dx + q\,dy - dz = 0$ and this may be expressed as the scalar product $(p\mathbf{i} + q\mathbf{j} - \mathbf{k})\cdot(\mathbf{i}\,dx + \mathbf{j}\,dy + \mathbf{k}\,dz) = 0$. As $d\mathbf{r} = \mathbf{i}\,dx + \mathbf{j}\,dy + \mathbf{k}\,dz$ lies in the tangent plane the required result follows.

8.6. ELEMENTS OF VECTOR FIELD THEORY

A *scalar field* is said to be defined in a region R of space, which may be a curve or a surface or a volume, when a real number is associated with each point of R. The association is given by a continuous *scalar point-function* $\varphi(x, y, z)$, single-valued and independent of a coordinate system, so that a number φ corresponds with any point (x, y, z) of R. For example, if $\varphi = x+2y-3z$ then $\varphi = 1$ is the number associated with the point $(2, 4, 3)$, $\varphi = 0$ with $(1, 1, 1)$, and so on. Distribution of temperature is a typical scalar field and this can be described graphically by a set of isothermal surfaces. Surfaces such as these, over which the scalar has a constant value, are called *level surfaces*; they cannot intersect because their point-function is defined to be single-valued.

A vector field is defined in a similar way by a continuous and single-valued vector point-function $\mathbf{v}(x, y, z)$, such as $\mathbf{v} = xy\mathbf{i} + xz\mathbf{j} - 2xyz\mathbf{k}$ which associates the vector $\mathbf{v} = 2\mathbf{i} - \mathbf{j} + 4\mathbf{k}$ with the point $(1, 2, -1)$. Such a field may be shown graphically by constructing a line which starts from, and in the same direction as, any one vector and is guided by the directions of successive vectors separated by very short distances. This technique is shown in Fig. 3.1 (p. 109) where an integral curve is sketched with the aid of a direction field. Starting from other vectors the process is repeated and a set of non-intersecting lines is formed which are called lines of flow, streamlines or *flux lines*. The distribution of velocity in a fluid is a typical vector field.

By means of vector analysis the features of fields can be described concisely. Three main concepts are used to describe the properties of fields; these are gradient, divergence and rotation. Each word will already have a fairly clear meaning in some contexts for the student. The purpose now is to formulate vector definitions of these terms. The symbolic forms of the definitions are simple and easily applied but, rather than quote these with little explanation, a reasoned analysis is given for each to ensure a sound understanding.

THE GRADIENT OF A SCALAR FIELD

The equation $\varphi(x, y, z) = C$ represents a surface in space, and when the parameter C takes a set of values a set of level

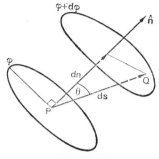

FIG. 8.17.

surfaces is defined. Figure 8.17 shows parts of two level surfaces φ and $\varphi + d\varphi$ separated by a very small distance dn along the normal $\hat{\mathbf{n}}$ to the surface φ at P. Q is any point on the surface $\varphi + d\varphi$ and its position vector relative to P is

$$d\mathbf{s} = \mathbf{i}\, dx + \mathbf{j}\, dy + \mathbf{k}\, dz.$$

Now

$$d\varphi = \frac{\partial \varphi}{\partial x}\, dx + \frac{\partial \varphi}{\partial y}\, dy + \frac{\partial \varphi}{\partial z}\, dz \qquad (6.1)$$

and the rate of change of φ at P in the direction $d\mathbf{s}$ is

$$
\begin{aligned}
\frac{d\varphi}{ds} &= \frac{\partial \varphi}{\partial x}\, \frac{dx}{ds} + \frac{\partial \varphi}{\partial y}\, \frac{dy}{ds} + \frac{\partial \varphi}{\partial z}\, \frac{dz}{ds} \\
&= \frac{\partial \varphi}{\partial x} \cos \alpha + \frac{\partial \varphi}{\partial y} \cos \beta + \frac{\partial \varphi}{\partial z} \cos \gamma,
\end{aligned}
\qquad (6.2)
$$

where $\cos \alpha$, $\cos \beta$, $\cos \gamma$ are the direction cosines of $d\mathbf{s}$. By (3.8) the far right-hand side of (6.2) may be interpreted as the scalar component of the vector

$$\mathbf{v} = \mathbf{i}\, \frac{\partial \varphi}{\partial x} + \mathbf{j}\, \frac{\partial \varphi}{\partial y} + \mathbf{k}\, \frac{\partial \varphi}{\partial z}$$

in the direction of $d\mathbf{s}$. This vector $\mathbf{v}$ is called the *gradient* of the scalar φ at the point (x, y, z) and is denoted by

$$\mathbf{v} = \mathrm{grad}\ \varphi = \nabla\varphi, \tag{6.3}$$

where ∇, termed nabla, is the vector differential operator

$$\left(\mathbf{i}\frac{\partial}{\partial x} + \mathbf{j}\frac{\partial}{\partial y} + \mathbf{k}\frac{\partial}{\partial z} \right).$$

Equation (6.1) may now be put in the form

$$d\varphi = (\mathrm{grad}\ \varphi)\cdot d\mathbf{s}$$

showing that if Q lies on the level surface containing P, so that $d\varphi = 0$, then the direction of grad φ is along $\hat{\mathbf{n}}$; as grad φ is independent of $d\mathbf{s}$ its direction is always along $\hat{\mathbf{n}}$. Thus

$$d\varphi = |\mathrm{grad}\ \varphi|\,\hat{\mathbf{n}}\cdot d\mathbf{s} = |\mathrm{grad}\ \varphi|\,ds \cos\theta$$

whence

$$\frac{d\varphi}{ds} = |\mathrm{grad}\ \varphi|\cos\theta,$$

showing that, since the numerical value of $d\varphi/ds$ is greatest when $d\mathbf{s}$ lies along $\hat{\mathbf{n}}$, $|\mathrm{grad}\ \varphi|$ is the greatest rate of change of φ at a given point.

The component of $\nabla\varphi$ in the direction of a vector $\mathbf{a}$ is $\hat{\mathbf{a}}\cdot\nabla\varphi$, called the *directional derivative* of φ in the direction of $\mathbf{a}$. The scalar product

$$\mathbf{a}\cdot\nabla\varphi = a_x\frac{\partial\varphi}{\partial x} + a_y\frac{\partial\varphi}{\partial y} + a_z\frac{\partial\varphi}{\partial z}$$

may also be expressed in the form

$$(\mathbf{a}\cdot\nabla)\varphi = \left(a_x\frac{\partial}{\partial x} + a_y\frac{\partial}{\partial y} + a_z\frac{\partial}{\partial z} \right)\varphi,$$

where $(\mathbf{a}\cdot\nabla)$ is treated as an operator acting on the scalar φ.

This operator may also act on a vector, but the resulting vector expression will contain nine terms in general.

Since the gradient exists at every point of a scalar field $\varphi(x, y, z)$ the operation $\nabla\varphi$ defines a vector field $\mathbf{v}(x, y, z)$ in all cases. However, the converse process of deriving a scalar field from a vector field is not always possible.

If φ is the potential in an electric field caused by static charges then the electric intensity $\mathbf{E}$ is normal to the equipotential surfaces $\varphi = C$ and the fields are related by

$$\mathbf{E} = -\operatorname{grad} \varphi,$$

the flux lines being the lines of force of $\mathbf{E}$.

Problems 8.6a

1. Show that $\nabla(\varphi\psi) = \varphi \nabla\psi + \psi \nabla\varphi$, where φ and ψ are scalar point-functions.

2. Show that (a) $\nabla r^n = nr^{n-1} \hat{\mathbf{r}}$, (b) $\nabla \log r = \hat{\mathbf{r}}/r$.

3. Obtain the directional derivative of $\varphi = xy - yz + xz$ at $(3, 2, 1)$ in the direction of $\overrightarrow{OP}$, where P is $(1, 2, 2)$, and state whether φ is increasing or decreasing in that direction.

4. The gradient of $\varphi = ax^2z - bxyz + cy^2$ at $(1, 1, 2)$ is required to have magnitude 6 and to be parallel to Ox. Evaluate the constants a, b, c.

5. Obtain the equation for the tangent plane to the surface of the sphere $x^2 + y^2 + z^2 = 9$ at $(2, 2, 1)$.

6. Given $\varphi = 2x^3y^2 + 3yz^2$, obtain grad φ and show that the point $(0, 1, 2)$ lies in the tangent plane which meets the surface $\varphi = 5$ at $(1, 1, 1)$.

7. Obtain a unit normal to the cone $z^2 = 9(x^2 + y^2)$ at $(0, 4, 12)$.

8. Show, by means of the gradient, that at $(0, a, b)$ the cylinders $x^2 + y^2 = a^2$ and $x^2 + z^2 = b^2$ intersect orthogonally.

9. Obtain the acute angle between the normals to the paraboloid $z = 4 - (x^2 + y^2)$ and the sphere $x^2 + y^2 + z^2 = 16$ at $(2, \sqrt{3}, -3)$.

10. Given $\mathbf{v} = 2(x + yz)\mathbf{i} + 2(xz - y)\mathbf{j} + (2xy + 1)\mathbf{k}$ derive a scalar function φ such that $\mathbf{v} = \operatorname{grad} \varphi$.

11. Show that if $\mathbf{a}$ and $\mathbf{b}$ are constant vectors then

$$r^5\mathbf{a} \cdot \operatorname{grad} \{\mathbf{b} \cdot \operatorname{grad} (1/r)\} = 3(\mathbf{a} \cdot \mathbf{r})(\mathbf{b} \cdot \mathbf{r}) - r^2(\mathbf{a} \cdot \mathbf{b}).$$

12. Show that the vector components of **v** which are normal and tangential to the surface $\varphi = C$ are respectively

$$\frac{(\mathbf{v} \cdot \operatorname{grad} \varphi) \operatorname{grad} \varphi}{(\operatorname{grad} \varphi)^2} \quad \text{and} \quad \frac{\operatorname{grad} \varphi \times (\mathbf{v} \times \operatorname{grad} \varphi)}{(\operatorname{grad} \varphi)^2}.$$

THE DIVERGENCE OF A VECTOR FIELD

The implication of divergence is most easily conceived by considering the behaviour of a fluid. The divergence at a point in a fluid, liquid or gas, is a measure of the rate per unit volume at which the fluid is flowing away from the point. A negative divergence is a convergence so the flow is towards the point. Physically divergence means either that the fluid is expanding or that fluid is being supplied by a source external to the field. Conversely convergence means a contraction or the presence of a sink through which the fluid is removed from the field. The flux lines diverge from a source and converge to a sink.

The *flux* of a vector **v** through a surface, usually taken to be small and plane, is defined as the scalar product of **v** with the vector area of the surface. If the surface is large and cannot be considered as an elemental area then the flux of the vector point-function **v** is defined by a surface integral.

A vector point-function $\mathbf{v}(x, y, z)$ is shown in Fig. 8.18 at a point P which is the centre of an elemental volume dV with

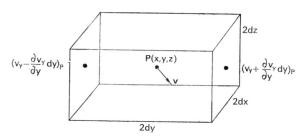

FIG. 8.18.

sides of length $2\,dx$, $2\,dy$ and $2\,dz$ parallel to the coordinate axes. In general the values of $\mathbf{v}$ in the vector field vary from point to point and the derivatives $\partial v_x/\partial x$, $\partial v_y/\partial y$, $\partial v_z/\partial z$ of the scalar components of $\mathbf{v}$ are not zero. The scalar component v_y of the vector which acts at the centre of the right-hand face of dV is approximately

$$\left(v_y + \frac{\partial v_y}{\partial y}\,dy\right)_P,$$

where P indicates that v_y and $\partial v_y/\partial y$ are evaluated at P, and at the centre of the left-hand face it is approximately

$$\left(v_y - \frac{\partial v_y}{\partial y}\,dy\right)_P.$$

As the two faces become vanishingly small the expressions for v_y at their central points may be taken to apply over all and the excess of flux leaving dV over that entering it in the positive y direction is

$$\left(v_y + \frac{\partial v_y}{\partial y}\,dy\right) 4\,dx\,dz - \left(v_y - \frac{\partial v_y}{\partial y}\,dy\right) 4\,dx\,dz = \frac{\partial v_y}{\partial y}\,dV.$$

It follows likewise that in the x and z directions the flux excess is $\dfrac{\partial v_x}{\partial x}\,dV$ and $\dfrac{\partial v_z}{\partial z}\,dV$, so that the total excess leaving dV is

$$\left(\frac{\partial v_x}{\partial x} + \frac{\partial v_y}{\partial y} + \frac{\partial v_z}{\partial z}\right)\,dV.$$

The *divergence* of a vector field $\mathbf{v}$, denoted by div $\mathbf{v}$, is defined as the quantity of this flux per unit volume and is therefore given by

$$\text{div } \mathbf{v} = \frac{\partial v_x}{\partial x} + \frac{\partial v_y}{\partial y} + \frac{\partial v_z}{\partial z}, \tag{6.4}$$

which is a scalar field. A field $\mathbf{v}$ for which div $\mathbf{v} = 0$ everywhere is said to be solenoidal.

The expression derived for div $\mathbf{v}$ may be generated by the operation $\nabla \cdot \mathbf{v}$, where ∇ is treated formally as a vector, thus

$$\nabla \cdot \mathbf{v} = \left(\mathbf{i} \frac{\partial}{\partial x} + \mathbf{j} \frac{\partial}{\partial y} + \mathbf{k} \frac{\partial}{\partial z} \right) \cdot (\mathbf{i} v_x + \mathbf{j} v_y + \mathbf{k} v_z)$$

$$= \frac{\partial v_x}{\partial x} + \frac{\partial v_y}{\partial y} + \frac{\partial v_z}{\partial z} \equiv \text{div } \mathbf{v}. \tag{6.5}$$

Problems 8.6b

1. Evaluate the divergence of $\mathbf{v} = 3x^2yz\mathbf{i} - 2xy^3\mathbf{j} + yz\mathbf{k}$ at $(1, 1, 2)$.

2. Find the value of c which makes $\mathbf{v} = (2cx - z)\mathbf{i} + y\mathbf{j} - (x - z)\mathbf{k}$ solenoidal.

3. Show that

    (a)   div $(\mathbf{u} + \mathbf{v}) = \text{div } \mathbf{u} + \text{div } \mathbf{v}$,

    (b)   div $(\varphi \mathbf{v}) = \varphi$ div $\mathbf{v} + \mathbf{v} \cdot \text{grad } \varphi$.

4. Show that (a) div $\mathbf{r} = 3$, (b) div $(\mathbf{r}/r^3) = 0$.

5. Find the directional derivative of div $\mathbf{v}$ at $(1, 2, 2)$ in the direction of the outer normal to the sphere $x^2 + y^2 + z^2 = 9$, where $\mathbf{v} = x^3y\mathbf{i} + y^3z\mathbf{j} + xz^3\mathbf{k}$.

THE ROTATION OF A VECTOR FIELD

Particles of a fluid may have two distinct types of motion. One is a pure translation by which fluid can enter a field and leave it. In this case the flux lines, or stream-lines, start at one point or set of points in the field and terminate elsewhere in the field. This feature of a field is its divergence. For the other type of motion the stream-lines form closed loops and the fluid has a movement which is vortical or rotational. Usually a fluid in motion has both streaming and rotational characteristics. For example the motion of water through a length of river is mainly streamline but small eddies are common and the presence of a whirlpool is possible. Again, the flow of water towards the plug-hole of a basin is convergent but it is often accompanied by a marked vortical motion, not to mention the noise which this induces.

If a vector field has the propensity to cause rotation then a particle placed in it will be moved in a loop. This simple idea is the basis for the definition of the rotation effect of a vector field now given.

In general the line integral of a vector point-function, to be defined in section 8.7, taken along any closed curve C in a vector field is not zero. The special case when the integral is

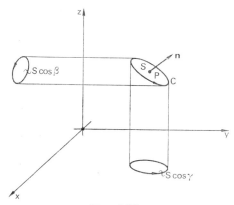

FIG. 8.19.

zero is of great importance and is discussed later. The rotation effect of a vector field is another vector field, the magnitude of which may be defined as the maximum line integral per unit area possible "at" a point in the original field.

Figure 8.19 shows a small plane area S with normal $\mathbf{n}$, which has direction cosines $\cos \alpha$, $\cos \beta$, $\cos \gamma$, and boundary C placed in a vector field $\mathbf{v}$. P is a fixed point (x_0, y_0, z_0) on S. The line integral of $\mathbf{v}$ round C is $\oint_C v_s \, ds$, where v_s is the scalar component of $\mathbf{v}$ along the tangent to C at any point and ds is the differential of arc s. Now

$$\oint_C v_s \, ds = \oint_C (v_x \, dx + v_y \, dy + v_z \, dz),$$

where the three scalar components of $\mathbf{v}$ are functions of (x, y, z). The value of $\mathbf{v}$ at any point on C may be related to $\mathbf{v}(x_0, y_0, z_0)$ at P approximately through the first few terms of a Taylor expansion; then the first term of the line integral becomes

$$\oint_C v_x \, dx \doteq \oint_C \left\{ (v_x)_0 + \left(\frac{\partial v_x}{\partial x} \right)_0 (x - x_0) \right.$$

$$\left. + \left(\frac{\partial v_x}{\partial y} \right)_0 (y - y_0) + \left(\frac{\partial v_x}{\partial z} \right)_0 (z - z_0) \right\} dx.$$

By properties of line integrals

$$\oint_C dx = \oint_C x \, dx = 0,$$

so that

$$\oint_C v_x \, dx \doteq \left(\frac{\partial v_x}{\partial y} \right)_0 \oint_C y \, dx + \left(\frac{\partial v_x}{\partial z} \right)_0 \oint_C z \, dx.$$

Example 2.2 of Chapter 7 shows that

$$\oint_C y \, dx = -S \cos \gamma$$

and it follows likewise that

$$\oint_C z \, dx = S \cos \beta,$$

the different signs arising from due regard to the senses in which the projections of C are described. A similar treatment of the second and third terms of the line integral gives the complete result

$$\oint_C v_s \, ds \doteq \left\{ \left(\frac{\partial v_z}{\partial y} - \frac{\partial v_y}{\partial z} \right)_0 \cos \alpha + \left(\frac{\partial v_x}{\partial z} - \frac{\partial v_z}{\partial x} \right)_0 \cos \beta \right.$$

$$\left. + \left(\frac{\partial v_y}{\partial x} - \frac{\partial v_x}{\partial y} \right)_0 \cos \gamma \right\} S.$$

Thus the line integral per unit area close to P tends to the value of the expression within curly brackets as S becomes smaller and C shrinks towards the point P, for it is easily shown that the contribution from the neglected terms of the Taylor expansion tends to zero. This expression is a maximum when S is so oriented about P that $\mathbf{n}$ is in the same direction as the vector

$$\left(\frac{\partial v_z}{\partial y} - \frac{\partial v_y}{\partial z}\right)\mathbf{i} + \left(\frac{\partial v_x}{\partial z} - \frac{\partial v_z}{\partial x}\right)\mathbf{j} + \left(\frac{\partial v_y}{\partial x} - \frac{\partial v_x}{\partial y}\right)\mathbf{k}. \quad (6.6)$$

This vector measures the rotational property of a vector field defined by the point-function $\mathbf{v}$. It is denoted by rot $\mathbf{v}$, but more usually by curl $\mathbf{v}$. A vector field is said to be *irrotational* if curl $\mathbf{v} \equiv \mathbf{0}$.

The expression derived for curl $\mathbf{v}$ may be generated by the operation $\nabla \times \mathbf{v}$ where, as with divergence, ∇ is treated as a vector. By the determinant form (4.13)

$$\nabla \times \mathbf{v} = \begin{vmatrix} \mathbf{i} & \mathbf{j} & \mathbf{k} \\ \partial/\partial x & \partial/\partial y & \partial/\partial z \\ v_x & v_y & v_z \end{vmatrix} \equiv \text{curl } \mathbf{v}. \quad (6.7)$$

THE OPERATOR ∇^2

The divergence of the gradient of a scalar φ, div grad φ, is

$$\text{div}\left(\mathbf{i}\,\frac{\partial \varphi}{\partial x} + \mathbf{j}\,\frac{\partial \varphi}{\partial y} + \mathbf{k}\,\frac{\partial \varphi}{\partial z}\right) = \frac{\partial^2 \varphi}{\partial x^2} + \frac{\partial^2 \varphi}{\partial y^2} + \frac{\partial^2 \varphi}{\partial z^2},$$

which may be expressed in operator form as

$$\left(\frac{\partial^2}{\partial x^2} + \frac{\partial^2}{\partial y^2} + \frac{\partial^2}{\partial z^2}\right)\varphi.$$

In terms of ∇

$$\text{div grad } \varphi = \nabla \cdot \nabla \varphi, \quad = \nabla^2 \varphi$$

by (4.2). Thus the operation div grad may be represented by ∇^2, which is called Laplace's operator or the Laplacian. In three dimensional rectangular Cartesian coordinates Laplace's equation is

$$\frac{\partial^2 \varphi}{\partial x^2} + \frac{\partial^2 \varphi}{\partial y^2} + \frac{\partial^2 \varphi}{\partial z^2}, \quad \equiv \nabla^2 \varphi, \quad = 0.$$

The operation of ∇^2 on a vector $\mathbf{v} = v_x \mathbf{i} + v_y \mathbf{j} + v_z \mathbf{k}$ is performed on the scalar components of $\mathbf{v}$, so that

$$\nabla^2 \mathbf{v} = \nabla^2 v_x \mathbf{i} + \nabla^2 v_y \mathbf{j} + \nabla^2 v_z \mathbf{k}.$$

An important relation is

$$\text{curl curl } \mathbf{v} = \text{grad div } \mathbf{v} - \nabla^2 \mathbf{v}$$

or

$$\nabla \times (\nabla \times \mathbf{v}) = \nabla(\nabla \cdot \mathbf{v}) - (\nabla \cdot \nabla)\mathbf{v}.$$

This has the same form as the vector triple product (4.16) when $\mathbf{b}$ is replaced by $\mathbf{a}$:

$$\mathbf{a} \times (\mathbf{a} \times \mathbf{c}) = \mathbf{a}(\mathbf{a} \cdot \mathbf{c}) - (\mathbf{a} \cdot \mathbf{a})\mathbf{c}.$$

Problems 8.6c

1. Obtain the curl of $\mathbf{v} = xz^2 \mathbf{i} - 3xyz \mathbf{j} + 2yz^3 \mathbf{k}$ at $(1, 1, -1)$.

2. Given that $\mathbf{v} = (x+ay+3z)\mathbf{i} + (x+2y+bz)\mathbf{j} + (cx+2y+3z)\mathbf{k}$ is irrotational, obtain the values of a, b and c.

3. Show that (a) curl grad $\varphi = \mathbf{0}$, (b) div curl $\mathbf{v} = 0$.

4. Show that

(a) curl $\varphi\mathbf{v} = \varphi$ curl $\mathbf{v} - \mathbf{v} \times \text{grad } \varphi$,

(b) div (φ curl $\mathbf{v}$) = (curl $\mathbf{v}$)·grad φ, [see 3(b) of Problems 8.6b]

(c) div ($\mathbf{u} \times \mathbf{v}$) = $\mathbf{v}$·curl $\mathbf{u} - \mathbf{u}$·curl $\mathbf{v}$.

5. Show that (a) curl $\mathbf{r} = \mathbf{0}$, (b) curl $r^n \mathbf{r} = \mathbf{0}$.

6. Show that if $\mathbf{a}$ is a constant vector then

(a) grad ($\mathbf{a} \cdot \mathbf{v}$) = ($\mathbf{a} \cdot \nabla$)$\mathbf{v} + \mathbf{a} \times \text{curl } \mathbf{v}$,

(b) curl ($\mathbf{a} \times \mathbf{v}$) = $\mathbf{a}$ div $\mathbf{v} - (\mathbf{a} \cdot \nabla)\mathbf{v}$.

7. Show that $2\mathbf{v} \times \text{curl } \mathbf{v} = \text{grad } \mathbf{v}^2 - 2(\mathbf{v} \cdot \nabla)\mathbf{v}$.

8. Obtain |curl curl **v**| at (2, 1, 2), where $\mathbf{v} = x^2yz\mathbf{i} + xy^2z\mathbf{j} + xyz^2\mathbf{k}$.

9. Show that

(a) div $(\varphi \text{ grad } \psi) = \varphi \nabla^2\psi + (\text{grad } \varphi) \cdot (\text{grad } \psi)$,
   [see 3(b) of Problems 8.6b],

(b) curl curl $\mathbf{v} = \text{grad div } \mathbf{v} - \nabla^2\mathbf{v}$.

10. Show that $\varphi = 1/r$ is a solution of Laplace's equation.

8.7. INTEGRALS OF VECTOR FUNCTIONS

ORDINARY INTEGRALS

The vector

$$\mathbf{v}(t) = v_x(t)\mathbf{i} + v_y(t)\mathbf{j} + v_z(t)\mathbf{k},$$

with components that are assumed to be continuous over a given range, depends on the single variable t. The process of deriving a vector **u**, say, whose derivative with respect to t equals **v** is called integration. Thus

$$\int \mathbf{v} \, dt, \quad = \mathbf{i} \int v_x \, dt + \mathbf{j} \int v_y \, dt + \mathbf{k} \int v_z \, dt,$$

$$= \int \frac{d\mathbf{u}}{dt} \, dt = \mathbf{u} + \mathbf{c},$$

where **c** is an arbitrary constant vector. The definite integral of **v** between limits $t = a$, b may be defined as the limit of a sum and is expressed as

$$\int_a^b \mathbf{v} \, dt = \mathbf{u}(t) \bigg|_a^b = \mathbf{u}(b) - \mathbf{u}(a).$$

Example 7.1.

Evaluate $\displaystyle\int_1^2 \mathbf{v} \, dt$, where $\mathbf{v} = 3t^2\mathbf{i} + 2t\mathbf{j} + \mathbf{k}$.

$$\int_1^2 \mathbf{v} \, dt = t^3\mathbf{i} + t^2\mathbf{j} + t\mathbf{k} \bigg|_1^2 = 7\mathbf{i} + 3\mathbf{j} + \mathbf{k}.$$

Problems 8.7a

1. A particle moves from rest at the origin at time $t = 0$ with an acceleration $4 \sin 2t\mathbf{i} + 2 \cos t\mathbf{j} - 6t\mathbf{k}$. Derive the velocity $\mathbf{v}$ and displacement $\mathbf{r}$ at any time.

2. Given that $\mathbf{u} = \sin t\mathbf{i} + \cos t\mathbf{j} + 2\mathbf{k}$, $\mathbf{v} = \sin t\mathbf{i} - \cos t\mathbf{j} - 2\mathbf{k}$ evaluate
(a) $\int_0^{\pi/2} (\mathbf{u} \cdot \mathbf{v}) \, dt$, (b) $\int_0^\pi (\mathbf{u} \times \mathbf{v}) \, dt$.

3. Obtain $\int \left(\mathbf{v} \times \dfrac{d^2\mathbf{v}}{dt^2} \right) dt$.

4. Determine the constant of integration if at $t = 0$

$$\int (\mathbf{u} \times \mathbf{v}) dt = \mathbf{u} \times \mathbf{v},$$

where $\mathbf{u} = t\mathbf{i} + \sin t\mathbf{j} + \sinh t\mathbf{k}$ and $\mathbf{v} = t^2\mathbf{i} + 4e^t\mathbf{j} + 4 \cos t\mathbf{k}$.

LINE INTEGRALS

The line integral of a vector point-function $\mathbf{v}(x, y, z)$ between points A, B on a curve C is the definite integral, with respect to arc length s, of the scalar component of $\mathbf{v}$ along the tangent to C. Let $\mathbf{\tau}$ be the unit tangent vector to C at any point; then the line integral may be expressed as

$$\int_C \mathbf{v} \cdot d\mathbf{s} = \int_C \mathbf{v} \cdot \mathbf{\tau} \, ds = \int_C \mathbf{v} \cdot \frac{d\mathbf{r}}{ds} \, ds = \int_C \mathbf{v} \cdot d\mathbf{r},$$

where $\mathbf{r}$ is the position vector of any point on C. The expansion of $\int_C \mathbf{v} \cdot d\mathbf{r}$ as

$$\int_C (v_x \, dx + v_y \, dy + v_z \, dz) \equiv \int_C (P \, dx + Q \, dy + R \, dz)$$

shows how the definition of a vector line integral leads to expressions already considered in Chapter 7.

The important special case when $\mathbf{v}$ is the gradient of a scalar function φ is now taken and the line integral is

$$\int_C \mathbf{v} \cdot d\mathbf{r} = \int_C \nabla\varphi \cdot d\mathbf{r} = \int_C \left(\frac{\partial\varphi}{\partial x}\, dx + \frac{\partial\varphi}{\partial y}\, dy + \frac{\partial\varphi}{\partial z}\, dz \right)$$

$$= \int_C d\varphi = \varphi_B - \varphi_A .$$

The integral is independent of the path C joining A to B and the condition $\mathbf{v} = \operatorname{grad} \varphi$, in vector terms, is equivalent to conditions (4.3) of Chapter 7. It also follows that $\oint_C \mathbf{v} \cdot d\mathbf{r} = 0$, provided $\mathbf{v}$ is single-valued.

Since the curl of a vector $\mathbf{v}$ is a point-function a compact form of its definition is

$$\hat{\mathbf{n}} \cdot \operatorname{curl} \mathbf{v} = \lim_{S \to 0} \frac{1}{S} \oint_C \mathbf{v} \cdot d\mathbf{s},$$

where $\hat{\mathbf{n}}$ is normal to any small area S in the field. This implies that when a vector field is such that all line integrals round closed curves are zero, the curl itself should be zero. A comparison of expression (6.6) for curl $\mathbf{v}$ with conditions (7.4.3) confirms this conclusion. When $\mathbf{v} = \operatorname{grad} \varphi$ the vector field $\mathbf{v}$ is said to be *conservative* and φ to be its scalar potential. The condition $\mathbf{v} = \operatorname{grad} \varphi$ implies and is implied by the condition curl $\mathbf{v} \equiv \mathbf{0}$.

Example 7.2.

The work done by a force field $\mathbf{F}$ in moving a particle along a path C is $W = \int_C \mathbf{F} \cdot d\mathbf{s}$. Evaluate W when $\mathbf{F} = e^x \mathbf{i} + e^{x^2} \mathbf{j}$ and C is the arc of the parabola $y = x^2$ between the origin and $(1, 1)$.

$$W = \int_C (e^x\, dx + e^{x^2}\, dy) = \int_0^1 (e^x + 2xe^{x^2})\, dx$$

$$= e^x + e^{x^2} \Big|_0^1 = 2(e-1).$$

Problems 8.7 b

1. A force field is defined by $\mathbf{F} = 2xz\mathbf{i} + 3y^2z\mathbf{j} - x^2y\mathbf{k}$. Evaluate the work done by $\mathbf{F}$ in moving a particle along the path $x = y = z$ from the origin to $(2, 2, 2)$.

2. Evaluate the line integral of $\mathbf{v} = y\mathbf{i} + z\mathbf{j} + x\mathbf{k}$ along the curve $x = t$, $y = t^2$, $z = t^3$ from the origin to the point $t = -1$.

3. Evaluate the line integral of $\mathbf{v} = \mathbf{r}$ along the circular helix $\mathbf{r} = a \cos t\mathbf{i} + a \sin t\mathbf{j} + ct\mathbf{k}$ from $t = 0$ to $t = 2\pi$.

4. Evaluate $\int_C \{xyz\,dx + (x+z)^2\,dy - 3xy\,dz\}$, where C is the arc of the curve $\mathbf{r} = t\mathbf{i} + t\mathbf{j} + t^2\mathbf{k}$ between the origin and $(-1, -1, 1)$.

5. Evaluate the work done by moving a particle once along the circle $x^2 + y^2 = 4$, in the x–y plane, in the force field

$$\mathbf{F} = (x+y-z)\mathbf{i} + (x-y+z)\mathbf{j} + (y-x+z)\mathbf{k}.$$

Evaluate $\oint_C \mathbf{v} \cdot d\mathbf{s}$ for the vector functions and closed paths defined in problems **6–10** below. The additional information given about some of the surfaces related to boundaries C is required in Problems 8.7c.

6. $\mathbf{v} = e^x\mathbf{i} + xy^2\mathbf{j} - (y+z)\mathbf{k}$ and C is the boundary of the square $x = \pm 1, y = \pm 1$ which lies in the plane $z = 3$, described clockwise when viewed from the origin.

7. $\mathbf{v} = xy\mathbf{i} - xyz\mathbf{j} + (x+y+z)\mathbf{k}$ and C is the boundary of the triangle, described clockwise when viewed from the origin, formed by the intersection of the plane $3x + 4y + 6z = 12$ with the planes $x = 0$, $y = 0$, $z = 0$.

8. $\mathbf{v} = (x^2 + y^2)\mathbf{i} - (2xy + z)\mathbf{j} + xe^z\mathbf{k}$ and C is the rim of the hemisphere $x^2 + y^2 + z^2 = a^2, 0 \leqslant z \leqslant a$, described counterclockwise when viewed from a point on the positive z axis.

9. $\mathbf{v} = z^2\mathbf{i} + x^2\mathbf{j} + y^2\mathbf{k}$ and C is the circle, described counterclockwise when viewed from the origin, formed by the intersection of the cone $z^2 = 4(x^2 + y^2)$ with the plane $z = 4$.

10. $\mathbf{v} = 2z^2\mathbf{i} + yz\mathbf{j} + y^2\mathbf{k}$ and C is the boundary, described clockwise when viewed from the origin, of the smaller section of the cylinder $x^2 + y^2 = 25$ cut off by the planes $z = 0$, $z = 2$ and $y = 3$.

Consider a curved surface S on which a vector point-function $\mathbf{v}$ is defined and let $\hat{\mathbf{n}}$ be a unit normal vector parallel to the normal at any point on S; $\hat{\mathbf{n}}$ is usually drawn outwards if S is closed and always away from the same side if S is not closed. Problems which involve open surfaces should state which side of the surface is to be considered. The component of $\mathbf{v}$ along the normal is $\mathbf{v} \cdot \hat{\mathbf{n}}$ and the flux of $\mathbf{v}$ through an element δS of S is $\mathbf{v} \cdot \hat{\mathbf{n}} \, \delta S$ approximately. The total flux of $\mathbf{v}$ through S is the limit, as $\delta S \to 0$, of the sum $\sum \mathbf{v} \cdot \hat{\mathbf{n}} \, \delta S$ applied to all elements of S and is called the surface integral of $\mathbf{v}$ over S. The surface integral of a vector function is expressed as

$$\iint_S \mathbf{v} \cdot d\mathbf{S} \quad \text{or as} \quad \iint_S \mathbf{v} \cdot \hat{\mathbf{n}} \, dS,$$

and it may be evaluated by projecting S on to one of the coordinate planes in the manner of Fig. 7.15 (p. 285). For the projection shown $\cos \gamma = \hat{\mathbf{n}} \cdot \mathbf{k}$ so that

$$\iint_S \mathbf{v} \cdot \hat{\mathbf{n}} \, dS = \iint_R \mathbf{v} \cdot \hat{\mathbf{n}} \, \frac{dx \, dy}{|\hat{\mathbf{n}} \cdot \mathbf{k}|}.$$

The surface integral of a scalar function φ is $\iint_S \varphi \, d\mathbf{S}$, which is a vector.

Since the divergence of a vector $\mathbf{v}$ is a point function a compact form of its definition is

$$\operatorname{div} \mathbf{v} = \lim_{V \to 0} \frac{1}{V} \iint_S \mathbf{v} \cdot d\mathbf{S},$$

where V is the volume enclosed by the surface S.

Example 7.3.

Evaluate $\iint\limits_{S} \mathbf{v} \cdot d\mathbf{S}$, where $\mathbf{v} = y\mathbf{i} + 2\mathbf{j} - 3x\mathbf{k}$ and S is that section of the plane $6x + 3y + 2z = 6$ within the positive octant (see Fig. 7.7, p. 263). Take the side of S on which $\hat{\mathbf{n}}$ is positive.

A unit normal to the surface $\varphi = 6x + 3y + 2z = 6$ is

$$\hat{\mathbf{n}} = \frac{\operatorname{grad} \varphi}{|\operatorname{grad} \varphi|} = (6\mathbf{i} + 3\mathbf{j} + 2\mathbf{k})/7,$$

which is constant over S because the surface is plane. Now $\mathbf{v} \cdot \hat{\mathbf{n}} = (6y + 6 - 6x)/7$ and, for projection on to the x–y plane, $|\hat{\mathbf{n}} \cdot \mathbf{k}| = 2/7$ so the surface integral becomes

$$3 \int_{0}^{1} \int_{0}^{2(1-x)} (y + 1 - x) \, dx \, dy = 3 \int_{0}^{1} \left[\frac{y^2}{2} + y - xy \right]_{0}^{2(1-x)} dx$$

$$= 12 \int_{0}^{1} (x - 1)^2 \, dx = 4(x - 1)^3 \Big|_{0}^{1} = 4.$$

Example 7.4.

Evaluate $\iint\limits_{S} \mathbf{v} \cdot d\mathbf{S}$, where $\mathbf{v} = 3z\mathbf{i} + 2xy\mathbf{j} + x\mathbf{k}$ and S is the outer surface of the cylinder $x^2 + y^2 = 4$ in the positive octant between $z = 0$ and $z = 3$.

In this case $\hat{\mathbf{n}} = (2x\mathbf{i} + 2y\mathbf{j})/4$, $\hat{\mathbf{n}} \cdot \mathbf{j} = y/2$ and $\mathbf{v} \cdot \hat{\mathbf{n}} = (3xz + 2xy^2)/2$. R is a rectangular region of the x–z plane and the surface integral is

$$\int_{0}^{3} dz \int_{0}^{2} \frac{(3xz + 2xy^2)}{y} \, dx,$$

from which y is eliminated through the equation of S to give

$$\int_0^3 dz \int_0^2 \left\{ \frac{3xz}{\sqrt{(4-x^2)}} + 2x\sqrt{(4-x^2)} \right\} dx$$

$$= \int_0^3 \left[-3z\sqrt{(4-x^2)} - \frac{2}{3}\sqrt{(4-x^2)^3} \right]_0^2 dz$$

$$= \frac{2}{3} \int_0^3 (9z+8)\, dz = 43.$$

Problems 8.7c

The meaning of the terms "upper" and "lower", which occur in some of these problems, is related to a z axis which is vertical.

1. Evaluate the integral in Example 7.3 by projecting S on to (a) the x–z plane, (b) the y–z plane.

2. Evaluate the integral in Example 7.4 by projecting S on to the y–z plane. Note that the use of this projection involves simpler integrals.

3. Evaluate $\iint_S \mathbf{v} \cdot d\mathbf{S}$, where $\mathbf{v} = xe^z\mathbf{j} + (x^2+z)\mathbf{k}$ and S is the lower surface of the parabolic cylinder $z = y^2$ within the circular cylinder $x^2 + y^2 = a^2$.

4. Evaluate $\iint_S \varphi\, d\mathbf{S}$, where $\varphi = 1/\sqrt{x^2+y^2}$ and S is the outer surface of the sphere $x^2 + y^2 + z^2 = a^2$ (a) in the positive octant, (b) above the plane $z = 0$.

5. Evaluate $\iint_S \varphi\, d\mathbf{S}$, where $\varphi = xyz$ and S is the upper surface of the parabolic cylinder $z = x(2-x)$ in the positive octant between the planes $y = 0$ and $y = 3$.

6–10. Consider **6–10** of Problems 8.7b. In each case obtain curl $\mathbf{v}$ and evaluate $\iint_S (\text{curl } \mathbf{v}) \cdot d\mathbf{S}$, where S is the defined open surface of which C is the boundary. The required side of S is defined by the sense in which C is described, this taken to be positive.

Evaluate $\iint_S \mathbf{v} \cdot d\mathbf{S}$ for the vector functions and outsides of the closed surfaces defined in **11–15** below.

11. $\mathbf{v} = x^2y\mathbf{i} - y^2z\mathbf{j} + z^2x\mathbf{k}$ and S is the surface of the cube bounded by the planes $x = 0$, $y = 0$, $z = 0$, $x = 2$, $y = 2$, $z = 2$.

12. $\mathbf{v} = (2x+y)\mathbf{i} + (3y-2z)\mathbf{j} + (2z+x)\mathbf{k}$ and S is the surface of the tetrahedron bounded by the planes $x+2y+4z = 8$, $x = 0$, $y = 0$, $z = 0$.

13. $\mathbf{v} = \mathbf{r}$ and S is the surface of the sphere with radius a and centre at the origin.

14. $\mathbf{v} = xyz\mathbf{i} + x(x^2+z^2)\mathbf{j} + (x+y+xz)\mathbf{k}$ and S is the surface formed by the cylinder $x^2+z^2 = 4$ and the planes $y = 0$, $y = 3$. Consider carefully the relative signs of $|z|$ and z.

15. $\mathbf{v} = \mathbf{r}$ and S is the surface formed in the positive octant by the paraboloid $z = 9-(x^2+y^2)$ and the planes $3x+3y+z = 9$, $x = 0$, $y = 0$, $z = 0$.

VOLUME INTEGRALS

A vector function $\mathbf{v}$ or a scalar function φ may be integrated through a volume V and, since an element of volume is not represented by a vector, the integrals take the forms

$$\iiint_V \mathbf{v}\, dV \quad \text{and} \quad \iiint_V \varphi\, dV.$$

These are evaluated simply as triple integrals, but when the integrand is a vector each of its components is treated as the integrand of a triple integral and the result is a vector.

Problems 8.7d

1–5. Consider **11–15** of Problems 8.7c. In each case obtain div $\mathbf{v}$ and evaluate $\iiint_V \text{div } \mathbf{v}\, dV$, where V is the region enclosed by the surface S.

6. Evaluate (a) $\iiint_V \mathbf{r}\, dV$, (b) $\iiint_V |\mathbf{r}|\, dV$, where V is the region enclosed by the sphere $x^2+y^2+z^2 = a^2$. Try to anticipate the result of (a) by simple considerations.

8.8. INTEGRAL THEOREMS IN VECTOR FORM

Gauss's theorem (7.12.2) is also known as the *divergence theorem* and, in vector form, it states that if V is the volume enclosed by a surface S and $\mathbf{v}$ is a continuous vector function of position with continuous first partial derivatives then

$$\iiint\limits_V \operatorname{div} \mathbf{v} \, dV = \iint\limits_S \mathbf{v} \cdot \hat{\mathbf{n}} \, dS,$$

where $\hat{\mathbf{n}}$ is the outward drawn unit normal to S.

A comparison of the solutions to **11–15** of Problems 8.7c with those to **1–5** of Problems 8.7d illustrates the validity of this theorem. These problems indicate that the evaluation of the triple integral may often be easier than the evaluation of the double integral.

In vector form Stokes's theorem (7.12.3) states that if $\mathbf{v}$ is a continuous vector function of position, with continuous first partial derivatives, defined on an open surface S and on its boundary C then

$$\iint\limits_S (\operatorname{curl} \mathbf{v}) \cdot \hat{\mathbf{n}} \, dS = \oint\limits_C \mathbf{v} \cdot d\mathbf{s},$$

where $\hat{\mathbf{n}}$ is the unit normal to the required side of S and C is described in the related positive sense.

A comparison of the solutions to **6–10** of Problems 8.7b with those to **6–10** of Problems 8.7c illustrates the validity of this theorem.

Answers to Problems

CHAPTER 1

Problems 1.2

1. Equality. **2.** Alternately unequal and equal.

3. $\begin{vmatrix} -5 & -7 \\ -4 & 3 \end{vmatrix}.$ **4.** $\begin{vmatrix} 6 & 5 \\ 9 & 4 \end{vmatrix},$ $\begin{vmatrix} 1 & 3 \\ 8 & 4 \end{vmatrix},$ $\begin{vmatrix} 1 & 2 \\ 7 & 6 \end{vmatrix}.$

5. 95 in every case. **6.** 0 in each case.

Problems 1.3

1. 59. **2.** -20. **3.** 0. **4.** 1. **5.** 2576.

Problems 1.4

1. -54. **2.** 52. **5.** 387. **6.** 312.

7. $6a+7$. **13.** $8abcd$. **14.** (a) 0. **15.** $(a-b)(b-c)(c-a)$.

17. (a) 4; (b) 3; (c) 2. **18.** 2. **19.** $-5, 4$. **20.** $-3, 12$.

Problems 1.5

1. $x = 7,$ $y = -5$. **2.** No solution. Parallel lines.

3. $x = 3t+2, y = t$. **4.** $x = 15,$ $y = -4$.

5. $x = 1,$ $y = 2,$ $z = 3$. **6.** $5x = -6t,$ $5y = 10-t,$ $z = t$.

7. $x = 3,$ $y = -5,$ $z = 7$. **8.** No solution. Parallel planes.

9. $x = 2-4t, y = t,$ $z = 0$. **10.** $w = -1, x = 0, y = 1, z = 2$.

Problems 1.6

1. $i_1 = \frac{13}{8},$ $i_2 = \frac{1}{4},$ $i_3 = -\frac{7}{8};$ $v_1 = \frac{11}{8},$ $v_2 = \frac{9}{8}$.

2. $I_1 = (R_2+R_3)v/\Delta,$ $I_2 = R_2v/\Delta,$ $\Delta = (R_1+R_2+R_4)(R_2+R_3)-R_2^2$.

$i_1 = \frac{35}{31},$ $i_2 = \frac{21}{31},$ $i_3 = \frac{14}{31}$.

3. $(R_1+R_2+R_5)I_1-R_2I_2-R_5I_3 \quad = v,$
$R_2I_1-(R_2+R_3+R_4)I_2+R_4I_3 \quad = 0,$
$R_5I_1+R_4I_2-(R_4+R_5+R_6)I_3 \quad = 0. \qquad v=1110.$

4. $(G_1+G_2+G_3)v_1-G_3v_2-G_1v_3 \quad = G_1v_a,$
$-G_3v_1+(G_3+G_4+G_6)v_2-G_6v_3 = G_6v_b,$
$G_1v_1+G_6v_2-(G_1+G_5+G_6)v_3 \quad = G_1v_a+G_6v_b.$
$v_1=\frac{30}{101}, \quad v_2=\frac{12}{101}, \quad v_3=-\frac{96}{101}.$

Problems 1.7

1. Zero solution only. **2.** $x=-3t,\quad y=2t,\quad z=t.$
3. $19x=4t,\quad 38y=29t,\quad z=t.$ **4.** $43x=7t,\quad 43y=11t,\quad z=t.$
5. $x=3s-5t,\quad y=s,\quad z=t.$ **6.** $3x=2t,\quad y=t,\quad z=0.$

Problems 1.8

1. $5(1st)+3(2nd)=2(3rd).$
2. $c=6;\quad 7(1st)+5(2nd)-2(3rd)=(4th).$

Problems 1.9

1. $\lambda=8;\quad x=2t-3,\quad y=t;\quad$ case (7).
$\lambda=18;\quad x=\frac{1}{5},\quad y=\frac{3}{5};\quad$ case (1).
2. $\lambda=6;$ inconsistent; case (6).
$\lambda=11;\quad x=\frac{4}{5},\quad y=\frac{12}{5};\quad$ case (1).

Problems 1.11

1. $(s-t)(t-u)(u-s).$ **2.** 360. **3.** $\begin{vmatrix} 2ab & ac & bc \\ bc & ab+c^2 & a^2 \\ ac & b^2 & ab+c^2 \end{vmatrix}.$

4. $e^x\{(x^2+4x+2)\log x+(2x+3)\}+2\sin 2x.$ **5.** 1.

CHAPTER 2

Problems 2.1

2. (a) $\begin{pmatrix} -10 & 7 & 7 \\ 8 & 15 & -18 \end{pmatrix}$, (b) Not defined, (c) $\begin{pmatrix} -10 & 8 \\ 7 & 15 \\ 7 & -18 \end{pmatrix}$,

(d) $\begin{pmatrix} 11 & 4 & -14 \\ 11 & 15 & 0 \end{pmatrix}$,   (e) $[0]_{23}$.   **3.** Yes to each.

4. (a) $\begin{pmatrix} 6 & -1 & 9 \\ 8 & 4 & 10 \\ 1 & 8 & 0 \end{pmatrix}$;   (b) $\begin{pmatrix} 2 & -1 & 0 \\ 0 & 2 & 4 \\ 0 & 0 & -6 \end{pmatrix} + \begin{pmatrix} 0 & 0 & 0 \\ 6 & 0 & 0 \\ 5 & -2 & 0 \end{pmatrix}$ or an

equivalent form; (c) $\begin{pmatrix} 1 & -\frac{3}{2} & 0 \\ -\frac{3}{2} & 1 & 5 \\ 0 & 5 & 9 \end{pmatrix} + \begin{pmatrix} 0 & \frac{3}{2} & 2 \\ -\frac{3}{2} & 0 & -1 \\ -2 & 1 & 0 \end{pmatrix}$;

(d) $3\mathbf{I} + \begin{pmatrix} 0 & -1 & 2 \\ 0 & 0 & 8 \\ 0 & 0 & 0 \end{pmatrix} + \begin{pmatrix} 0 & 0 & 0 \\ 3 & 0 & 0 \\ 3 & 4 & 0 \end{pmatrix}$ or an equivalent form;

(e) $\begin{pmatrix} -1 & 0 & 0 \\ 5 & 1 & 0 \\ -9 & 2 & 5 \end{pmatrix}$.   **5.** $a = 1$; $b = 2$; $c = -3$.

Problems 2.2

1. $\mathbf{Ac}, \mathbf{cA}$. $\mathbf{A}^2 = \begin{pmatrix} -1 & 8 \\ -4 & 7 \end{pmatrix}$, $\mathbf{AB} = \begin{pmatrix} 2 & 1 & 5 \\ -2 & -6 & 5 \end{pmatrix}$, $\mathbf{Ad} = \begin{pmatrix} 1 \\ -11 \end{pmatrix}$,

$\mathbf{B'A} = \begin{pmatrix} 2 & 4 \\ 4 & 3 \\ -1 & 8 \end{pmatrix}$, $\mathbf{BB'} = \begin{pmatrix} 14 & -1 \\ -1 & 5 \end{pmatrix}$, $\mathbf{Bc'} = \begin{pmatrix} 9 \\ -5 \end{pmatrix}$, $\mathbf{B'd} = \begin{pmatrix} 10 \\ 17 \\ 1 \end{pmatrix}$.

$\mathbf{cB'} = (9 - 5)$, $\mathbf{cc'} = 21$, $\mathbf{c'd'} = \begin{pmatrix} 20 & -8 \\ 5 & -2 \\ -10 & 4 \end{pmatrix}$, $\mathbf{d'A} = (7\ 4)$,

$\mathbf{d'B} = (10\ 17\ 1)$, $\mathbf{dc} = \begin{pmatrix} 20 & 5 & -10 \\ -8 & -2 & 4 \end{pmatrix}$, $\mathbf{dd'} = \begin{pmatrix} 25 & -10 \\ -10 & 4 \end{pmatrix}$.

2. $\begin{pmatrix} -11 & 57 & 40 \\ -8 & 24 & 16 \\ -21 & 39 & 24 \end{pmatrix}$. **3.** $\begin{pmatrix} 8 & 8 & 13 \\ 8 & 5 & 8 \\ 13 & 8 & 8 \end{pmatrix}$. **4.** $\begin{aligned} x_1 &= -5z_1 - 5z_2, \\ x_2 &= 9z_1 - 4z_2. \end{aligned}$

5. Given, no change, rotation through π, rotation through $\pi/2$, reflection in y_2 axis, rotation through θ.

7. $\mathbf{A}^2 + \mathbf{AB} + \mathbf{BA} + \mathbf{B}^2$.

Problems 2.3

1. $\begin{pmatrix} 5 & -2 \\ -7 & 3 \end{pmatrix}$, $\frac{1}{21}\begin{pmatrix} 12 & -1 & -3 \\ 0 & 7 & 0 \\ -3 & 2 & 6 \end{pmatrix}$, not defined, $\begin{pmatrix} 0 & 0 & 1 \\ 0 & 1 & 0 \\ 1 & 0 & 0 \end{pmatrix}$,

$\begin{pmatrix} \cos\theta & -\sin\theta \\ \sin\theta & \cos\theta \end{pmatrix}$.   **2.** $x = 5$, $y = 6$, $z = 18$.

3. $3x = \ \ X - \ Y + \ Z$,
$9y = \ 5X - 2Y - \ Z$,
$9z = \ \ X + 5Y - 2Z$.

Problems 2.5

1. $3x = z$, $3y = 2z$.   **2.** $w = z$, $2x = -11z$, $2y = z$.

3. $22w = 23z - 2y$, $22x = -32y - 105z$.

4. $w = -z$, $x = 0$, $y = 0$.

5. Zero solution only.   **6.** $v = -z$, $2w = -z$, $x = 0$, $2y = z$.

Problems 2.6

1. $x = -z - 2$, $y = 2$.   **2.** $4x = 7z - 29$, $4y = 51 - 9z$.

3. Unique solution $x = 4$, $y = 5$.

4. No solution.   **5.** $x = 2y - z + 1$.

6. $25w = 2(71 - 7z)$, $25x = 8(3 - z)$, $25y = 47z - 91$.

7. $55w = 2(9 - 32z)$, $55x = 31(z - 2)$, $5y = 2(2 - z)$.

8. $11w = 11(3 - y) - 49z$, $11x = 11 - 17z$.

9. No solution.   **10.** Unique solution $x = -1$, $y = 1$, $z = 2$.

Problems 2.7

1. (a) $i_1 = 58$, $v_1 = 183$;   (b) $i_2 = -38$, $v_2 = 11$.

2. $R_1(R_2 + R_3 + R_4) + R_3(R_2 - R_4) = 0$. $R_2 = -1$. R_2 is a source of power and the value of R_4 is arbitrary.

Problems 2.8

1. (a) $\begin{pmatrix} 0 & 0 & 2 & -3 \\ 0 & 0 & -1 & 2 \\ 2 & -1 & -13 & 16 \\ 5 & -2 & -31 & 38 \end{pmatrix}$;   (b) $\begin{pmatrix} -101 & 22 & 3 & 1 \\ -24 & 5 & 1 & 0 \\ 4 & -1 & 0 & 0 \\ 1 & 0 & 0 & 0 \end{pmatrix}$.

2. $\begin{bmatrix} 0 & 0 & 0 & 7 & -6 & 15 \\ 0 & 0 & 0 & -3 & 3 & -7 \\ 0 & 0 & 0 & 5 & -5 & 12 \\ 14 & 2 & -7 & -57 & 43 & -112 \\ -8 & -1 & 4 & 33 & -25 & 65 \\ 3 & 1 & -2 & -8 & 5 & -14 \end{bmatrix}$.

Problems 2.9a

1. -1, $\{1 \ -1\}/\sqrt{2}$;   2, $\{1 \ -4\}/\sqrt{17}$.   **2.** 1, $\{1 \ 0\}$.

3. 0, any vector of order 2×1.   **4.** 0, $\{0 \ 1\}$; 4, $\{1 \ 0\}$.

5. -1, $\{1 \ 0\}$;   1, $\{1 \ 1\}/\sqrt{2}$.   **6.** 3, any vector of order 2×1.

7. 2, $\{5 \ 2\}/\sqrt{29}$.   **8.** -5, $\{1 \ 3\}/\sqrt{10}$;   5, $\{2 \ 1\}/\sqrt{5}$.

9. 1, $\{1 \ 0 \ 0\}$;   2, $\{0 \ 1 \ 0\}$;   3, $\{0 \ 0 \ 1\}$.

10. -1, $\{1 \ 6 \ -1\}/\sqrt{38}$;   2, $\{2 \ 6 \ 1\}/\sqrt{41}$;   4, $\{4 \ 4 \ 1\}/\sqrt{33}$.

11. -1, $\{0 \ 1 \ -1\}/\sqrt{2}$;   2, $\{1 \ -1 \ 0\}/\sqrt{2}$;   3, $\{1 \ -1 \ 1\}/\sqrt{3}$.

12. 2, $\{1 \ 0 \ 0\}$, $\{0 \ 0 \ 1\}$;   5, $\{0 \ 1 \ 0\}$.

13. 0, $\{2 \ 2 \ -1\}/3$;   1, $\{3 \ 0 \ -2\}/\sqrt{13}$, $\{3 \ 1 \ -2\}/\sqrt{14}$.

14. 0, $\{0 \ -1 \ 3\}/\sqrt{10}$;   1, $\{1 \ 4 \ -12\}/\sqrt{161}$.

15. -10, $\{0 \ 0 \ 31 \ -8\}$; -7, $\{0 \ 21 \ 35 \ -19\}$;

     3, $\{1170 \ 468 \ 360 \ -407\}$;   21, $\{0 \ 0 \ 0 \ 1\}$. (The first three

     vectors are not normalised.)

22. $\{1 \ 1 \ 2\}$, $\begin{pmatrix} 3 & 1 & 1 \\ 1 & 3 & 1 \\ 1 & 1 & 5 \end{pmatrix}$.

24. $\frac{1}{2}\begin{pmatrix} -3 & 1 & 1 \\ -1 & 1 & 1 \\ -5 & 1 & 3 \end{pmatrix}$. -1, $\{1 \ 0 \ 1\}$ 1, $\{1 \ 2 \ 3\}$;   2, $\{1 \ 3 \ 1\}$.

    -1, 1, $\frac{1}{2}$;   same eigenvectors.

Problems 2.9b

2. -9, $\{-1 \ 1 \ 4\}$;   9, $\{4 \ 0 \ 1\}$, $\{1 \ 17 \ -4\}$.

Problems 2.10

1. $\{1 \ 1 \ 0\}$, $\begin{pmatrix} 2 & 2 & 0 \\ 2 & 2 & 0 \\ 0 & 0 & 1 \end{pmatrix}$.   **3.** $\frac{1}{4}\begin{pmatrix} 3 & -1 \\ 5 & -3 \end{pmatrix}$.

4. 0, $\{-1 \ 2 \ 2\}$;   1, $\{2 \ 0 \ -3\}$, $\{2 \ 1 \ -3\}$.   **5.** $\frac{1}{5}\begin{pmatrix} -7 & -6 \\ 16 & 13 \end{pmatrix}$.

6. $\frac{1}{6}\begin{pmatrix}(8-2^{p+1}) & (-2+2^{p+1}) & (-2+2^{p+1}) \\ (6-3\times2^{p+1}) & (3\times2^{p+1}) & (-6+3\times2^{p+1}) \\ (2-2^{p+1}) & (-2+2^{p+1}) & (4+2^{p+1})\end{pmatrix}$,

$\frac{1}{4}\begin{pmatrix}5 & -1 & -1 \\ 3 & 1 & -3 \\ 1 & -1 & 3\end{pmatrix}$.

7. $-6X^2-3Y^2+9Z^2$; $\quad 3X = 2x+y+2z$, $\quad 3Y = x+2y-2z$,
$3Z = -2x+2y+z$.

8. $4X^2+9Y^2+36Z^2 = 36$; $\quad 3, 2, 1$; $\quad x = z$, $\quad y = 0$, $\quad x = -z$.

9. $X = (y+2z)/\sqrt{5}$, $\quad Y = (5x+2y-z)/\sqrt{30}$,
$Z = (x-2y+z)/\sqrt{6}$;

$\frac{1}{7}\begin{pmatrix}6 & 2 & -1 \\ 2 & 3 & 2 \\ -1 & 2 & 6\end{pmatrix}$.

10. (a) $y_1 = 5(A\cosh t+B\sinh t)+C\cosh 5t+D\sinh 5t$,
$\quad\quad y_2 = -19(A\cosh t+B\sinh t)+C\cosh 5t+D\sinh 5t$.

(b) $y_1 = 2(A\cosh 2t+B\sinh 2t)+C\cosh\sqrt{7}t+D\sinh\sqrt{7}t$,
$\quad\quad y_2 = 5(A\cosh 2t+B\sinh 2t)+C\cosh\sqrt{7}t+D\sinh\sqrt{7}t$.

CHAPTER 3

Problems 3.3

1. (a) 3, 1; (b) 1, 1; (c) 2, 1; (d) 2, 3; (e) 1, 2.

3. $y = 2x+C$.

Problems 3.4

3. It is the solution $y = 2(x-1)$ of the equation and is the asymptote to the set of integral curves $y = 2(x-1)+Ce^{-x}$.

Problems 3.5

1. $(2, -9/8)$.

Problems 3.6

1. $y'' = 0$.   **3.** (a) $y'' - y = 0$;   (b) $(y' - y)y' = e^x$;
(c) $y'' + y = 0$;   (d) $(y' - 2)y' = y - x$;   (e) $y'' - 4y' + 4y = 0$.
4. (a) $y = \sinh x$;   (b) $y = e^x - 1$;   (c) $y = \sin x$;
(d) $4(y + 4) = (x + 4)^2$;   (e) $y = xe^{2x}$.
5. A singular solution.

Problems 3.7a

1. $\sin y = Cx^2$.   **2.** $e^{x+y} \cos x = C$.   **3.** $(1 + x^2)y^2 = C$.
4. $xy(\log x)^2 = C$.   **5.** $x^2 y \sin^2 x = C$.   **6.** $e^y = C - 2(x + 1)e^{-x}$.

Problems 3.7b

1. $2a \tan^{-1}(y/x) + \log(x^2 + y^2) = C$.   **2.** $y + x = Cx^4(y + 3x)$.
3. $e^{y/x} = Cx^6(y + 2x)^2$.   **4.** $xe^{\cot(y/x)} = C$.
5. $y^2 = Cx^3 \exp\{(x^2 - y^2)/xy\}$.   **6.** $xy = Ce^{x/y}$.

Problems 3.7c

1. $y = 3 + (2 - x) \log C(x - 2)$.   **2.** $(2y + x - 4)^2 = C(y + x - 3)^3$.
3. $3y = C + x + 5 \log(x - 2y - 4)$.   **4.** $(y - x + 1)(y + x + 2)^2 = C$.
5. $\log(y + 1) + 2 \tan^{-1} \dfrac{(y + 1)}{(x + 1)} = C$.

Problems 3.7d

1. $e^{x+y} \cos x = C$.   **2.** $y^2 \log x = C$.   **3.** $x \cos y + y \sin x = C$.
4. $x^2 + y^3 + \sin 2x + \cos 3y - y^3 e^{2x} = C$.   **5.** $y = C(x + 3)^2 - 4$.

Problems 3.7e

1. $x^2 e^{xy} = C$.   **2.** $(x^2 + y^2)e^{x \sin x} = C$.   **3.** $y \sqrt{(1 + x^2)} = C$.
4. $xy(\log x)^2 = C$.   **5.** $x^2 y \sin^2 x = C$.

Problems 3.7f

1. $3y = (x^3 + C) \cot x$.   **2.** $yxe^{3x} = 2 \sinh x + C$.
3. $y^2 = (x + C)^2/(x^2 + 1)$.   **4.** $y^2 = 4(x + C)^2/(x^2 + 3x + 2)^3$.
5. $y = x \log x + Cx^2$.   **6.** $x^2 e^{xy} = C$.

Problems 3.7g

1. $y^2 = Ce^x - x^2$.   **2.** $(x^2 + y^2)e^{x \sin x} = C$.

Problems 3.7h

1. $ye^{xy} = C$.

2. $y = (\cosh^{-1} x + C)/\sqrt{(x^2 - 1)}$.

3. $y + \log C(y^2 + 1) = (x^2 - 1)/x$.

4. $xye^{-xy} \cos(y/x) = C$.

5. $x^4 + 4xy^3 - y^4 = C$.

6. $y = x + 1/x(x^2 + C)$.

7. $(x + 2y + 3)^5 (x - y)^3 = C$.

8. $(x + y - 1)ye^x = C$.

9. $y = 4x^2/(4x^2 + C)^{2/3}$.

10. $y = C \cos x - x$.

CHAPTER 4

Problems 4.2a

1. $y = Ae^{-3x} + Be^{2x}$; $A = 1$, $B = 0$.

2. $y = Ae^x + Be^{4x}$; $A = -1$, $B = 1$.

3. $y = Ae^{-2x/3} + Be^{x/2}$; $A = 3$, $B = 0$.

4. $y = Ae^{-2x} + Be^{2x}$; $A = B = 1$, $y = 2 \cosh 2x$.

5. $y = Ae^{-3x} + B$; $A = -1$, $B = 3$.

Problems 4.2b

1. $y = e^{-2x}(A \cos 3x + B \sin 3x)$; $A = 4$, $B = 5$.

2. $y = e^x(A \cos x + B \sin x)$; $A = 2$, $B = 0$.

3. $y = e^{-3x/2}\left(A \cos \dfrac{\sqrt{3}}{2} x + B \sin \dfrac{\sqrt{3}}{2} x\right)$; $A = 0$, $B = 2\sqrt{3}$.

4. $y = e^{-3x}(A \cos 2x + B \sin 2x)$; $A = 6$, $B = 8$.

5. $y = A \cos 4x + B \sin 4x$; $A = 1$, $B = \frac{1}{4}$.

Problems 4.2c

1. $y = (Ax + B)e^{-3x}$; $A = 4$, $B = 1$.

2. $y = (Ax + B)e^{-3x/2}$; $A = 4$, $B = 3$.

3. $y = (Ax + B)e^x$; $A = B = 2$.

4. $y = (Ax + B)e^{5x}$; $A = -4$, $B = 1$.

5. $y = (At + B)e^{-Rt/2L}$; $A = 1$, $B = 0$.

Problems 4.2d

1. $y = A + Be^x + Ce^{2x}$.   **2.** $y = A + (Bx + C)e^{3x}$.

3. $y = (A + Bx + Cx^2)e^x$.   **4.** $y = e^{-2x}(A \cos 3x + B \sin 3x) + Ce^{2x}$.

5. $y = Ax + B + (Cx + D)e^{3x}$.

Problems 4.4a

1. e^{-x}.   **2.** e^{-x}.   **3.** e^x.   **4.** xe^{2x}.   **5.** $e^{3x} + x^2 e^{2x}$.

6. $x(e^{2x} - e^{x/3})/5$.   **7.** $x \sinh 3x$.   **8.** xe^{2x}.   **9.** $x(e^x - e^{2x} + e^{3x})$.

10. $x^3 e^{-2x}$.

Problems 4.4b

1. $(3 \cos x + \sin x)/10$.   **2.** $(2 \cos x + \sin x)/10$.   **3.** $\frac{1}{18} \sin 3x$.

4. $\sin x - \cos x$.   **5.** $2(4 \cos 2x - 15 \sin 2x)/241$.

6. $x \sin 2x$.   **7.** $\sin 2x$.   **8.** $-x \sin 2x$.

9. $(3 \sinh 3x + 4 \cosh 3x)/21$.   **10.** $x \cosh x$.

Problems 4.4c

1. $x^2 - 2$.   **2.** $3x + 2$.   **3.** $-(8x^4 + 8x^3 + 6x^2 + 3x)/128$.

4. $(18x^2 - 30x + 19)/108$.   **5.** $-(x^4 + 12x^3 + 72x^2 + 240x + 360)$.

Problems 4.4d

1. $e^x(2 \sin x - 3 \cos x)$.   **2.** $e^x(x - 1)$.   **3.** $-xe^{4x} \cos 5x$.

4. $xe^{-x} \sin 6x$.   **5.** $x^4 e^x/12$.   **6.** $e^{-3x}(x^2 + 4x + 6)$.

7. $2(1 - x) \cos 2x + \sin 2x$.

8. $e^x\{(26x - 9) \cos x + (39x - 46) \sin x\}/169$.

9. $e^{2x}\{(1 - 3x) \cos 3x + \sin 3x\}/54$.

10. $xe^{4x}\{3x \cos x + (2x^2 - 3) \sin x\}/12$.

Problems 4.4e

1. $y = Ae^x + Be^{2x} + 2xe^x(e^x + 2) + 1$; $A = 4$, $B = -5$.

2. $y = Ae^{-3x} + Be^{-2x} + 3(\cos 2x + 5 \sin 2x)/26 + 1$; $A = \frac{18}{13}$, $B = -\frac{3}{2}$.

3. $y = Ae^{-4x} + Be^x + xe^x(25x^2 + 60x + 51)/375$; $A = -B = \frac{2}{75}$.

4. $y = e^x(A \cos x + B \sin x) + 8(2 \cos x + \sin x) + \cos 2x + 2 \sin 2x + 15$;
$A = -32$, $B = 20$.

5. $y = (Ax + B)e^{-4x} + e^{-4x}(2x^2 - \sin 2x)$; $A = 2$, $B = 0$.

Problems 4.6

1. 2A; 0·693 sec. **3.** $i = \dfrac{E}{2} \sqrt{\dfrac{C}{L}} (\sin \omega t - \omega t e^{-\omega t})$.

4. $q = CV\left(1 + \dfrac{2t}{CR}\right)e^{-2t/CR}$, $i = -\dfrac{V}{L} te^{-Rt/2L}$. **5.** 0·00322 sec.

Problems 4.7

1. $x = \cos t + \sin t + 3 \sin 2t$; $\quad y = \sin t - \cos t - 3 \cos 2t$.

2. $x = e^t \cos t - t$; $\quad y = e^t \sin t + t^2 - 1$.

3. $x = 1 - 2 \cos t + \cos 2t$; $\quad y = -t + 5 \sin t - 2 \sin 2t$.

4. $x = 3(t^2 - 1) \sin t - 14t \cos t$; $\quad y = (1 - t^2) \cos t$.

5. $x = t^2 + \log t$; $\quad y = t^2 - 2t + e^t$.

Problems 4.8

1. $y = \{(\log x)^2 + A \log x + B\}x^2$. **2.** $y = \{A + x(B + \log x)\}x^2$.

3. $y = \{A + x(B\sqrt{x} - 1)\}/x^2$. **4.** $y = 1 + (A \log x + B)x + C/x$.

5. $y = A \log x + B + x^3/9$. **6.** $y = Ax^{1/3} + (B + \log x^{1/3})x^{2/3}$.

7. $y = \{A \cos (\log x) + B \sin (\log x) + \log x\}x$.

8. $y = \{A \cos (\log x) + (B + \log x) \sin (\log x)\}x$.

9. $y = \{A \cos (\log x) + B \sin (\log x)\}x^3 + (\log x^5)^2 + 6 \log x^5 + 13$.

10. $y = \dfrac{A}{(x-1)^3} + B(x-1) + \dfrac{(x-1)^2}{5} + \dfrac{(x-1)}{2} \log (x-1) - \dfrac{1}{3}$.

CHAPTER 5

Problems 5.2

1. $y = c_0\left(1 + \dfrac{x^2}{2!} + \dfrac{x^4}{4!} \ldots\right) + c_1\left(x + \dfrac{x^3}{3!} + \dfrac{x^5}{5!} \ldots\right)$

$= c_0 \cosh x + c_1 \sinh x$.

2. $y = c_0(1 - 2x^2 - \tfrac{8}{3}x^3 - 2x^4 \ldots) + c_1(x + 2x^2 + 2x^3 + \tfrac{4}{3}x^4 \ldots)$

$= \{(c_1 - 2c_0)x + c_0\}e^{2x}$.

3. $y = c_0\left(1 - x + \dfrac{x^2}{2!} - \dfrac{x^3}{3!} + \dfrac{x^4}{4!} \ldots\right) + \dfrac{x^2}{2!} - \dfrac{x^3}{3!} + \dfrac{x^4}{4!} \ldots$

$= (c_0 + 1)e^{-x} + x - 1.$

4. $y = c_0\left(1 + x - \dfrac{x^2}{2!} - \dfrac{5}{3!} x^3 + \dfrac{13}{4!} x^4 + \dfrac{101}{5!} x^5 \ldots\right) = \dfrac{c_0 e^x}{(x^2 + 1)}.$

5. $y = c_0\left(1 - \dfrac{x^2}{2!} + \dfrac{x^4}{4!} \ldots\right) + x\left(1 - \dfrac{x^2}{2!} + \dfrac{x^4}{4!} \ldots\right) = (c_0 + x)\cos x.$

Problems 5.3

1. $\pm 2.$ **2.** None.   **3.** $(2n+1)\pi/2.$

4. $0.$ **5.** $n\pi/2 \quad (n \neq 0).$ **6.** $0, \pm 1.$

Problems 5.4

1. $y = 1 + \dfrac{x^3}{3!} - \dfrac{x^4}{4!} - \dfrac{2}{5!} x^5 + \dfrac{4}{6!} x^6 \ldots.$

2. $y = x + \dfrac{5}{3!} x^3 + \dfrac{101}{5!} x^5 + \dfrac{4241}{7!} x^7 \ldots.$

3. $y = 2x + \dfrac{6}{3!} x^3 - \dfrac{30}{5!} x^5 + \dfrac{630}{7!} x^7 \ldots.$

4. $y = 1 + 6(x-1) + \tfrac{15}{2}(x-1)^2 + \tfrac{5}{2}(x-1)^3.$

5. $y = 1 - \dfrac{(x-2)^3}{3!} + \dfrac{4}{6!} (x-2)^6 - \dfrac{28}{9!} (x-2)^9 \ldots.$

Problems 5.5

1. $y = c_0\left(1 + x^2 + \dfrac{x^4}{12}\right) + c_1\left(x + \dfrac{x^3}{4} + \dfrac{x^5}{160} \ldots\right).$

2. $y = c_0 + c_1 \sum\limits_{n=0}^{\infty} \dfrac{(-1)^n}{(2n+1)} x^{2n+1} = c_0 + c_1 \tan^{-1} x \qquad (|x| < 1).$

3. $y = c_0\left(1 + \dfrac{x^2}{2}\right) + c_1 x.$

4. $y = c_0(1 + 2x^2) + c_1\left(x - \dfrac{x^3}{3} - \dfrac{x^5}{6} \ldots\right).$

5. $y = c_0\left(1 - \dfrac{x^2}{2} + \dfrac{x^4}{24} - \dfrac{x^6}{80} \ldots\right) + c_1 x.$

6. $y = c_0(1 + x^2 + x^4 \ldots) + c_1(x + x^3 + x^5 \ldots) = \dfrac{c_0 + c_1 x}{(1 - x^2)} \quad (|x|) < 1).$

7. $y = c_0\left(1 - \dfrac{9}{2}x^2 + \dfrac{105}{8}x^4 - \dfrac{539}{16}x^6 \ldots\right) + c_1\left(x - \dfrac{10}{3}x^3 + 9x^5 \ldots\right).$

8. $y = c_0\left(1 + \dfrac{x^2}{2!} + \dfrac{x^3}{3!} + \dfrac{x^4}{4!} + \dfrac{x^5}{5!} \ldots\right) + c_1 x = c_0 e^x + (c_1 - c_0)x.$

9. $y = c_0\left(1 - \dfrac{3}{2}x^2 + \dfrac{x^3}{6} + \dfrac{3}{8}x^4 \ldots\right) + c_1\left(x - \dfrac{x^3}{2} + \dfrac{x^4}{12} \ldots\right).$

10. $y = c_0\left(1 - \dfrac{x^4}{4!} + \dfrac{5}{8!}x^8 \ldots\right) + c_1\left(x - \dfrac{2}{5!}x^5 + \dfrac{12}{9!}x^9 \ldots\right)$

$\qquad + c_2\left(x^2 - \dfrac{x^6}{5!} + \dfrac{42}{10!}x^{10} \ldots\right).$

Problems 5.6

1. $y = c_0 e^x + b_0\sqrt{x}\left(1 + \dfrac{2}{3}x + \dfrac{(2x)^2}{15} + \dfrac{(2x)^3}{105} \ldots\right).$

2. $y = c_0\left(1 + \dfrac{x}{2} + \dfrac{3x^2}{20} + \dfrac{x^3}{30} \ldots\right) + b_0 x^{-3}\left(1 + \dfrac{x}{2}\right).$

3. $y = c_0 x\left(1 - \dfrac{x}{3} + \dfrac{x^2}{3.4} - \dfrac{x^3}{3.4.5} \ldots\right) + \dfrac{b_0}{x}(1 - x)$

$\qquad = \dfrac{2c_0}{x}(e^{-x} + x - 1) + \dfrac{b_0}{x}(1 - x).$

4. $y = \dfrac{c_0}{x} + \dfrac{c_1}{(1 - x)}.$

5. $y = c_0\left(1 - \dfrac{x}{2!} + \dfrac{x^2}{4!} - \dfrac{x^3}{6!} \ldots\right) + b_0\sqrt{x}\left(1 - \dfrac{x}{3!} + \dfrac{x^2}{5!} - \dfrac{x^3}{7!} \ldots\right)$

$\qquad = c_0\cos\sqrt{x} + b_0\sin\sqrt{x}.$

6. $y = \dfrac{c_0}{x^2} + \dfrac{c_2}{6}(3x^2 - 8x + 6).$

7. $y = \dfrac{c_0}{\sqrt{x}}\left(1 - \dfrac{x^2}{2^2\times2!} + \dfrac{x^4}{2^4\times4!} - \dfrac{x^6}{2^6\times6!} \ldots\right) + \dfrac{c_1}{\sqrt{x}}$

$\qquad \times\left(x - \dfrac{x^3}{2^2\times3!} + \dfrac{x^5}{2^4\times5!} \ldots\right) = \left(c_0\cos\dfrac{x}{2} + 2c_1\sin\dfrac{x}{2}\right)\Big/\sqrt{x}.$

8. $y = c_0 e^x + b_0\sqrt{x}.$

Problems 5.7

1. $y = \{Ax + B(x\log x + 1)\}/(1 - x)^2.$

2. $y = \{A + B(x - \log x)\}/(1 - x)^2.$

3. $y = e^x(A + B\log x).$ **4.** $y = (x + 1)(A + B\log x).$

5. $y = x\{A + B(\log x + x)\}/(x+1)^2.$

6. $y = Ay_1 + By_2;$ $y_1 = 1 - 2x + \frac{3}{2}x^2 - \frac{2}{3}x^3 \dots ;$

$y_2 = y_1 \log x + 3x - \frac{13}{4}x^2 + \frac{31}{18}x^3 \dots .$

Problems 5.9

1. $y = AJ_0(x) + BY_0(x).$ **2.** $y = A'J_\nu(ax) + B'J_{-\nu}(ax).$

3. $y = x^{-\nu}\{AJ_\nu(x) + BJ_{-\nu}(x)\}.$ **4.** $y = \sqrt{x}\{AJ_0(x) + BY_0(x)\}.$

5. $y = \sqrt{x}\{AJ_1(2\sqrt{x}) + BY_1(2\sqrt{x})\}.$

6. $y = \sqrt{x}\left\{AJ_{1/3}\left(\frac{2}{3}\sqrt{\frac{x^3}{3}}\right) + BJ_{-1/3}\left(\frac{2}{3}\sqrt{\frac{x^3}{3}}\right)\right\}.$

7. $y = \sqrt{x^3}\{AJ_{5/4}(x^2) + BJ_{-5/4}(x^2)\}.$

8. $y = (A\cos x + B\sin x)/\sqrt{x}.$ Exact for $\nu = \pm\frac{1}{2}.$

$y = A'J_{1/2}(x) + B'J_{-1/2}(x).$

CHAPTER 6

Problems 6.3

1. $(cs^2 + bs + 2a)/s^3.$ **2.** $(a^2s^2 + 2as + 2)/s^3.$ **3.** $e^b/(s-a).$

4. $(s+3)/(s^2+9).$ **5.** $2/(s^2+16).$ **6.** $4/s(s^2+4).$

7. $(s\cos b - a\sin b)/(s^2+a^2).$ **8.** $1/(s-1).$ **9.** $1/(s-1)^2.$

10. $(s-a)/\{(s-a)^2+b^2\}.$ **11.** $e^{2t}.$ **12.** $t(t+2).$

13. $(3\sin 2t - 2\sin 3t)/6.$ **14.** $\cosh 2t - \cos t.$ **15.** $1 + e^{-t}.$

16. $\{\cos(5t/2) - \sin(5t/2)\}/2.$ **17.** $e^{4t/3}.$

18. $\cos 2\sqrt{2}t - \sqrt{2}\sin 2\sqrt{2}t.$

Problems 6.4

1. $t^3e^{-t}.$ **2.** $t^{n-1}e^{at}.$ **3.** $e^{3t} + 4e^{-2t}.$ **4.** $e^{4t}\cos 2t.$

5. $e^t(3\cosh 2t + 4\sinh 2t).$ **6.** $e^{-3t}(2\cos t - \sin t).$

7. $e^{-3t/2}(6 - 5t)/2.$ **8.** $e^{-2t}(3t^2 - 8t + 2)/2.$

9. $e^{-3t/4}(12\cos 2\sqrt{3}t - 5\sqrt{3}\sin 2\sqrt{3}t)/16.$

10. $e^{2t}(1 + 4\cos 4t).$

Problems 6.6

1. $y = 1 + t$.   **2.** $2y = e^{2t} - 1$.   **3.** $2y = e^{3t} - e^t$.

4. $9y = e^{2t} + (15t + 8)e^{-t}$.   **5.** $12y = e^t + 8e^{-2t} - 9e^{-3t}$.

6. $y = e^{-t}(2\cos 2t - \sin 2t)$.

7. $10y = 5e^{-t} - 6e^{-2t} + \cos t + 3\sin t$.

8. $10y = 3(e^{3t} - \cos t) + \sin t$.

9. $x = e^{6t}\sin t$, $y = e^{6t}(\sin t - \cos t)$.

10. $2x = -2 + t^2 + e^{-t} + \cos t + 3\sin t$;
$4y = -2 + e^{-t} + \cos t + 3\sin t$.

Problems 6.7

1. $2y = e^t - 2e^{2t} + e^{3t}$.   **2.** $450y = 2(15t + 8)e^{-2t} - 25e^t + 9e^{3t}$.

3. $4y = 2t^2 - 2t + 1 - e^{-2t}$.   **4.** $6y = -1 + 3e^t - 3e^{2t} + e^{3t}$.

5. $72y = 16e^{4t} + 9e^{5t} - 12te^t - 25e^t$.   **8.** $y = e^{2t} + \sin 2t - 2t\cos 2t$.

Problems 6.8

1. $y = e^{-2t}$.   **2.** $2y = t^2 - 4t + 2$.   **3.** $y = 2J_0(3t)$.

4. $y = 3e^{-2t}J_0(t)$.   **5.** $y = -2e^{-t}\sin t$.   **6.** $y = (t + 1)e^{-t} - 1$.

Problems 6.9

1. $-e^{-s}(e^{-s} - 1)^2/s$.   **2.** $A/s(1 + e^{as})$.   **3.** $(e^{-2s} + s - 1)/s^2$.

4. $A\{1/as + 1/(1 - e^{as})\}/s$.   **5.** $(1/s^2)\tanh(as/2)$.   **6.** $H(t - \pi)$.

7. $2H(t - 2) - 6H(t - 3)$.   **8.** $H(t) + H(t - 1) + H(t - 2) + \ldots$.

9. As for **8**.   **10.** $tH(t - a)$.

Problems 6.10

1. $(t - 4)^2 e^{t-4} H(t - 4)$.   **2.** $(\sin 3t)H(t - 2\pi/3)$.

3. $-e^{-2(t-\pi)}(\cos t + \sin t) H(t - \pi)$.

4. $\{e^{-3(t-2)}\cosh 2(t - 2)\} H(t - 2)$.   **5.** (a) $\sin 2t + \cos 2t$; (b) 2.

6. $y = 1 - (1 + \cos \pi t) H(t - 1)$.

7. $y = 1 - \cos 2t + \sin 2t - (1 - \cos 2t) H(t - \pi)$.

8. $i = 1 - (t + 1)e^{-t} - \{1 - te^{-(t-1)}\} H(t - 1)$.

9. $y = t - \sin t - (t + \sin t - \pi) H(t - \pi)$.

10. $y = (\sin t - t \cos t)/2 + \{1 - \sin t + \frac{1}{2}\left(t - \frac{\pi}{2}\right)\cos t\}\, H\left(t - \frac{\pi}{2}\right).$

11. $y = 1 + 5t - e^{2t}(\cos 4t + \frac{3}{4}\sin 4t) - [1 + 5(t-a) - e^{2(t-a)}\{\cos 4(t-a)$
$\qquad + \frac{3}{4}\sin 4(t-a)\}]\, H(t-a).$

12. $y = 0;\quad y = \{2(t-\pi) + 3 - 4e^{t-\pi} + e^{2(t-\pi)}\}/4;$
$\qquad y = t + 3(1-\pi)/2 - e^{t-\pi}(1+e^{-\pi}) + e^{2(t-\pi)}(1+e^{-2\pi})/4.$

Problems 6.11

3. $\{Aa \coth (\pi s/2a)\}/(s^2+a^2).$ **4.** $\{e^{2(1-s)\pi} - 1\}/(1-s)(1-e^{-2\pi s}).$

5. $y = A(t - \sin t)/a - A[\{1 - \cos (t-a)\}\, H(t-a)$
$\qquad + \{1 - \cos (t-2a)\}\, H(t-2a) + \ldots].$

6. $y = (-1)^n A\left[\left\{at - \pi(n - \left| \sin \frac{n\pi}{2} \right|)\right\}\cos at - \sin at\right]\Big/2a^2.$

CHAPTER 7

Problems 7.3

1. (a) $\frac{17}{12}$; (b) $\frac{9}{7}$.   **2.** 1.   **3.** $(3e^4 - 23)/3$.   **4.** 14.

5. $\frac{28}{3}$.   **6.** 6.   **7.** 0.   **8.** π^4.

9. $-2^{11}/5$.   **10.** -8π.   **11.** 0.   **12** (a) 2π, (b) 0.

Problems 7.4

1. 56.   **2.** $k = 1$; 1.   **3.** e in every case.   **4.** 0.

Problems 7.6

1. 6.   **2.** 9.   **3.** 1.   **4.** $-\frac{1}{70}$.

5. $\frac{164}{3}$.   **6.** 0.   **7.** 0.   **8.** 54.

9. $-2^8/5$.   **10.** $\frac{3}{8}a^2 \log 3$, 8.   **11.** $\frac{128}{3}$.   **12.** $\frac{16}{15}$.

13. $\frac{5}{12}$.   **14.** $16a^3/3$.   **15.** $\frac{64}{3}$.   **16.** $\pi/128$.

17. $\frac{8}{5}$.   **18.** $3(4\pi - 3\sqrt{3})/4$.   **19.** π.

20. $\cosh 4 - \frac{1}{4}\sinh 4$, 20·5.

Problems 7.7

1. 6, 9, 1, $-\frac{1}{70}$, $\frac{164}{3}$.   **2.** $e-1$.   **3.** $2\log 2 - 1$.

4. $\frac{5}{2} + \log \left(\frac{4}{27}\right)$.   **6.** $\frac{155}{84}$.

Problems 7.8

1. 6.   **2.** 0.   **3.** $\frac{1}{6}$.   **4.** a^5.   **5.** $e^2 - 19/3$.

6. $3^7/8$.   **7.** $64\sqrt{2}/15$.   **8.** $2(4e - 9)$.

Problems 7.9

1. $\frac{8}{3}$.   **2.** $(e - 1)/2$.   **3.** $(e^2 - 1)/4e$.   **4.** $e^2 - 1$.

5. $2/\pi$.   **6.** $(\cosh 1 - 1)/2$, 0·272.   **7.** $1/6$.

8. $2\{\log(1 + \sqrt{2}) + (1 - \sqrt{2})\}$.   **9.** $2/3(6!)$.

10. $(e - 1)/6$.

Problems 7.10

1. $-\frac{32}{3}$.   **2.** $\frac{9}{2}$.   **3.** $2\pi a^4$.   **4.** π.

5. $4\pi a$.   **6.** $8\pi(\sqrt{2} - 1)/3$.   **7.** $\pi(2\log 2 - 1)/2$.   **8.** π.

9. $1 - 5/2e$.   **10.** $6(3\pi - 4)$.   **11.** $9\log(1 + \sqrt{2})$.

12. $2\sqrt{2}\pi \log 2$.   **13.** $\pi a^2/8$.   **14.** 372π.   **16.** 7π.   **17.** $4\pi \log 2$.

18. $\pi/6$.   **19.** $2\{3\pi + 4(5 - 4\sqrt{2})\}/9$.   **20.** 256π.

Problems 7.11

1. $dS = a^2 \sin\theta \, d\theta \, d\varphi$, $4\pi a^2$.   **2.** $a(\log 2/\sqrt{3})\log(2 + \sqrt{3})$.   **3.** $\frac{21}{4}$.

4. $\pi(17\log 17 - 16)/16$.   **5.** $a^2(8a^3 + 5a^2b + 135b^2 + 60ab - 30ab^2)/60$.

6. 0.   **7.** $3^7/2^6$, $3^6/140$.   **8.** $4\pi a^3/3$.   **9.** $\frac{1}{6}$.   **10.** $1 + 3(1 - e^2)/8e$.

Problems 7.12

1. 1, $(3e^4 - 23)/3$, 14, -8π, 0.   **2.** $\frac{1}{6}$, $1 + 3(1 - e^2)/8e$, 0.   **3.** 0.

4. $16 + \pi$.   **5.** 0.

CHAPTER 8

Problems 8.3

1. (a) 5; (b) 7.   **2.** $\mathbf{a} - \mathbf{b} + \mathbf{c}$.   **4.** (a) $3\sqrt{6}$; (b) $2\sqrt{29}$; (c) $\sqrt{106}$.

5. $(3\mathbf{i} + 6\mathbf{j} - 2\mathbf{k})/7$.   **6.** $\mathbf{a} = \mathbf{b} + \mathbf{c}$; $\sqrt{42}/2$, $\sqrt{186}/2$, $\sqrt{222}/2$.

7. $4\sqrt{3}\{-(\mathbf{i} + \mathbf{j} + \mathbf{k})/\sqrt{3}\}$.   **8.** $(3, -4, -2)/\sqrt{29}$.   **9.** 2, $64°7'$.

10. (a) $\mathbf{r} = \mathbf{i} + \mathbf{j} + (1 + t)\mathbf{k}$; (b) $\mathbf{r} = (1 + t)(\mathbf{i} + \mathbf{j} + \mathbf{k})$;

    (c) $\mathbf{r} = (1 + 3t)\mathbf{i} + (1 - 2t)(\mathbf{j} + \mathbf{k})$.

Problems 8.4a

1. $-6/\sqrt{59}$.　**3.** $78°13'$.　**4.** $\pm(3i-5k)/\sqrt{34}$.　**5.** $(\pm\sqrt{3}j+k)/2$.

8. (a) 4; (b) 17.　**9.** 49.　**10.** $x+2y+3z = 16,\ 16/\sqrt{14}$.

Problems 8.4b

1. $\pm(4i+5j+7k);\ 2(4i+5j+7k)$.　**2.** $2\sqrt{6}$.　**3.** $\pm(6i-2j+3k)/7$.

5. $\pm(8i-7j+3k)/\sqrt{122}$.　**6.** $a\times b+b\times c+c\times a = 0$.

8. (a) $4(-11i+4j+k)$;　(b) $5(-i+2j+5k)$;　(c) -90;　(d) -120.

Problems 8.4c

1. 3.　**2.** $d\cdot(b\times c)/a\cdot(b\times c)$.　**3.** $(c\times b+sa)/a\cdot c$.

4. $c-(c\cdot d)b/(b\cdot d)$.　**5.** $\{s^2b+(a\cdot b)a+s(a\times b)\}/s(s^2+a^2)$.

Problems 8.6a

3. 3, increasing.　**4.** $a = b = c = 3$.　**5.** $2x+2y+z = 9$.

7. $\pm(3j-k)/\sqrt{10}$.　**9.** $\cos^{-1}(11/4\sqrt{29})$.　**10.** $x^2+2xyz-y^2+z+C$.

Problems 8.6b

1. 7.　**2.** -1.　**5.** 42.

Problems 8.6c

1. $i-2j+3k$.　**2.** $a = 1;\ b = 2;\ c = 3$.　**8.** $8\sqrt{6}$.

Problems 8.7a

1. $v = 2(1-\cos 2t)i+2\sin tj-3t^2k$,

$r = (2t-\sin 2t)i+2(1-\cos t)j-t^3k$.

2. (a) -2π; (b) $8j$.　**3.** $v\times\dfrac{dv}{dt}+c$.　**4.** $2(i+j+3k)$.

Problems 8.7b

1. $\frac{40}{3}$.　**2.** $\frac{1}{60}$.　**3.** $2c^2\pi^2$.　**4.** $-\frac{26}{15}$.　**5.** 0.　**6.** $\frac{4}{3}$.

7. -9.　**8.** 0.　**9.** 0.　**10.** 64.

Problems 8.7c

1. (a) 4; (b) 4.　**2.** 43.　**3.** $-\pi a^4/2$.

4. (a) $a\left(i+j+\dfrac{\pi}{2}k\right)$; (b) $2\pi ak$.　**5.** $6(2i+5k)/5$.

6. $\frac{4}{3}$.   **7.** -9.   **8.** 0.   **9.** 0.   **10.** 64.   **11.** 16.

12. $\frac{224}{3}$.   **13.** $4\pi a^3$.   **14.** 0.   **15.** $81(3\pi - 4)/8$.

Problems 8.7d

1. 16.   **2.** $\frac{224}{3}$.   **3.** $4\pi a^3$.   **4.** 0.   **5.** $81(3\pi - 4)/8$.

6. (a) 0; (b) πa^4.

Index